WHAT CAUSED THE BIG BANG?

VIBS

Volume 115

A volume in
Philosophy and Religion (PAR)
Kenneth A. Bryson, Editor

WHAT CAUSED THE BIG BANG?

Rem B. Edwards

Rodopi

Amsterdam – New York, NY 2001

The paper on which this book is printed meets the requirements of "ISO 9706:1994, Information and documentation - Paper for documents - Requirements for permanence".

ISBN: 90-420-1407-5

Printed in The Netherlands

In memory of Curry T. Haynes

Professor of Biology, Emory at Oxford,

Who helped me to understand and love living things

CONTENTS

EDITORIAL FOREWORD

The appearance of Rem B. Edwards's *What Caused the Big Bang?* marks the introduction of the first title in the Philosophy and Religion (PAR) special series. I cannot imagine a better way to introduce the special series to the academic world than to do so through the thought of a frequently cited scholar. This is PAR's first book, Edwards's sixteenth.

Edwards's erudition is everywhere in evidence as he devours the pages of Big Bang literature, separating fact from fancy, the examined from the unexamined. Socrates would recognize his sting as belonging to the most energetic of gadflies, unrelenting, pestering those who would readily ascribe the origin of the universe to anything less than disciplined reason requires. What caused the Big Bang? Now that the 15 billion-year-old cosmic dust has settled, several likely explanations emerge from the cosmic broth. But not all explanations are proven equal, as Edwards amply demonstrates: Steady State and Plasma Cosmologies; Antecedent Universe Cosmologies; Big Fizz and Big Divide Quantum Cosmologies; Quantum Observership Cosmology; Big Accident Quantum Cosmology; Atheistic Anthropic Cosmology; the Final Anthropic Principle–each view contains fatal flaws.

Edwards's thesis that God caused the Big Bang follows a detailed deconstruction of alternate models showing their weakness: where and how they commit fallacies. The burden of proof now falls squarely on the shoulders of those who do not accept the claim that God caused the Big Bang. Critics must point to the deficiencies in Edwards's argument and defend the superiority of their own view. This is a hard sell, given the breadth and depth of his work. But if God created the universe, what is our place in it? Who is God, why did God create, is God responsible for the suffering of innocent victims, and since the universe is contingent, does God sustain creation? Like all good philosophy, Edwards's answer to questions raises more questions!

In my own work on death and immortality, the mysterious nature of the nothing has long beckoned forth, inviting me to visit the nurturing intelligibilities it incloses. In discussions on death and dying, I find useful the distinction between the absence of something and the removal of ground in which the possibility of this absence arises. For instance, is death the absence of life or is it the removal of the possibility in which the possibility of absence arises? The simple answer is that it is both. The complex answer is that one distinction (ontological) raises the question of what death might be like to the dead (if *post mortem* states exist), while the other (epistemic) addresses the ordinary-language view of death as absence of life. The investigation into the ontological character of death (death as such), then is conducted from the perspective of the nothing as reversal in the possibility of temporal existence. Death is a return to the conditions that existed before the Big Bang. In part, my thesis depends on the existence of a state in the likeness of the nothing. Edwards's *What Caused the Big Bang?* provides solid evidence and confirms my own belief that God is at work in this domain.

If God caused the Big Bang, then, the universe had a beginning. It might not have had a beginning in time (the universe could be eternal), but it must have had a beginning in the order of existence (thereby providing an answer to the question, "Why is there something rather than nothing?"). If matter has a beginning, science cannot reach that far. The laws of the universe only become applicable at Planck Time and Planck Space or length; at 10^{-43rd} of a second, the size of the universe was 10^{-33rd} centimeter in diameter (see pp. 98–99 of this volume). This is as close to the moment of creation as science can get. So how can the Big Bang have a cause? Philosophy and/or religion take over at that point. Edwards's inquiry reveals that the gap separating the before and the after of existence (beyond scientific measurement) is not nothing at all since it is pregnant with the divine laws and patterns of existence. How else would the universe know to open the first act of existence in a scene of well-orchestrated expansion and contraction? Planets could not have formed in the absence of laws and patterns. The existence of the law implies structure. And structure points beyond contingency to the existence of a Necessary Being–or God at work in the *ex nihilo*.

Edwards's book is powerful and timely. His cogent analysis of quantum physics provides at least one indubitable truth that cannot be deconstructed–God exists! The current crisis in Ethics is due to the excesses of relativism. Once we accepted Hume's invitation to skepticism, Heidegger's critique of the Absolute, Nietzsche's death-of-God movement, and the genetic secularization of our species, nothing special was left to unite us. We found ourselves doing moral theory in the absence of a unified ethical vision of our common origin, nature, and destiny. Edwards's book provides the ontological grounding required for a fresh start. It should be required reading, not only where physics is taught, but whenever Philosophy and Religion matter.

Kenneth A. Bryson
Editor, Philosophy and Religion
University College of Cape Breton
Sydney, Nova Scotia, Canada
January 2001

PREFACE

In 1988, Stephen W. Hawking wrote,

> Up to now, most scientists have been too occupied with the development of new theories that describe what the universe is to ask the question why? On the other hand, the people whose business it is to ask why, the philosophers, have not been able to keep up with the advance of scientific theories.[1]

My professional training as a philosopher has contributed immensely to my preparation for writing this book; and I have done my best to try to understand those scientific theories which have a direct bearing on my central question: What caused the Big Bang? I hope that my inquiry brings together successfully both the what and the why of the origin of the universe. This topic has long fascinated me, and I have read and thought extensively about it. My considerable reading about the Big Bang, my background in Process Philosophy, with its emphasis on uniting philosophy and science, and my training, teaching, and writing in the philosophy of religion have all helped to prepare me for this enterprise.

Almost everyone is curious about the origin of the universe; and my intended audience is philosophers, theologians, scientists, and all inquisitive persons who wonder how and why it all began. I agree with George Smoot that there is a "deep public interest in understanding the origin of the universe and our place in it;"[2] so this book is written for the average literate person, not just for professionals. In places, however, the subject matter is difficult. To quote Hawking again:

> ...If we do discover a complete theory, it should in time be understandable in broad principle by everyone, not just a few scientists. Then we shall all, philosophers, scientists, and just ordinary people, be able to take part in the discussion of why it is that we and the universe exist. If we find the answer to that, it would be the ultimate triumph of human reason–for then we would know the mind of God.[3]

With Smoot and Hawking, I share this ultimate goal and address this broad audience. My contribution will be to ask and try to answer philosophical questions of scientists, who are usually at least as naive about philosophy as philosophers are about science. I also want to show ordinary people what the best scientific minds are saying about the origin of the universe, and how to think critically and philosophically about their theories. In thinking about the ultimate origin of the universe, we are in the borderlands between science, philosophy, and religion.

I will try to present the major answers that contemporary scientific cosmologists are giving to: What caused the Big Bang? For many persons, this question has a simple and direct answer: God did it. When the Belgian astronomer/priest George Lemaître first understood the evidence for the Big Bang, he was probably convinced personally that science had discovered the moment when God created the world;[4] but he tended to keep his views on how religion relates to astronomy to himself. In a 1951 address, Pope Pius XII claimed that "True science discovers God in an ever-increasing degree–as though God were waiting behind every door opened by science....Science has provided proof of the beginning of time....Hence, creation took place in time. Therefore, there is a Creator; therefore, God exists."[5]

Those who believe that things are so easy will be surprised to learn that most contemporary scientific cosmologists are doing their best to avoid the hypothesis that God created or caused the Big Bang. Most presuppose a Naturalistic metaphysics, according to which the universe has been around in some form from eternity with no conceivable dependence on Deity. Many scientists believe that the Big Bang, which initiated our cosmic epoch, was caused by an antecedently existing universe, not by God's creative activity. Some scientific cosmologists try to avoid God by maintaining that the Big Bang had no cause at all. Coming chapters will survey both secular and religious accounts of cosmic origins and evaluate them on their own merits.

Chapter One of this book reviews the overwhelming evidence that convinces most scientists today that our universe began with a Big Bang somewhere between eight and twenty billion years ago; and it charts the course of the evolution of the universe from an initiating Big Bang to where we are today. It explores the possibility that science cannot answer the question of ultimate origins because the topic lies beyond the proper bounds of legitimate science. Good science involves both theory and empirical confirmation, but many contemporary scientific cosmologists are producing only content-less theories about what caused the Big Bang. Scientific Cosmological Agnostics deny that our question is properly scientific and indicate that the pseudoscièntific theories which try to answer it cannot be verified, even indirectly. Only experience can separate actualities from abstruse possibilities, reality from wild speculation; yet we have no experience of worlds creating other worlds.

In Chapter Two, Naturalistic Humanistic theories of reality (metaphysics), of knowledge (epistemology), and of human origins (humanistic anthropology) and well-being (ethics and axiology) are introduced, subjected to thorough philosophical analysis and criticism, and shown definitively to be untenable. All the non-theistic answers to: What caused the Big Bang? examined in Chapters Three through Nine are deeply embedded within an indefensible Naturalistic Humanistic philosophical outlook.

Chapter Three explores significant challenges to the idea that our universe originated in a Big Bang. Steady State Cosmology, developed by Fred Hoyle and his associates in the 1940s and 1950s, affirms that the universe is uncreated and that it maintains its present general appearance from eternity; so there was no Big Bang. The Hubble expansion of the universe results from the continuous creation of hydrogen atoms out of nothing by matter. New atoms fill in the blanks left by the Hubble expansion of the universe. Steady State Cosmology is not alone in affirming that the physical universe is spatially and temporally infinite. In *The Big Bang Never Happened*,[6] Eric Lerner offers seemingly powerful objections to the evidence that convinces most scientists that the Big Bang really happened, and he presents his own Plasma Cosmology which, when all is said and done, relies upon a local Mini-Bang to explain what is happening in the finite part of the infinite universe that is observable to us. Decisive objections to his position are developed.

In Chapter Four, two versions of the theory that our universe was created by the collapse of an antecedently existing universe are discussed. George Gamow thought that a shrinking universe infinitely preceded our own in a time before our time and finally collapsed in a Big Crunch. It then rebounded, and our resulting universe will expand forever. We exist in a life-supporting phase of the endless rebound period. Where Gamow's Cosmology postulates only one contraction, one crunch, one Bang, and one rebound, Oscillation Cosmology conjures up an infinite number of antecedent universes, each of which began in a Big Bang, expanded to a maximal state, recontracted, then renewed the whole process with another Big Bang. Oscillationists propose that an influx of energy from an antecedently existing universe caused the Big Bang and our resulting cosmos, but the position is fatally flawed, as this chapter shows.

Quantum theory has powerfully influenced cosmological speculation since the early 1980s. Chapters Five through Eight explore a variety of Quantum Cosmologies, each of which has its own peculiar answer to the question of cosmic origins.

Big Fizz Cosmology covered in Chapter Five says that our Big Bang was created when energy bubbles formed through spontaneous quantum fluctuations in the womb of an antecedently existing Superspacetime or Mother Spacetime. Infinitely many bubbles form spontaneously to make infinitely many universes, which co-exist within Mother Spacetime. Our bubble inflated fifteen billion or so years ago, so here we are! Big Divide Cosmology says that every universe sub-divides itself into infinitely many universes at every turn of events, so we are here for a brief moment within a universe that looks like it began in a Big Bang, but it really began only a fraction of a second ago when an antecedent universe sub-divided to actualize all possibilities. But these cosmologies are utterly implausible, as demonstrated.

Quantum Observership examined in Chapter Six emphasizes the important role that some interpreters of quantum mechanics assign to scientific observers, measurers, and experimenters. It maintains that the indefinite and indeterminate domain of quantum events takes on definiteness and determinateness only when observers view it. Evidence for the Big Bang exists only as human observers find it–and thereby create it. So what created the Big Bang? We did! (But we didn't, as the concluding critique proves.)

Big Accident Cosmology contends, as explained in Chapter Seven, that the question of what caused the Big Bang presupposes something that supposedly is not true, namely that everything which comes into being has a cause. Quantum physics denies this, we are told, and discloses that our universe originated out of nothing, was caused by nothing, exists for no purpose, and is nothing. Nothing caused the Big Bang. It is so easy for nothing to cause nothing! Just why our universe *did not* originate this way is carefully explained.

Chapter Eight considers Atheistic Anthropic Cosmology. Many recent scientific cosmologists note that our universe is exceptionally fine-tuned for the creation of life, including intelligent forms of life. Tiny changes in any of the initial conditions, constants, and laws of nature would have resulted in a universe inhospitable to life. For every successful way of creating a life-supporting universe, there are infinitely many futile ways to get it wrong. Lifeless universes are infinitely probable, and life-supporting universes are infinitely improbable. Why, then, do we live in a life-supporting universe? The Anthropic Principle says that we live in a life-supporting universe because we are here, that is, because if the universe were not life-supporting, we would not be here to ask questions about it. Even atheists do not deny the remarkable life-supporting design of our universe, but they think that they can account for this without having to appeal to God. Atheistic Anthropic Cosmology explains that if infinitely many worlds exist, as many Quantum Cosmologies profess, then universes as rare as our own will just happen occasionally. Given an infinite number of shoes, one will fit now and then by pure chance. The metaphysical Principle of Plenitude, that all possibilities must be actual somewhere, guarantees the existence of an infinite number of universes. For Atheistic Quantum Cosmologists, the Principle of Plenitude is the ultimate cause of our Big Bang plus infinitely many other universes, very few of which life-sustaining. The innumerable flaws of Atheistic Anthropic Cosmology are spelled out in detail.

Chapter Nine deals with the bizarre claims made by the Final Anthropic Principle, according to which our universe and an infinite number of others will ultimately coalesce into a single omniscient and omnipotent Omega Point that will be God. God does not now exist and did not create the world; but the world, which began without God, now exists and will ultimately create God. Human life is meaningful because through our android descendants we can contribute to the development of the Omega Point by traveling in space and ultimately inhabiting our entire universe. The position borders on madness, as explained!

These atheistic theories and a few theistic accounts of what caused the Big Bang are explained and critically examined in significant depth in chapters to follow. When considered critically and seriously, much of the atheistic cosmological speculation being done by today's astronomers, astrophysicists, and other scientists is quite outlandish, as the preceding paragraphs suggest. Once this is fully understood and appreciated, the door is open to reexamine the possibility that God caused the Big Bang. Still, some ways of conceiving of the nature of God and of God's relations with the world are much more intelligible than others, and much easier to relate to the universe disclosed to us in Big Bang Cosmology.

Two quite different concepts of God are examined in Chapter Ten. Classical Theism is committed to the absolute changelessness of God in every conceivable respect. Process Theism, by contrast, affirms that God is indeed changeless in certain desirable respects but is in process in other highly desirable respects. It is desirable both that God be changelessly good and that God's experiences change as God interacts with created worlds and their creatures as they come in to being in spacetime and history. A comprehensive but modified Process Theology best reconciles science and religion. Chapter Ten also discusses several senses in which God may be said to "exist" and develops and justifies several changes in Process Theism that seem desirable, upon examination, if it is to be rationally and religiously appealing.

Chapter Eleven presents a revitalized Biopic Teleological Argument for the existence of God, based upon massive evidence for the fine tuning of the universe for life, as disclosed by contemporary scientific cosmology. Note that when masculine pronouns are used occasionally in reference to God in this chapter and elsewhere, this is done merely from convention and for economy or convenience of expression; but it in no way implies that God is masculine in any intelligible or defensible sense.

Chapter Twelve further develops the case for Theism with a refurbished Cosmological Argument from Contingency for the existence of God, again based upon what contemporary physics and astrophysics have revealed about the radically contingent nature of physical reality.

To my knowledge, no existing book covers and critically examines philosophically all the major options for explaining the origin of the Big Bang. The astute debate between William L. Craig and Quentin Smith in their *Theism, Atheism, and Big Bang Cosmology*[7] focuses almost exclusively on the Standard Model of the Big Bang, with its initial singularity, and on the quantum Big Accident option; but it neglects all the other theories of origin explored here. M. A. Corey's *God and the New Cosmology: The Anthropic Design Argument*[8] deals mainly with the teleological but not in depth with the cosmological argument for God's existence. As endnotes for each following chapter will indicate, numerous books and articles examine and defend one particular theory or another. Yet, no previous book takes a hard philosophical look at all the basic options presented here while critically examining the Naturalistic assumptions that underlie the non-theistic scientific (or pseudoscientific) cosmologies covered in Chapters Three through Nine. Many cosmologists emphasize scientific data and theories. While not neglecting these, I also introduce relevant philosophical questions, analysis, and theories.

I wish to express special appreciation to The University of Tennessee for the Faculty Development Grant which it awarded to me to work on this project, and to my former Department Heads, George Brenkert and Kathy Bohstedt, for their substantial support for my efforts. Kenneth A. Bryson, editor of the Religion and Philosophy special series of the Value Inquiry Book Series, was immensely helpful in spotting defects that I was not able to see, but final responsibility for imperfections in the book rests with me. And many, many thanks to my wife Louise who managed to tolerate my existence while I was so deeply immersed in the project of figuring out the universe.

ACKNOWLEDGMENTS

I thank the following publishers for their kind permission to reprint or paraphrase copyrighted materials, especially Barry Whitney, editor of *Process Studies*, which previously published my discussion of "How Process Theology Can Affirm Creation *Ex Nihilo*," 29:1 (Spring-Summer 2000), pp. 77–96 that appears here as a section of Chapter Ten. Short quotations and references in this volume should fall within the realm of "fair use," but for permission to use more detailed material, I thank the following:

Cambridge University Press: James Cornell, ed., *Bubbles, Voids, and Bumps in Time: The New Cosmology*, 1989.

Clarendon Press, a division of Oxford University Press: William L. Craig and Quentin Smith, *Theism, Atheism, and Big Bang Cosmology*, 1993.

Mind, an Oxford University Press publication: Ian Hacking, "The Inverse Gambler's Fallacy," 96 (1987), pp. 331–340.

Open Court Publishing Company: Charles Hartshorne, *Creative Synthesis and Philosophic Method*, 1970.

Oxford University Press: John D. Barrow and Frank J. Tipler, *The Anthropic Principle*, 1989.

Process and Faith: John B. Cobb, Jr., "Is God Creator *Ex Nihilo*?" Web site, July-August, 1999.

Prometheus Books: Kai Nielsen, *Naturalism without Foundations*, 1996; Victor J. Stenger, *The Unconscious Quantum: Metaphysics in Modern Physics and Cosmology*, 1995.

Random House, Alfred A. Knopf: Eric J. Lerner, *The Big Bang Never Happened*, 1991.

Random House, Bantam Dell Publishing Group: Stephen W. Hawking, *A Brief History of Time*, 1988, 1996.

Sky & Telescope: Cheryl J. Beatty and Richard T. Fienberg, "Participatory Cosmology: The Big Bang Challenge" (March 1994), pp. 20–22; "Extrasolar Planet Seen Transiting Its Star" (February 2000), pp. 16–17; "The Great Attractor's Hidden Heart" (December 1999), p. 12; Robert Jastrow, "What Are the Chances for Life?" (June 1997), p. 63; Alan M. MacRobert, "Beyond the Big Bang" (March 1983), pp. 211–213; Sten Oldenwald, "Spacetime: The Final

Frontier" (February 1996), pp 24–29; "The Primordial Soup: A Recipe for Life" (February 1998), p. 20; Joshua Roth and Joel R. Primack, "Cosmology: All Sewn Up or Coming Apart at the Seams" (January 1996), pp. 20–26; Gary H. Sanders and David Beckett, "LIGO: An Antenna Tuned to the Songs of Gravity" (October 2000), pp. 41–48.

Yale University Divinity School: Robert Jastrow, "Science and the Creation," in *Creation* (a special issue of *Reflection*), edited by Thomas Schattauer, 1980.

Writers House LLC: Stephen W. Hawking, *A Brief History of Time*, 1988, 1996.

Zygon, a Blackwell publication: Joel Friedman, "The Natural God: A God Even an Atheist Can Believe In" (September 1986), pp. 369–388; Fred W. Hallberg, "Barrow and Tipler's Anthropic Cosmological Principle" (June 1988), pp. 139–157; Frank J. Tipler, "The Omega Point as *Eschaton:* Answers to Pannenberg's Questions for Scientists" (June 1989), pp. 217–253; Patricia A. Williams, "Christianity and Evolutionary Ethics: Sketch Toward a Reconciliation" (June 1996), pp. 253–268.

One

SCIENTIFIC COSMOLOGY AND THE BIG BANG

Greek philosophers assumed that the world, the universe, the cosmos, or nature as a whole, existed in some form from eternity, that is, infinitely into the past, and that the basic stuff of the universe is uncreated, everlasting, self-sufficient, and indestructible. The official Christian view, by contrast, was and is that the world was created by God out of nothing (*ex nihilo*) at some point in the finite past. Naturalistic cosmologists through the centuries have sided with the Greek view and affirmed the everlasting self-existence and self-sufficiency of nature as a whole. Traditionally, each view was affirmed solely as a matter of dogma and blind faith, though Naturalists consistently intimated that science confirms their metaphysics. Up to the present century, however, no convincing evidence was available to resolve the dispute. Has this now changed?

In the twentieth century, cosmology tried to become a science, something more than mere speculation and dogma; and up to a point it succeeded. Twentieth-century cosmologists produced a plethora of astonishing discoveries about the universe as a whole.[1] The most amazing is that the world of nature as we know it came into being at a definite point in the finite past. Our universe has not existed forever after all; it was created between 8 and 20 billion years ago. Do not conclude too hastily that God created it, or even that the origin of the world had a cause. Controversies about what, if anything, caused the Big Bang constitute the main subject matter of this book. We cannot address our central concern–What caused the Big Bang?–until we first survey evidence for the Big Bang and this theory's story of the evolution of the universe. Later, some significant challenges to Big Bang Cosmology will be confronted.

The astronomer Fred Hoyle, who did not favor the view, coined the apt pejorative phrase "Big Bang" for the theory that our universe began with the greatest thermonuclear explosion of all time in the finite but far distant past. In 1993, *Sky and Telescope* magazine ran a contest designed to find a better name for it, but after reading over 13,000 entries, some quite imaginative, the panel of prestigious judges decided that "Big Bang" is the best name after all. The winner of the contest was Fred Hoyle! As expressed in *Sky and Telescope*, "And the winner is...Nobody. Or Fred Hoyle, really, though he didn't submit 'the Big Bang' as an official entry last summer, having coined the term 44 years ago." Some imaginative losers were: "Hubble Bubble," "Planck Point," "Bertha D. Universe," "What Happens If I Press This Button?" and "You're Never Going to Get It All Back In There Again."[2]

As the theory developed in the twentieth century, differing estimates were given about how long ago the Big Bang happened. In the 1920s and 1930s,

Edwin Hubble greatly overestimated the expansion rate of the universe; and this resulted in a significant miscalculation of the age of the universe–first at a billion years, and later at around two billion years. In 1952, George Gamow, an early popularizer of the view, published his *The Creation of the Universe.* Gamow calculated that the Big Bang occurred between 1.7 and 3.4 billion years ago.[3] In 1958, Werner Heisenberg's *Physics and Philosophy* placed the origin of the universe at about four billion years ago.[4] During the 1950s, available estimates of the age of the earth and its solar system showed them to be older than the universe, so something had to give!

Periodic improvements in methods for estimating cosmic distances and velocities bring about more accurate estimates of the time of the origin of the universe; but uncertainty lingers. The work of Allan Sandage and W. A. Baade in the late 1950s and early 1960s immensely expanded time/distance estimates in astronomy. In 1978, Robert Jastrow estimated the age of the universe to be around 20 billion years, give or take a few billion years one way or the other;[5] and Frank Tippler uses that figure in his 1994 book titled *The Physics of Immortality.*[6] Today, most cosmologists accept 15 billion years as a workable compromise between the 8 to 20 billion year estimates generated by the inexact dating techniques and calculations available to today's astronomers.

Much more detailed observations and exact measurements by the Chandra X-ray Observatory launched in 1999, and the Hubble Space Telescope (deployed in 1990, repaired during 1993, and upgraded in 1997 and 1999), will yield a more precise time for the origin of the universe. At first, as many questions were raised as answered. Late in 1994, some data from Hubble observations suggested that the universe may be only 8 to 12 billion years old, in which case the age of some stars seems to be greater than that of the universe–perhaps twice as great! Still, most available data indicates a 10 to 18 billion-year-age.[7] Astronomers scrambled to reconcile conflicting data, either by reintroducing an Einsteinian cosmological constant, by correcting the Hubble constant, by looking for ways to lower the estimated age of the stars, by developing more accurate measurements of celestial distances, or by greatly reducing estimates of the critical density or mass/energy in the universe.[8] In early 1997 and thereafter, new and more accurate measurements of celestial distances were made public that harmonize the age of the universe with its stellar and galactic contents.[9] With the availability of much more accurate measurements, by early 1999 something close to a fifteen billion-year-old cosmos seems very plausible.[10] I will use the convenient 15 billion-year figure for the age of the universe. Exact timing is not essential for identifying the ultimate cause of the Big Bang; what really matters is the well-founded scientific consensus that our world came into being somewhere between 8 and 20 billion years ago. In a later discussion of Big Accident Cosmology, the mass/density of the universe will be important for deciding whether the universe actually had a cause.

According to Big Bang Cosmology, how did the universe evolve after its beginning, and what evidences support this sensational view of cosmic origins?

1. The Evolution of the Universe

Contemporary scientific cosmologists apply the concept of evolutionary development to the cosmos as a whole, not just to living things on earth. They are convinced that they can trace the broad outlines of cosmic evolution back to the beginning, around fifteen billion years ago, and that we can know more about the origin of the universe than about today's weather and earthquakes. The Standard Hot Big Bang Model prevailed between 1965 and 1980, when it was modified to take account of inflation and quantum effects. According to the Standard Model and a few widely publicized modifications, the universe as we know it developed as follows.

A. The Initial Singularity

Those of us who take pride in our common sense would be skeptical if told that in the distant past all of the material comprising the building in which we now live or work was once compacted into a ball no bigger than a basketball or a baseball. We would be astounded if further informed that at one time in the past, all of the matter/energy comprising the entire universe was compacted into a space no bigger than a baseball or a golf ball. Once, space itself was even smaller than that. We would be utterly incredulous to learn that in the beginning the totality of spacetime and all its contents were the size of a penny, a pinhead, or even a point. Yet, this is precisely what the Standard Model tells us, not as a matter of baseless dogma or mere speculation, but as grounded presumably in hard evidence. The historical development of Big Bang theory in the twentieth century will concern us only minimally. That story is well told by many other authors,[11] but science cannot be divorced entirely from its history.

In the beginning, at T = 0, there was a singularity, according to the Standard Big Bang Model; and in this initial singularity, all the mass/energy of creation was compressed infinitely to zero size. The initial singularity had the volume of an Euclidean point, which has position but no magnitude at all; but, since no space existed in which it could assume a position, the initial singularity had neither position nor magnitude. In singularities, including the initial one and those that may exist in black holes and collapsed neutron stars, mass/energy is infinitely curved, infinitely hot, infinitely dense, and infinitely small.[12] For the universe as a whole, at T = 0, the being of the world was as close to non-being as it could get! Perhaps, as some suggest, it just was non-being.

Between 1965 and 1970, two eminent British astrophysicists, Roger Penrose and Stephen Hawking, collaborated to prove that our universe began in a singularity, assuming that classical general relativity theory is correct.[13] In his

1988 best selling book, *A Brief History of Time*, Hawking reported that he had changed his mind; he wrote that he was "trying to convince other physicists that there was in fact no singularity at the beginning of the universe," that "It can disappear once quantum effects are taken into account."[14] Like Hawking, many other contemporary cosmologists are greatly troubled by the initial singularity, as we will see in Chapter Four.

B. Planck Time and Space

Quantum Cosmologists trace the origin of the universe back to a tiny fraction of a second after the initiating explosion emerged from near nothingness, but they reject the initial singularity. Momentous developments occurred in the first milliseconds after creation, including the arrival of time, space, and all the laws of physics.

Planck Time is how long it takes a photon of light, which obviously travels at the speed of light, to traverse a unit of Planck Length or Space. In contemporary Quantum Cosmology, time and space are physically inseparable; and both have minimal quantum energy units, below which the very concepts of time and space have no meaning. These minimal units are called Planck Time and Planck Length or Space; at 10^{-43} ($1/10^{43}$) of a second, the size of the universe was 10^{-33} ($1/10^{33}$) of a centimeter in diameter in volume. At those dimensions, Planck Density can also be calculated (10^{90} kilograms per cubic centimeter). The universe had to be at least that big and that old to exist at all. In Standard Big Bang Cosmology, by contrast, time and space can be condensed all the way down to a pointlike singularity; but Quantum Cosmologists think that physical time and space have no meaningful referent at less than Planck dimensions,[15] and that it makes no sense to ask what happened *before* that in lesser quantities of spacetime. Yet, we will see, the question of "before that" is almost irresistible. Before that was a singularity, or the laws of quantum physics, or a vacuum fluctuation, or an antecedent universe, or Divine creation *ex nihilo*–or something.

Scientific cosmologists agree that all laws of physics break down or become totally inapplicable either at the initial singularity or at Planck Time. This includes the First Law of Thermodynamics, according to which energy is neither created nor destroyed. Laws of nature are employed in reasoning backward in time scientifically. If natural laws break down totally at an initial singularity or at the Planck Wall, nothing before that can be known scientifically because nothing remains to guide our extrapolations. As John A. Wheeler put it, "There never has been a law of physics that did not demand 'space' and 'time' for its statement....With the collapse of space and time the framework falls down for everything one ever called a law of physics."[16] The exact point of breakdown is a major controversy. If what Wheeler says is true, and if Planck spacetime is the point of ultimate collapse, extrapolating back to an initial singularity is impossible. Wheeler once did so, as we will see in discussing Oscillation Cosmology in Chapter Four; but this extrapolation assumes that the laws of physics continue to hold earlier than they can be known to apply, if Plank dimensions really are the ultimate limits of scientific knowledge.

C. Inflation

Several versions of Big Bang Cosmology were formulated early in the twentieth century, initially by Willem de Sitter in 1917 and Alexander Friedmann in 1922.[17] Other interpretations were developed in the 1920s and 30s by George Lemaître, Arthur Eddington, and James Jeans. In the late 1940s George Gamow and his associates, Ralph Alper and Robert Herman, made major contributions. The Standard Model of the Big Bang based on relativity physics dominated astrophysics between 1965 and 1980. It affirms that the early universe emerged from a singularity and thereafter expanded at a fairly uniform rate.

No model published prior to 1980 made a place for inflation. What is inflation? Some scientific cosmologists now believe that when the universe was around 10^{-35} of a second old, it underwent a short but spectacular period of exponentially rapid expansion from Planck dimensions to roughly the size of a grapefruit.[18] This lasted until 10^{-33} of a second after creation. Then the universe returned somehow (a great mystery) to the more leisurely and uniform pace of expansion predicted by the Standard Model. Alan H. Guth, who proposed the idea in 1980,[19] called this rapid expansion "inflation." Today, many cosmologists accept inflation, though not in the exact form originally proposed by Guth, not even Guth himself. Significant modifications of the inflationary scenario were proposed by Andreas Albrecht, Paul Steinhardt, and the Russian cosmologist Andrej Linde.[20] Guth now acknowledges that inflation is more of a family of theories than a single theory of cosmic origins and recognizes more than fifty varieties.[21] Some cosmologists are concerned that inflation may be so elusive that it could never be falsified, thus casting doubt upon its status as a scientific hypothesis. Although there is no direct empirical evidence for it, inflation is still widely accepted as a viable explanatory hypothesis because it is aesthetically appealing and provides plausible solutions to the horizon, magnetic monopole, and flatness problems. However, inflation theory may create the very problems to which it is supposedly the solution!

The *horizon problem* is the puzzle about why the observed universe is so homogenous and isotropic in structure and content, that is, why the universe is observed everywhere to be so remarkably uniform in temperature, background radiation, and large scale distributions of matter, despite the minor irregularities introduced by intermittent concentrations of matter in planets, stars, gaseous clouds, galaxies, and supergalaxies. Inflation suggests that the irregularities occurred when the universe was larger than the Standard Model could allow, and that the homogeneities were established when the universe was smaller than the Standard Model permits, thus enabling diverse parts of an initially tiny universe to influence one another causally and bring about the universe's observed uniformity. According to later versions of the theory, the universe did not begin homogeneously; but the chaotic perturbations and inhomogeneities that might have existed prior to the inflationary epoch were smoothed out by inflation itself. After inflation ended, additional fluke deviations or quantum fluctuations occurred; and these eventually carved out the stars and galaxies.[22] Of course, inflation is not needed if the universe began homogeneously and isotropically, that is, if the uniformities were given as initial conditions.

The problem of *magnetic monopoles* is logically implied by Grand Unification Theories (GUTs), according to which four basic physical forces were unified as one force in the earliest trillionths of a second of creation. GUTs imply that massive numbers of large physical particles should exist with either a south or north pole, but not both. Such particles would be radically different from and incompatible with the existence of all physical particles identifiable today, which have both poles. According to Guth's first inflationary scenario, all the magnetic monopoles were exterminated during the inflationary era. Now he is convinced that this would not happen; and a new scenario is accepted because inflation's prediction of magnetic monopoles is "at odds with observation." In inflationary models, the production of magnetic monopoles is either avoided by delaying the genesis of material particles during a brief period of rapid supercooling that released the energy that caused inflation,[23] or they are so dispersed by inflation itself that we just can't find them. None of this has been verified directly by observation; but supposedly the inflationary hypothesis is justified by its fruits, by its power to resolve otherwise intractable problems. Yet, it seems to create the very problems it solves, which otherwise don't exist! Inflation theory says that the universe should be rife with magnetic monopoles, but not one has ever been seen, and inflation theory tries to explain why nonexistent entities don't exist!

The *flatness problem* is the mystery over why space may be so little curved. It is closely related to the question of whether the universe contains so little mass/energy that it will continue to expand forever, or so much of it that gravity will eventually halt its expansion and bring about a contraction phase during which the universe slowly shrinks toward an ultimate Big Crunch. All the early versions of inflation predicted that the universe was blown out nearly flat at the beginning and that it is either just barely closed or perfectly balanced between open and closed. If these predictions cannot be sustained and confirmed, inflation is in deep trouble. As explained later, our universe looks very much like an open one, and inflation theorists are now scurrying (without great success) to develop models of inflation compatible with an open universe.[24]

The issues of homogeneity, inhomogeneity, and the ultimate fate of the universe will be considered again later. Alan Guth believes that inflation is close to gaining universal acceptance, but it actually has many serious critics. Cosmologists like Roger Penrose,[25] A. Karel Velan,[26] William L. Craig,[27] and many other critics[28] have serious doubts about inflation. Profound questions about it remain unanswered. Since scientists cannot repeat, verify, or falsify inflation, should it be taken seriously as a scientific hypothesis? Is it instead just metaphysics at its worst? Once inflation got underway, what brought it to a halt? How do we know that the universe was not just created by God or something else with initial conditions at the outset that insured general uniformity, the absence of magnetic monopoles, and whatever the curvature of space happens to be? Is inflation just a clever ruse for avoiding God? As Joseph Silk indicates, "Initial conditions are an alternative to inflation. Inflation is really a way of trying to erase arbitrary initial conditions. And it hasn't succeeded."[29] By assuming arbitrarily that the conditions for which it wishes to account were not present as initial conditions for the universe, inflation theory creates all the problems that it tries to resolve.

Is the universe really as flat, as uncurved, as most theories of inflation imply? If so, does inflation really explain this flatness? If wrinkles and irregularities existed in early spacetime, why didn't inflation just blow them out into bigger wrinkles and irregularities? More seriously, how can inflation's usual presupposition that Omega (critical density) equals at least 1, that the universe is either closed or ultimately balanced between open and closed, be reconciled with empirical findings of only a tenth to a thirtieth or so of the mass/energy required to balance or to halt and reverse the expansion process? Peter Coles and George Ellis indicate that, though controversial, "An open [not a closed] universe might arise from inflation;" and they caution that "We do not in fact have any proof that inflation ever took place...."[30] Considerations strongly favoring the openness of the universe will be presented later, especially in Chapter Three. At any rate, with or without inflation, the universe began about fifteen billion years ago, give or take a few billion here or there. With or without inflation, What caused the Big Bang? is a very good question.

D. Evolution of Physical Forces, Particles, and the Laws of Nature

During the remainder of the first minutes of creation, extremely important things happened that brought about the universe recognizable to us today, or at least to today's astrophysicists. During the first one to four minutes of time, the universe continued to expand and cool very rapidly; the four basic forces of physics separated from one another; the most elemental physical particles were formed; and the laws of nature came into being.

According to Grand Unification Theory (GUT), which aspires to be a theory of everything, the four basic forces of nature originally were only one force that divided into four at around 10^{-39} of a second. These physical forces are: 1. gravity, the force that holds together large-scale objects like those perceptible by our senses; 2. the strong nuclear force that holds protons and neutrons together in atomic nuclei; it is immensely more powerful than gravity, but operates only at short distances within the nuclei of atoms; 3. the electromagnetic force that holds electrons in orbit around atomic nuclei; and 4. the weak force that causes the slow decay of radioactive elements like uranium, ionium, radium, and radon. All of these forces are presumed to be mediated by messenger particles like gluons, gravitons, and W^+, W^-, and Z^0. The last two of these elemental physical forces have now been combined into an "electroweak force;" but gravity is especially resistant to unification. At present, Grand Unification has not been confirmed; it is a hope, not a fact, an aspiration, not an accomplishment. When and if Grand Unification is achieved, this will be additional powerful evidence for the Big Bang.

Sub-atomic physical particles first emerged from the matterless energy of preceding phases of creation at around 10^{-6} of a second, but nothing as complex as atoms and chemical elements then existed. A dense unstable soup of photons plus sub-atomic matter and anti-matter particles evolved from pure energy. In particle accelerators, when particles are generated from energy, an equal number of matter and antimatter particles are always created. For this reason, astrophysicists believe that for a short time equal numbers of quarks and anti-quarks, neutrons and neutrinos, electrons and positrons, protons and antiprotons, and a vast plethora of particles and anti-particles were created in dense thermonuclear reactions in the very early universe. None of these particles and antiparticles endured for very long. When they collided, they annihilated one another, released more energy, and generated additional radiation and sub-atomic particles. By the end of the first thirty minutes, by some process mysterious even to astrophysicists, the symmetry of matter and antimatter in the universe was broken. Somehow matter came to predominate over antimatter; and thus we live in a world of matter, not antimatter.

Natural laws describe general features of structures and processes in the world of public, perceptible spacetime. Laws of nature that apply to the four basic physical forces and to diverse sub-atomic, atomic, chemical, biological, psychological, and social entities came into being along with the realities whose patterns they depict. Through the centuries, the precise kind of reality possessed by natural laws has been much debated. A once popular view, now outmoded, affirmed that natural laws are powers or forces that compel things to be what they are and do what they do. The dominant view today is that laws of nature are merely statistical patterns that describe but do not forcefully compel or restrain natural structures and processes.[31] Natural laws have no causal efficacy in themselves; they do not cause things to exist; they neither force things to do what they do nor to be what they are; and they do not impose external restraints. At best, natural laws are only formal causes, not efficient causes. They do not even exist prior to the things that they "regulate" or "govern." They tell us nothing about the behavior of a single entity like a single atom or molecule; they merely summarize the average behavior of actual things in large groups. They do not describe the activities of individuals with absolute precision. Individuals come first; and their existence, natures, activities, and habits gradually engender recognizable statistical patterns. Knowledge of these general configurations by intelligent beings comes much later. Laws of nature merely describe how things behave on the average, as we discern them. Until entities exists in quantity, their configurations and the laws that portray their functional patterns do not exist, except perhaps as possibilities or exemplars in the mind of God. The real significance of natural laws is not that they are efficient causes but rather that they permit us to reason from the observed to the unobserved. Our abilities to predict and control future events and to understand past events depend upon this reasoning. Natural laws are only formal and probabilistic, not efficient and exact, causes of groups of physical events. They do not pre-exist these events to impose anything upon them.

Natural laws operate within given physical conditions. Big Bang Cosmology prompts astrophysicists to inquire about the initial conditions of the universe. What kind of habituated or habit-forming stuff was given originally for the laws of nature to describe? Attempts to dispense with initial conditions by converting or reducing them to natural laws seem doomed to failure. The quantity of mass/energy in the universe, extremely low entropy or disorder, no more than four basic physical forces, their relative strengths, permissible kinds of physical particles, the directionality of time, and other conditions resulting in regular or lawful patterns of events were simply given at the beginning of creation. These primordial conditions spawned numerical constants of nature like the relatively unchanging numbers associated with the four basic forces, Planck's constant, charges for electrons and protons, definite masses for different kinds of physical particles, the rate of Hubble expansion, and the speed of light. Physical constants are extremely useful to us in predicting past and future courses of events. Natural laws merely describe the formal patterns of processes and realities in nature; but in themselves they are devoid of energy or power.

E. From a Universe of Radiation to a Gaseous Universe

The nuclei of hydrogen and helium atoms were formed when the universe was about one second old; but for the first 700,000 years, the cosmos, composed of turbulent radiant energy, was almost without form and void. Stable atoms could not form during the first part of this radiation era because electrons were constantly being knocked out of place from their orbits around protons. The "primordial fireball," as it is often called, was at first completely dark and fireless for many hundreds of thousands of years. After around 300,000 years, atoms began to form as the universe further expanded and cooled. Until then, darkness was upon the face of the deep; but as atoms formed, more space was created, and photons were set free to illuminate the universe for the first time; and suddenly, after 300,000 years or so of darkness, there was light. The all-pervasive microwave background that we detect today reflects these processes and originated during this era of cosmic evolution. Finally, after 700,000 to a million years, the universe became a dense gaseous cloud consisting of about 75 percent hydrogen and 25 percent helium, with traces of deuterium and lithium. This gaseous plasma continued to expand and cool; but it contained no heavy elements. No quasars, galaxies, stars, or planets graced the skies.

F. Creation of Quasars, Galaxies, Stars, Solar Systems, Heavier Elements

When the gaseous universe was a billion or so years of age, two powerful forces and an initial irregularity or fluke fluctuations finally separated and produced massive objects in the heavens. The two forces were (1) the kinetic expansive energy of the original cosmic explosion that continued to dissipate the hydrogen and helium gases and (2) the opposing attractive force of gravity that slowly assembled huge masses of hydrogen and helium gasses to form the stuff of millions of quasars and supergalaxies, plus billions of galaxies and their innumerable stars.

Either irregularities were built into the universe as initial conditions, or quantum flukes explain why, despite gravity, the gaseous universe did not expand indefinitely as a homogeneous gaseous plasma. Most astrophysicists today think that quantum physics accounts for these flukes. Just where quantum physics begins to apply to the early universe is highly problematic. Some think it doesn't apply at all because they interpret quantum physics only phenomenologically or epistemologically as informing us only about how quantum conditions appear to us, not about how they really are or were.

As better explained in Chapter Six, quantum physics is interpreted in this book as applying realistically and ontologically to the actual structure of the physical world. Some ontological realists think that quantum physics applies at or very near the beginning of creation, perhaps at or immediately following the very earliest Planck or inflationary moments; others think that it begins to apply only when quantum fields of energy or tiny individuated quantum objects like electrons and photons made their first appearance. In any event, minuscule quantum fluctuations occurred at some point in the very early universe, and these significantly affected the density and distribution of existing mass/energy in the later universe. Eventually these perturbations had massive cumulative effects. Persisting and spreading for a billion years or more, the initially small effects of very early sub-microscopic vacillations caused huge clouds of gas to separate. Then gravity pulled their ingredients together to form quasars, galaxies, and stars.

Although only .002 percent of all existing hydrogen is heavy hydrogen or deuterium, that tiny amount is indispensable for igniting all stellar furnaces. Gravity alone is too weak to condense and heat the hydrogen in evolving stars to the temperature required to ignite the process of nuclear fusion that drives all stars and suns; but gravity can and did produce the lower densities and temperatures required to initiate deuterium fusion. Deuterium fusion then inaugurated the fusion of hydrogen into helium, and flaming galaxies, quasars, stars, and suns graced the skies many hundreds of thousand of years after the initiation of the Big Bang.

G. Formation of Our Sun, Its Planets, and Life on Earth

Aristotle and Ptolemy believed mistakenly that the heavens were absolutely perfect in their changelessness. They and their ancient and medieval disciples could see only the deceptive stability of stars in our Milky Way and a few "wandering stars," the planets. Unknown to the ancients, many other galaxies of stars exist, and so did earlier generations of stars. Eventually, some of these stars, the giant supernovas, exploded and filled the heavens with stardust.

Five billion or so years ago, ten billion years after the onset of the primordial Big Bang explosion, gravity assembled enough stellar debris to form our sun and its planets, including our earth. Our sun presently has enough nuclear fuel to burn for another five billion years, so it is a middle-aged star. The earth existed for about one to one and a half billion years before the most primitive forms of life appeared. Don't ask how! No one really knows! All the oxygen, carbon, nitrogen, iron, and other heavy elements that constitute our bodies and those of all living things were forged earlier in stellar furnaces. No life could have formed anywhere during the first generation of stars composed almost entirely of hydrogen and helium. Supernovas first had to manufacture the heavier elements, then explode to scatter their stuff of life into the cosmos. The process is cumulative, so later and later supernovas consist of more and more metals and other heavier elements. Our sun, all its planets, and we ourselves were fabricated by gravity out of the rubble of numerous supernova explosions. As William Fowler said, "Each one of us and all of us are truly and literally a little bit of stardust."[32] This may explain our fascination with and Immanuel Kant's being filled with awe by "the starry heavens above." We are all stuff of their stuff.

Assuming a 15 billion-year-old universe, about three and a half to four billion years ago, eleven and a half to twelve billion years after the Big Bang was initiated, microscopically small forms of life first appeared on earth. Much later, after more than 2,500,000,000 years of evolution, our earliest upright-walking hominid ancestors, members of the species *Australopithecus anamensis*, first emerged in Africa around 400 to 300 million years ago, more than 14.5 billion years after the Big Bang. They and their descendants like *Australopithecus afarensis* and *Homo erectus* flourished for well over a million years. *Homo erectus* migrated from Africa into Asia and Europe. Exactly when the earliest members of our own hominid species, *Homo sapiens*, descended from them and first appeared is a matter of great controversy; but it happened somewhere between 500,000 and 100,000 years ago, most probably around 200,000 years ago. We are newcomers in creation.

2. Evidences for the Big Bang

What evidences support the Big Bang account of the origin and development of our cosmos? *Why* do scientific cosmologists believe that our universe began in a cataclysmic thermonuclear explosion? Evidence accumulated slowly for most of the twentieth century. Today it is so overwhelming that almost all contemporary astronomers, astrophysicists, and scientists are convinced that our world origi-nated in an astronomical explosion at some point in the finite past. Its present structure evolved from a primordial fireball. Only a very few cosmologists like Eric Lerner are unconvinced. His serious challenge to Big Bang Cosmology will be discussed in Chapter Three.

Explosions familiar to us take place within a pre-existing spacetime system, and their debris scatters into that system; but the initial Big Bang was very different from familiar bomb and dynamite blasts. The Big Bang created spacetime itself. Space is not sheer nothingness, as common sense tends to assume; it has its own physical properties. It is a primordial physical medium that has its own energy density and texture. It is grainy, granular, or foamy; and it is elastic, temporally expanding, but capable of being shrunk, bent, knotted, and warped. All physical things, including explosions, are manifestations of it. The following converging lines of evidence imply that the basic spatiotemporally extended stuff of our universe originated in a thermonuclear explosion around fifteen billion years ago.

A. Receding Galaxies and the Redshift

Only since the early 1920s have we known with assurance of the existence of other galaxies, profuse with stars. Before powerful telescopes were built early in this century, astronomers could not see any individual stars beyond our own galaxy. They could see a few nebulae, but they could not confirm that they are composed of stars or that they exist beyond our Milky Way. With the naked eye, we can only see individual stars in our own galaxy, the Milky Way with its hundred billion suns. We can also see a few dim nebulae, like the nearby Andromeda Nebula, whose composition was unknown before the 1920s.

With his feeble telescope, Galileo could see the moons orbiting Jupiter. He was the first to see that the Milky Way is composed of individual stars, but he could not see any single stars within the hundred and twenty five billion or more galaxies that lie beyond our own Milky Way. By the middle of the eighteenth century, telescopes were powerful enough to see many nebulae; and in 1784, Charles Messier published a list of 103 bright clusters and nebulae, some of which turned out to be extragalactic.[33] Thomas Wright conjectured as early as

1750 that the spiral nebulae are clusters of stars. In 1850, Baron Alexander von Humboldt called them "island universes."[34] Without empirical verification, most astronomers accepted this view from Messier's day to late in the nineteenth century, when a few influential astronomers shifted to the position that they are only clouds of gas located within the Milky Way.[35]

Edwin Hubble first confirmed the island universes hypothesis. Beginning early in the twentieth century, astronomers employed much more powerful telescopes to probe the distant universe. In 1924, Edwin Hubble, using the 100-inch telescope at Mount Wilson Observatory in California, confirmed that the nebulae contain individual stars and that giant galaxies of stars exist beyond the Milky Way, our local galaxy. Hubble developed methods for measuring cosmic distances based upon the luminosity of "standard candles" in the sky, but we now know that his original computations greatly underestimated these distances and the corresponding age of the universe. He proved, nevertheless, that the universe is immensely larger and richer in contents than most people had ever dreamed it to be.

Edwin Hubble and Milton Humason, his associate, made and reported numerous observations that confirm another startling truth about the galaxies, the redshift; but they were not the first to see and discuss it. V. M. Slipher at the Lowell Observatory noticed the redshift in 1912 while observing and charting a small number of spiral nebulae, but he did not realize that they are clusters of stars and stardust, and that their redshift evidences an expanding universe.

What is the redshift, and what is its significance for the origin of our universe? In the middle of the nineteenth century, Christian Doppler discovered some important truths about light emitted by moving objects. If light coming from an object moving toward an observer is filtered through a spectroscope, its wavelength shortens and shifts toward the blue end of the spectrum. If the object is moving away from the observer, its wavelength lengthens and shifts toward the red end of the spectrum. Radar guns used today by traffic police make use of this "Doppler effect." Light waves coming from distant objects moving rapidly away from us are stretched toward the red end of the spectrum; and the further away and faster these objects are, the redder the shift. The degree of this shift is proportional to their speed and distance from the observer.

Hubble and Humason examined numerous galaxies and found in most cases that they manifest the "Doppler effect," the redshift; and they inferred that most observable galaxies in the universe are moving away from us. This was, as they recognized, the first hard empirical evidence that the universe expands as time marches on. The redshift of the galaxies does not result from their moving through space like projectiles, but from the expansion of space itself.

Galactic redshift is extremely important evidence for Big Bang Cosmology. If the spatiotemporal universe expands as it moves into the future, then earlier and earlier in the past it must have been more and more compacted or concentrated. The calculable rate of cosmic expansion, expressed in Hubble's

1929 law of uniform expansion, implies a relatively uniform rate of contraction, applied retroactively. Calculations reveal that at some point in the finite past, the universe was compacted to zero, an initial singularity; but this implication did not become clear for some time. Much later in the century, quantum physicists discerned that all laws of nature break down or become inapplicable at less than Planck time. If so, compaction below Planck dimensions to an even earlier initial singularity cannot be inferred after all; but that story will be told later.

Hubble found that although most galaxies manifest the redshift, a few do not. Our nearest neighbors, the Andromeda galaxy, the Magellanic Clouds, and twenty five or so other small galaxies nearby, do not exhibit the redshift because they are sufficiently close to be gravitationally bound to our Milky Way. In 1994, astronomers discovered a large, nearby, and previously unknown galaxy hidden behind our Milky Way; and there may be others.[36] The Andromeda galaxy, with a blueshift, is actually heading towards us and will collide with our Milky Way in about five billion years. This should not cause alarm, however; neither we nor our descendants will be around to see it; and it may only pass through the Milky Way with few collisions between vastly separated stars.

B. Hubble's Law of Uniform Expansion

In 1924 Edwin Hubble established that most galaxies are moving away from us; and by 1929 he realized that they do so in a lawlike manner. They move away from us and from one another at a uniform and calculable rate, like raisins in a rising loaf of bread, or dots on the surface of an inflating balloon. According to Hubble's law of uniform expansion, in a homogeneous universe, galaxies move away from us and from one another at speeds proportional to their distance.

Improved contemporary estimates of cosmic distances and velocities differ significantly from the results of Hubble's initial computations. We now know that the galaxies will double their distances from one another and from us in less than ten billion years. Before late 1997 or early 1998 astronomers assumed that the *rate* of cosmic expansion is being slowed by gravity, so it was much more rapid in the distant past when cosmic distances were smaller, and it will be much slower in the far distant future when the size of the universe is significantly greater. It now appears, as explained in Chapter Three, that the rate is accelerating, not slowing. This rate of expansion, known as the "Hubble constant," remains nearly the same during the lifetime of any human astronomer; but from a cosmic perspective, the pace of intergalactic distancing is anything but constant. Very recent work pins the Hubble constant down to between 60 and 75 kilometers per second for each megaparsec (3.26 million light years) of distance.[37]

The reverse of expansion is contraction. Hubble's law is significant for Big Bang Cosmology because it implies that in the finite but far distant past, all the mass/energy in the universe, including all matter in all the galaxies, was compressed together and belonged to one exploding primordial fireball. Even earlier, it was compacted either to zero, an initial singularity, or to Planck or other finite dimensions. If the laws of nature collapse at or beyond the Planck Wall, nothing smaller or earlier could exist as a part of our universe.

C. The First and Second Laws of Thermodynamics

The law of increasing entropy affirms that order and energy are constantly being lost or dissipated in closed systems that are not drawing and renewing them from elsewhere. Loss of order and energy is irreversible and time-asymmetrical. Applied to the whole universe as a closed system, this law indicates that disorder increases universally, and perhaps that overall energy concentrations decrease, as the cosmos expands and cools over time. It also implies that the energy and order of the universe were more and more concentrated and organized earlier and earlier in time. The universe began in a state of low entropy, and if it were infinitely old it now would be in a state of maximum entropy. However, since it is not infinitely disordered, it cannot be infinitely old. At some point in the finite past, about fifteen billion years ago, the universe was totally concentrated into a singularity or to Planck or other minute dimensions.

The *First Law of Thermodynamics* affirms that energy is conserved, that it can be neither created nor destroyed, that it can only be transformed into other types. This well verified law of nature implies nothing metaphysical about the ultimate origin of the mass/energy of the universe. It does not imply that mass/energy existed everlastingly and necessarily throughout an infinite past, though it is occasionally given this metaphysical interpretation.[38] Standard Big Bang Cosmology affirms that all the mass/energy of creation came into being with its laws about fifteen billion years ago. As a law of science, the First Law of Thermodynamics says simply that *once created,* energy is conserved and that we know of no physical way that we can destroy it. Superficially, the First Law of Thermodynamics, which says that mass/energy cannot be created, appears to conflict with Big Bang Cosmology, which says that all the mass/energy of the universe was created around fifteen billion years ago. However, the empirical or scientific claim that "The amount of energy in the universe is constant" should not be confused with the metaphysical claim that "The amount of energy in the universe is necessary, uncreated, self-sufficient, and everlasting." If our universe was actually preceded by an earlier universe that gravitationally collapsed to a singularity, all previous conservation laws would have collapsed and terminated as it crunched to spatiotemporal nothingness. As Charles Misner, Kip Thorne, and John A. Wheeler say,

> Of all principles of physics, the laws of conservation of charge, lepton number, baryon number, mass, and angular momentum are among the

> most firmly established. Yet, with gravitational collapse the content of these conservation laws also collapses. The established is disestablished.[39]

Conservation laws are really not incompatible with Standard Big Bang Cosmology because the conservation of mass/energy, like all other laws of nature, begins with and does not antedate the Big Bang.

The *Second Law of Thermodynamics* applies fully only to closed systems that do not draw energy and order from other systems. The growing bodies of living things, including living human beings, concentrate order and energy; but this does not violate the Second Law because all living things eat, drink, respire, photosynthesize (if plants), and draw their sustenance from outside themselves. They are not completely closed systems, so the Second Law does not apply fully to them. Considered apart from the energy that they draw from elsewhere, the law does apply. Animal bodies constantly lose and expend ordered energy, which is why they require renewal meal after meal, breath after breath. If not fed, they starve. Plants also lose ordered energy and renew it through photosynthesis and by absorbing nutrients through their foliage and roots.

The concept of "entropy" may involve subtle ambiguities. Roger Penrose notes that "Entropy is a concept that may be banded about in a totally cavalier fashion!"[40] He suggests that there may be something subjective about the "order" involved in low entropy, that "Various observers' aesthetic judgments might well get involved in what they deem to be 'order', rather than disorder."[41]

Increasing entropy is decreasing order, so to understand it we must advance and comprehend a concept of order. Contemporary discussions of entropy give many different accounts of what counts as *order*. Astrophysicists identify order with the undifferentiated homogeneity of mass/energy in the earliest universe, with matter that has not yet been converted into pure or unavailable energy, with an intense concentration of energy, with the relatively low ratio between the number of photons and the number of other particles in the universe, with the ability to recover information from antecedent states of affairs, with energy states not pervaded by destructive ripples, with complex environments that can support intelligent forms of life, and perhaps with all of these. But can all of the above be consistently combined? Measured by the standard of being life-supporting, the earliest universe was chaos, high entropy, great disorder, "without form and void." Yet, it was low in entropy when order is identified with the original concentrated and homogeneous soup of radiant energy or sub-atomic particles that ultimately produced an available-energy-environment supportive of and usable by intelligent life. Whether the very early universe is characterized as ordered or chaotic may be a matter of wording and emphasis. George Gamow opted for chaos or high entropy when he wrote that "In the distant past our universe was considerably less differentiated and complex than it is now and the state of matter at that time could be accurately described by the classical concept of 'primordial chaos.'"[42] This is really not incompatible with Roger Penrose's emphasis, discussed later, on low entropy as an initial condition of the universe because Penrose employs different concepts of order and entropy.

Entropy is often conceived as a decrease in the concentration of both *energy* and *order*, but these are two very different things. In the expansion phase of our universe, they normally hang together, but not necessarily, not always, not in the contraction phase if one is to occur, and not in contemporary regions of gravitational collapse. Neutron stars are so massive and dense that gravity has fused their electrons and protons into neutrons. In black holes or collapsing neutron stars or universes, energy concentrates but *dis*order increases. Neither black holes nor collapsing universes violate the *First* Law of Thermodynamics, for they are not losing energy absolutely. Spacetime and energy are continuously constricted and concentrated in them while life-supportive order, available information, usable energy, and recoverable information deteriorate, and destructive ripples increase, so most astrophysicists now believe; but the issue is contested.

With respect to energy, black holes as well as collapsing stars and universes actually do violate the Second Law of Thermodynamics, the law of increasing entropy, for energy concentrates in them; but with respect to life supporting, ripple-free usable energy and available information, they do not violate this law, for these elements of order decrease within them. Concerning order, but not energy, Roger Penrose affirms that "The second law will hold sway just as much inside a black hole as it does elsewhere."[43] By this, Penrose means that human existence in black holes and collapsing universes would be utterly impossible because the kind of order that could support us would be lost and lacking.

Penrose thinks that no living space traveler could survive being sucked into a black hole; the violence and disorder would be unendurable.[44] Reflecting on recollapsing universes, Stephen W. Hawking agrees and confesses,

> At first, I believed that disorder would decrease when the universe recollapsed. This was because I thought that the universe had to return to a smooth and ordered state when it became small again. This would mean that the contracting phase would be like the time reverse of the expanding phase. People in the contracting phase would live their lives backwards: they would die before they were born and get younger as the universe contracted.[45]

After reading a brilliant article by Don Page titled "Will Entropy Decrease if the Universe Recollapses?"[46] Hawking decided that he had made a mistake. He came to believe that "disorder would in fact continue to increase during the contraction" and that "Conditions in the contracting phase would not be suitable for the existence of intelligent beings...."[47] On this view, in black holes and collapsing stars and universes, energy concentrates; but life-supporting order, available information, and ripple-free usable energy do not. In his work on radiating black holes, Hawking contends that the wave functions of all objects sucked into black holes, and thus all information about them, would be lost irretrievably. Entropy in this sense would consist in the loss of wave function probabilities and information about them. Yet, some astrophysicists are not fully convinced and argue that entropy would decrease (meaning presumably that both energy and order increase) in a collapsing universe,[48] and that information initially lost into a black hole might be recovered from the radiation it slowly emits.[49] This issue will arise again in later discussions.

D. Inferences from Einstein's Theory of Relativity

In 1905, Albert Einstein developed his *special* theory of relativity. It denied the reality of absolute Newtonian time and space, which were always and everywhere the same. It affirmed that space and time are inextricably united; no timeless space or spaceless time can exist. Their apparent constancy depends on the speed at which observers and their immediately surrounding spacetime frames are traveling. The speed of light is Einstein's only constant; observers always find it to be the same no matter where they are or how fast they are moving; but this implies the relativity of Newton's absolute constants, space and time.

In 1915 Einstein developed, and in 1917 he published, his *general* theory of relativity. It was a new theory of gravity and curved non-Newtonian space. Einstein regarded them as identical. General relativity affirms that space itself, and the path of light waves and other particles in terms of which we measure space, are curved or distorted in the presence of omnipresent gravitational fields. Light travels in curved paths in such fields. Space is more than a homogeneous empty form; it is something real in itself that can be concentrated, stretched, shrunk, bent, and straightened. The curvature of space just consists of gravitational fields that vary in intensity with variations in mass.

To Einstein's surprise and dismay, when combined with Hubble's observations of the redshift, his relativity field equations indicated that the universe is actually finite but expanding with unbounded potential, and that it had a beginning in time. When he realized this, Einstein was horrified! He mistakenly assumed that a finite universe must have an absolute center, and this was prohibited by his special theory of relativity. The Copernican Principle affirms that there are no privileged positions and observers in a relativity universe.

Einstein was both a theoretical physicist and a philosophical metaphysician. When he first discovered his general theory of relativity, he believed with Spinoza in a universe that is static, uniform, infinite, and eternal. Robert Jastrow tells us that, "When Einstein came to New York in 1921 a rabbi sent him a telegram asking, 'Do you believe in God?' and Einstein replied, 'I believe in Spinoza's God, who reveals himself in the orderly harmony of what exists.'"[50]

Most Theists have a suspicion that Spinoza's God is no God at all. Spinoza's God is Nature, and Nature is his God. Spinoza did not deify nature; he naturalized deity. To believe in Spinoza's God is to affirm that the universe is penetrable to reason, that determinism reigns inexorably, that "God does not play dice" (as Einstein put it), and that the world is spatially infinite, eternal, necessary, unchanging, and static. Initially, Einstein accepted this metaphysics, believing that reality is timeless and changeless, and that human distinctions between past, present, and future are illusions. Einstein once wrote that "Space and time are not conditions in which we live, but modes in which we think."[51]

Jastrow asks: "Why did Einstein object to the idea of a beginning?" and he answers:

> I think it is plausible that he did not believe in God the Creator. A beginning presupposes an agent that set in motion the events, which we call the explosion of the universe. That was anathema to Einstein. He believed in Spinoza's God, who created order and harmony in the universe and is revealed in equations like Einstein's relativity equation, but he did not believe in a personal God or God the Creator.[52]

Given his metaphysical inclinations, it is small wonder that Einstein reacted with powerful negative emotions when Willem de Sitter, a Dutch astronomer, showed him in correspondence during 1917 that his relativity equations have non-static solutions which imply an expanding universe that originated in the finite past. According to Jastrow, Einstein wrote back, "This circumstance irritates me" and "To admit such possibilities seems senseless."[53] Notice, says Jastrow, the emotionally loaded language! When Alexander Friedman, a Russian mathematician, proved that Einstein's rejection of an expanding universe was based on a mistake in calculation, Einstein first ignored him, then attempted to prove him wrong, and finally published a confession of his error.[54] After Edwin Hubble decisively confirmed the redshift of the galaxies, Einstein refused to accept an expanding universe until he traveled from Germany to California in 1931 to see for himself. Then theory yielded to facts.

Einstein finally realized that his preferred static but finite universe would collapse under the weight of its own gravity; so he postulated a "cosmological term" as an equal repulsive force that would prevent this collapse. Later, he confessed that this was the greatest mistake of his career. It prevented him from discovering the Big Bang.

What is the nature of this repulsive force that might so nicely counterbalance gravitational attraction? Einstein had no answer. Big Bang Cosmologists know that a repulsive force does operate in the universe–the kinetic energy left over from the primordial explosion; but whether there is an additional repulsive force, a Cosmological Constant, is still being debated. For billions of years, the repulsive force of energy released by the Big Bang has exceeded the cohesive force of gravity. It may or may not do so always. The ultimate destiny of the universe hangs in the balance. Stay tuned for breaking developments!

E. Nucleosynthesis of Hydrogen, Helium, Deuterium, and Heavier Elements

In contemporary cosmology, astronomy merges with sub-atomic particle physics. Astrophysicists now believe that the smallest things in the universe can tell us something significant about the universe as a whole, and *vice versa*. Research with particle accelerators discloses both the laws of physics and the conditions under which stable atomic elements were created out of sub-atomic particle soup in the very early universe. Alchemists' dreams of converting one element into another come true in nuclear physics, but only under extremely exotic conditions. Particle physicists recognize that some states and processes predicted by their theories were realized only under the extreme conditions of density and temperature of the Big Bang, and in the stars as nuclear furnaces.

Conditions were right in the very early universe for the nucleosynthesis of hydrogen into helium, but not for producing the heavier elements. Big Bang Cosmology predicts that the universe consists of about 75 percent hydrogen and 24 percent helium. Spectroscopic examination of countless stars and galaxies confirms this prediction, which is powerful evidence for Big Bang Cosmology.

The remaining heavier elements, quite abundant in our local terrestrial environment, constitute only about one percent of the stuff of the universe. In 1950, George Gamow argued that the heavy elements were created very shortly after the initial explosion as the universe expanded and cooled. However, research with nuclear reactions and "atom smashers" shows that conditions in the earliest universe could generate only the prevailing hydrogen and helium abundances. Also, if all of the radioactively unstable elements had originated in the Big Bang, they would have completely decayed by now; but they have not. The primordial fireball expanded and cooled too quickly to produce the heavier elements. Aeons of "stellar cooking time" were required to convert hydrogen and helium into heavier elements like oxygen, carbon, nitrogen, manganese, iron, and radon. The bodies of living things as we know them consist, in part, of elements synthesized only by nuclear reactions in the stars. The heavy elements were dispersed through the cosmos only when giant supernovas exploded. Their debris forms the planets of our solar system and others like them throughout the universe. This account of the origin and distribution of the heavy elements in and through the explosion of supernovas was dramatically confirmed in 1987 when, for the first time in the modern world, astronomers actually observed a supernova explosion in the relatively nearby Large Magellanic Cloud of stars.[55] Late in 1992, astronomers detected another supernova explosion almost five billion light years away. The Hubble Space Telescope has since located many more in far away galaxies, and astronomers use them as "standard candles" for measuring vast cosmic distances.

Stellar cooking could not and did not produce the quantities of deuterium (heavy hydrogen) and helium that exist in the universe. Only the extreme temperatures, densities, and pressures that existed during the earliest phases of the Big Bang can account for them. Stars generate some helium at a rate that accounts for only 2 percent of all the helium in the universe. The Big Bang is required for all the rest. The Big Bang directly generated hydrogen, helium, and traces of deuterium and lithium. Irregularities in the primordial fireball eventually separated vast clouds of gas, which gravity then condensed into quasars and stars; and the stars later cooked up the heavy elements. Everything observable came out of the Big Bang, one way or another.

F. The Cosmic Microwave Background

Today's astronomers can "see" the universe in many ways. They view it through powerful optical telescopes; they listen to its radio waves; and they detect its radiant x-rays, gamma rays, ultraviolet rays, and infrared rays. These types of radiant energy, coming to us from local and non-local regions of the universe, can now be converted to visual images. Radio waves furnish some of the most powerful evidence available for the Big Bang.

Prior to 1965, many scientific cosmologists, especially those attracted to Steady State Cosmology (to be examined later) still had doubts about Big Bang Cosmology. A remarkable achievement of modern astronomy occurred in 1965, providing crucial evidence for the Big Bang. In 1965, two relatively new Ph.D.s, Arno Penzias and Robert Wilson, worked for Bell Laboratories in New Jersey on projects in radio astronomy. They tried to eliminate all extraneous sources of radio static from their instruments so they could obtain the purest possible signals from outer space, but they found that some noises could not be eradicated. These noises resembled the static we have all heard between stations on AM radio bands and the familiar snow between existing television channels; in fact about one percent of the photons that cause snow on TV belong to the microwave background that Penzias and Wilson discovered. To eliminate this radio hiss, they even shooed away pigeons and cleaned their droppings from their antenna. Nothing worked. No matter what they did, their radio telescope picked up rumbling noises from every direction. They did not realize at first that they had stumbled upon the still enduring echoes of the Big Bang.

Earlier cosmologists had predicted that a microwave background continued to exist as a remnant of the primeval explosion that created the universe. Some actively searched for it. As early as 1947, George Gamow and his associates predicted its existence at around 5 degrees Kelvin, which was off by two degrees, yet quite good for its day; but these predictions were ignored and forgotten. In the early 1960s, scientists in Russia and the United States again predicted its existence and calculated its temperature to be about 3 degrees Kelvin, the currently accepted figure. At Princeton University, astrophysicists under the leadership of Robert Dicke and P. J. E. Peebles were building a radio telescope to search for this microwave background; but Penzias and Wilson found it before the Princeton group completed construction. After Penzias and Wilson contacted the Princeton group, Dicke and Peebles visited their radioscope to listen. They realized that Penzias and Wilson had found what they were looking for, a cosmic radio hiss coming from everywhere in space. Most cosmologists now agree that the microwave background consists of continuing reverberations from the Big Bang. In 1978, Penzias and Wilson received a Nobel Prize for their discovery.

G. The Dark Sky at Night

The Big Bang theory of the origin of the universe is accepted for its explanatory power. It accounts for a vast amount of data that no other theory can explain–the receding galaxies and their redshift, Hubble's law of uniform expansion, the ongoing dissipation of energy and order within the world, Einstein's relativity theorems, the origin of the chemical elements, and the cosmic microwave background. It even accounts for something very commonplace that has puzzled astronomers since at least the middle of the eighteenth century: it explains why the sky is dark at night. Most of us have never wondered; the sun goes down; and that explains it. But if we live in an infinite universe, there is a puzzle, formulated first by the German physician Heinrich Olbers in 1826, and thereafter called "Olbers' Paradox."

Olbers knew perfectly well that the sun sets at night, but he wondered why there is not enough starlight to make the night sky as bright as day, or even as bright as the sun, assuming as Olbers did under the influence of Sir Isaac Newton, that the rest of the static but endless universe is as richly populated by stars as our visible universe. Imagine being at the center of a sphere surrounded by all the stars we can see. If we double the size of that visible sphere, as we might do with a more powerful telescope, stars will be only a fourth as bright, as predicted by Newton's inverse square law; but there will be four times as many of them to negate this effect. This doubling process could and should go on to infinity in an infinite universe until it encompasses an infinite number of stars. Light from each of these has had an infinite amount of time to reach us, so a star should shine in every last niche in the heavens, and the night sky should be as bright as the sun. Yet it is not. Why? Olbers thought that a gaseous medium filled the sky and blocked out most of the starlight; but this will not work. Given enough time, as Hermann Bondi showed, these gases would heat up and eventually radiate as much energy as they absorb.[56] And infinity is enough time!

Many attempts have been made to resolve Olbers' Paradox, but the most successful is provided by the Big Bang. Except for a few discussed in Chapter

Three, today's cosmologists accept the Big Bang and agree that it provides the ultimate solution to Olbers' Paradox. Big Bang theory says that neither time nor space are infinite. Finite spacetime can contain only a finite number of stars that shine only for a finite amount of time. Stars come and go; they are born from cosmic debris, burn for a few billion years, exhaust their nuclear fuel, and die. The spatiotemporal finitude of the universe explains why most of the sky is dark at night and counts also as powerful evidence against the naturalistic claim that our universe is infinite in both space and time.

Additional evidence for the Big Bang origin of our universe will be presented in later chapters. To anticipate a bit, the *hydrogen cycle*, the gradual conversion of hydrogen into the heavier elements in supernova nucleosynthesis, could not have been going on forever, or there would be no hydrogen left in the universe; but there is. The same is true for all radioactive elements; if they have been losing electrons through nuclear decay for an infinite amount of time, no radioactive elements would remain today; but they do. The ripples in the microwave background discovered by the COBE satellite confirm the basic Big Bang scenario for galaxy and supergalaxy formation.

We have now reviewed the primary scientific evidence for Big Bang Cosmology; in Chapter Three we will consider putative evidence against it. If Big Bang Cosmology does not survive under critical scrutiny, there is no point in exploring our central philosophical question: What caused the Big Bang? Can natural science answer this question? Cosmological Agnosticism says that it cannot. Naturalistic and theistic metaphysicians give philosophical and religious answers to our question of ultimate causes, but there may be no scientific answer.

3. Scientific Cosmological Agnosticism

Scientists now advance a variety of explanations for the origin and evolution of the universe; but agnosticism says that we really do not know the answers. General Cosmological Agnosticism does not deny that many cosmological puzzles can be resolved. It focuses on the central question of this book: What caused the Big Bang?; and it says that we do not know. We know only what happened after the Big Bang was inaugurated, and science can not tell us what caused the Big Bang. Coming chapters will examine naturalistic theories that purport to offer scientific answers; but they actually give only highly speculative and dubious philosophical accounts of the ultimate origin of the universe, without acknowledging their subtle shift from science to unverifiable metaphysics. Scientific Agnostics claim that a plausible scientific account of the cause of the Big Bang is not and never will be available.

Agnosticism in cosmology may be either scientific or philosophical. Scientific Agnosticism says that the cause of the Big Bang lies beyond the limits of scientific methodology and knowledge. Philosophical Agnosticism extends this outlook. It says that the cause of the Big Bang lies entirely beyond the limits of philosophical knowledge, beyond all human reason, broadly understood. The remainder of this book will accept Scientific Cosmological Agnosticism, but total Philosophical Agnosticism will be rejected. Standard Big Bang Cosmology is inherently agnostic because it recognizes that scientific methodology and knowledge break down completely at or near the onset of the Big Bang. It makes no attempt to go behind the initial singularity or the earliest Planck dimensions.

If an initial singularity existed at T = 0, there were no laws of nature, no space, no time, and no physical causation. Cosmologists disagree about whether the breakdown of the laws of nature occurs at an initial singularity or at Planck dimensions; but most agree that they become inapplicable at some point. We can extrapolate scientifically all the way back to where these laws break down completely; but we cannot go beyond that because we must appeal to these laws just to get that far. Observable processes in nature like the redshift of the galaxies, the Hubble expansion, nucleosynthesis of the elements, the microwave background, and the dark night sky, all imply a Big Bang origin for the universe. Extrapolating from known laws of nature, these processes can be traced backward in time; but calculations result in zero time and space and infinite energy after eight to twenty billion years. Scientific Cosmological Agnostics insist that science itself cannot take us beyond and before that to a space beyond space, and a time before time began.

The fact that a world-creating Big Bang erupted billions of years ago is scientifically well established. The laws of nature can take us back to an initial singularity, but they can take us no further for at that point all physical laws break down completely and become inapplicable. The zero spacetime and infinite energy of the initial singularity indicate methodological failure as well as a cosmological beginning. Science cannot transcend its own limitations. With no laws of nature, spacetime, physical causation, or empirical data to work with, no further scientific extrapolations are possible. At an initial singularity, the ultimate limits of scientific knowledge are reached, the Scientific Cosmological Agnostic insists; and What caused the Big Bang? has no scientific answer. We cannot know what would make an initial singularity explode because it lies beyond the limits of all known laws of physics.

If a collapsing universe could shrink below Planck dimensions, the same agnosticism follows if all laws of nature become inapplicable at or below the Planck Wall. Later we will see that Quantum Cosmologists deny that anything can exist on the nether side of this Wall; if a sub-Planck collapsed universe existed, we could have no scientific knowledge of it because physical laws and causation appear for the first time at Planck dimensions. Many cosmologists are convinced that scientific knowledge terminates at the Planck Wall, that the Planck Wall is a blank wall. If the laws of nature only become applicable when the universe was 10^{-43} seconds old and 10^{-33} centimeters in diameter, then these dimensions represent the ultimate limits of scientific knowledge. What caused the appearance fifteen billion years ago of a universe with minimal Planck numbers? What caused the Big Bang? There are no scientific answers if, looking backward, scientific knowledge ends before these questions arise.

Theologians and philosophers may address the question of ultimate origins; but science has no answer, Scientific Cosmological Agnostics insists. Those scientists and naturalistic metaphysicians who affirm on allegedly scientific grounds that "Nature is eternal" must recant. Even "Every event has a natural cause" must be relinquished, for science ultimately reaches back to a point of origin beyond which no natural physical causes can be found. Natural causes are those that operate within our system of spacetime; but scientists generally agree that space, time, natural laws and physical causation began with the Big Bang and did not exist prior to either Plank dimensions or an initial singularity. Numerous prominent cosmologists and scientists today are Scientific Agnostics on the question of the cause of the Big Bang.[57] In 1980, Robert Jastrow began his essay on "Science and the Creation," with these words:

> I should like to say at the start that I am an agnostic in religious matters. I am fascinated, however, by some recent developments in astronomy–partly because of their religious implications and partly because of the peculiar reactions of my colleagues. In a nutshell, the astronomers, studying the universe through their telescopes, have been forced to the conclusion that the world had a beginning. Scientists have always felt more comfortable with the idea of a universe that has existed forever because their thinking is permeated with the idea of cause and effect; they believe that every event that takes place in the world can be explained in a rational way as the consequence of some previous event. If there is a religion in science, this statement can be regarded as its main article of faith. But the latest astronomical results indicate that at some point in the past the chain of cause and effect terminates abruptly. An important event occurred–the origin of the world–for which there is no known cause or explanation within the realm of science.[58]

Jastrow, convinced that science does not know what caused the Big Bang, insists that "We cannot find out what caused that beginning" because the scientist

> comes to a blank wall where the big bang occurred. The cosmic explosion, the birth of the universe, is an effect for which he cannot find the cause. Some might say that if he cannot find it today, he will find it tomorrow; and we will read about it in the *New York Times* when Walter Sullivan gets around to it. This, however, is one finding in science that seems likely never to succumb to scientific investigation because in the first moments of the universe's existence the temperature and the pressure were infinitely

> high, which means that all relics of a pre-creation universe that might have given us a clue to some natural forces that conspired to bring about the explosive moment we call the big bang are gone. All of the evidence the scientist could examine to explain this cosmic holocaust has been melted down and destroyed in the fiery heat of that first moment itself. That is why it seems to me and a few other people that this is a blank wall, a curtain covering the mystery of creation never to be raised by human minds, at least in the foreseeable future.
>
> This brings us to a very interesting pass. The world has come into being as a product of forces that are today, and very likely forever, outside the reach of scientific inquiry. These forces do not fit into the present body of natural forces–gravity, electricity, nuclear forces–and, being outside the realm of nature, as the scientists understand it, they must therefore, properly be termed supernatural. In this statement cosmologist and astronomer finally come face to face with the theologian, who has always thought that what one might call a supernatural force, a creative force, has been responsible for the origin of this world.[59]

If our system of nature is the only cosmos, the whole of physical reality, and if it was brought into being by causes that transcend nature, these causes must be classified as "supernatural." Were the supernatural causes of the Big Bang Divine? Not necessarily. In the next few chapters, we will consider metaphysical Naturalism and a variety of prestigious but pseudoscientific naturalistic cosmologists who contend that the transcendent causes of our universe were not Divine. Yet, if Scientific Cosmological Agnosticism is correct, their own metaphysics is no more verifiable or scientific than that of the theologians.

Standard Big Bang Cosmology affirms that scientific knowledge breaks down at the beginning of *our* cosmos, so there can be no scientific knowledge of what caused the Big Bang. The question must be turned over to philosophers and theologians. Can they do any better? In particular, can philosophical Naturalism or Theism do any better? We shall see in the following pages. In his 1978 book, *God and the Astronomers*, Robert Jastrow first takes us on a scientific journey through the evolution of the universe all the way back to the primordial cosmic fireball. Then, speaking from the vantage point of Standard Big Bang Cosmology, he identifies the final step that science takes in its cosmological inquiry into origins.

> Now we would like to pursue that inquiry farther back in time, but the barrier to further progress seems insurmountable. It is not a matter of another year, another decade of work, another measurement, or another theory; at this moment it seems as though science will never be able to raise the curtain on the mystery of creation. For the scientist who has lived by his faith in the power of reason, the story ends like a bad dream. He has

> scaled the mountains of ignorance; he is about to conquer the highest peak; as he pulls himself over the final rock, he is greeted by a band of theologians who have been sitting there for centuries.[60]

Jastrow first wrote his book before the 1980s, when inflation and quantum physics were added to the Big Bang theory of origins. Chapters Four through Eight will consider more carefully whether Quantum Cosmologies can overcome the apparent limits of scientific knowledge by introducing quantum laws and effects at the beginning of our cosmic epoch.

In sum, Standard Big Bang Cosmology is agnostic on the question of ultimate cosmic or supercosmic causes. As a scientific theory, it deals with the evolution of the universe from either the initial singularity, or from Planck time, to today. It incorporates all empirical evidence that supports the theory; but it offers no scientific answer to: What caused the Big Bang? It turns this question over to agnostics, philosophers, and theologians. Can they do any better? We will see. Robert Jastrow suggests that this is "the most interesting question of all" about which to speculate.[61] But it is speculation, not natural science.

Two

HUMANISTIC NATURALISM

Humanistic Naturalists should be strongly inclined to reject Big Bang Cosmology. They should be horrified by its development, for they are committed to a philosophical outlook which appears, at least at first, to be completely refuted by Big Bang theory. Humanistic Naturalism had its heyday during the early and mid twentieth century, but the view is as old as some of the ancient Greek and Roman philosophers. Prominent twentieth century philosophers who are identified by themselves or others as Humanistic Naturalists were George Santayana, John Dewey, Morris Cohen, Sterling Lamprecht, Roy W. Sellars, John H. Randall, Jr., Sidney Hook, Ernest Nagel, Corliss Lamont, Bertrand Russell, Samuel Alexander, J. B. Pratt, William P. Montague, Paul Kurtz, Kai Nielsen, Daniel C. Dennett, and many others. Among recent naturalists, Nielsen mentions A. J. Ayer, C. I. Lewis, W. V. O. Quine, Donald Davidson, Richard Rorty, Hilary Putnam, P. F. Strawson, Donald Davidson, David Armstrong, and J. J. C. Smart.[1] Many prominent scientists like Carl Sagan are or have been Humanistic Naturalists, and we will examine some of their positions in later chapters.

Members of this philosophical family tend to share a common metaphysical, methodological, ethical, and anthropological outlook, though they do not completely agree with or perfectly resemble one another in every respect. Humanistic Naturalists subscribe to most if not all of the following philosophical doctrines; but individual Naturalists may reject, de-emphasize, or ignore a few of these family traits. As less and less of these traits are affirmed, the legitimacy of calling a position "Naturalism" becomes more and more doubtful. Because most natural scientists regard themselves as Naturalists, David Griffin tries to reconcile science and religion with what he calls a "naturalistic theism," by dropping almost everything Naturalists have ever meant by the term. His "minimal naturalism" retains only metaphysical trait D below, and he modifies it significantly by making divine causation a regular part of all natural causation.[2] Naturalists are likely to regard this as a purely verbal victory, but Griffin also launches more fundamental and substantive attacks on Naturalism's "scientific" status. So will the following pages, even with respect to D below.

Humanistic Naturalists tend to believe:

A. Only nature exists; the supernatural does not exist.
B. Nature as a whole has no purposes, values, or traits of personality.
C. The most general features of nature like time, space, and the basic physical stuff within them exist infinitely, eternally, and necessarily.
D. All events have natural causes; there are no supernatural causes.
E. Scientific method is the only legitimate method for discovering truth.
F. A Humanistic ethics and "philosophy of man" are adequate.[3]

The first four of these claims are metaphysical, and the fifth is methodological. By "metaphysical" claims, I mean those pertaining to the most universal or fundamental features of reality, the traditional meaning of the term. Unlike Kai Nielsen, who calls only *a priori* versions of such claims "metaphysical,"[4] I recognize both a priori and empirical approaches to such claims. Naturalists do not avoid metaphysics just because they profess to be empiricists. Humanistic Naturalists try to combine the fifth methodological claim with the sixth ethical and anthropological thesis. The first five of these have the most obvious importance and direct relevance to Big Bang Cosmology.

1. Family Traits of Humanistic Naturalism

Historically, the philosophical outlook of Humanistic Naturalism was developed expressly as an alternative to Theistic Supernaturalism, which takes just the opposite position on every point. Consider first how Naturalists themselves have expressed their fundamental beliefs.

A. Nature as All Existence

Humanism believes that nature or the universe makes up the totality of existence and is completely self-operating according to natural law, with no need for a God or gods to keep it functioning. Corliss Lamont[5]

Nature in which all interactions exist. John Dewey[6]

We find insufficient evidence for belief in the existence of a supernatural; it is either meaningless or irrelevant to the question of the survival and fulfillment of the human race. As nontheists, we begin with humans not God, nature not deity. Nature may indeed be broader and deeper than we now know; any new discoveries, however, will but enlarge our knowledge of the natural. Humanist Manifesto II[7]

What, then, are the controlling principles of naturalism? Essentially those of science: the beliefs that nature is an all-inclusive, spatiotemporal system and that everything which exists and acts in it is a part of this system. Roy W. Sellars[8]

The Cosmos is all that is or ever was or ever will be. Carl Sagan[9]

This first humanist principle, the rejection of the supernatural worldview, is shared with materialism and naturalism. Paul Kurtz[10]

> *There is nothing beyond nature. There is no supernatural reality, spiritual beings, or any purely mental realities.* Kai Nielsen[11]

The claim that "Only nature exists; the supernatural does not exist" is essential to being a Naturalist. Doubters about this may be skeptics, agnostics, or positivists; but they are not Naturalists. This claim invites the question: What is nature? Sometimes "nature" is conceived so broadly that it covers the whole of reality, in which case a real God who transcends our world would be an object in nature. "Nature" or "the universe" usually refers to our system of spacetime, but Frank J. Tipler defines "the universe"as "all that exists." Without the additional premise that our system of space time is all that exists, this definition implies that an existing transcendent God belongs to the universe. Tipler, for instance, insists that God is a natural entity and that theology is a branch of physics.[12] He actually wants to naturalize God and treat God as purely immanent in and ultimately produced by spacetime as we know it–in conjunction with infinitely many other spacetime universes that actualize all possibilities.

Philosophical Naturalists deliberately use "nature" in a more limited way to exclude even an immanent God, to say nothing of Heaven, Hell, Angels, and all other-worldly entities. Nature is all; nothing more exists. For atheistic Naturalists, especially in their debates with Theists, "nature" denotes this world, the visible universe in its totality; there is no other world; and no other-worldly entities are real. Nature, the cosmos, the totality of our public spatiotemporal universe, is the only reality.

The creative, transcendent, and eternal God of traditional western Theism supposedly caused nature, the totality of spacetime, to come into being. By definition, supernatural entities can only exist outside of and before our system of spacetime; but no such beings exist, Naturalists insist. We can only speak metaphorically at best, or unintelligibly at worst, they contend, of their existence, and no reliable scientific evidence supports belief in supernatural entities. "Before time" is a temporal metaphor; and "outside space" is a spatial metaphor; but these metaphors have no literal or intelligible extensional meaning or reference. Scientific method, they contend, does not and cannot verify the existence of other worlds or other-worldly entities, so nothing warrants belief in their existence.

B. Nature as Purposeless

> *Our world has been made by nature through the spontaneous and casual collision and the multifarious accidental, random and purposeless congregation and coalescence of atoms.* Lucretius[13]

> *[Naturalism] excludes cosmic purpose, a meaningful totality, and any variation of the Platonic form of the good.* Roy W. Sellars[14]

This cosmos, unbounded in space and infinite in time, consists fundamentally of a constantly changing system of matter and energy, and is neutral in regard to man's well-being and values. Corliss Lamont[15]

To a naturalist, evidence for purpose, needs, organization, and ends in nature, is discovered in the behavior of specific things and organisms. No reference to the purpose of the whole is empirically relevant to the purposes he discovers by natural observation and experiment. Sidney Hook[16]

Humanism asserts that the nature of the universe depicted by modern science makes unacceptable any supernatural or cosmic guarantees of human values. Humanist Manifesto I[17]

Nature for the humanist is blind to human purposes and indifferent to human ideals. Paul Kurtz[18]

Most if not all Naturalists insist that "Nature as a whole has no purposes, values, or traits of personality." They hold that no valuational, personal, or psychological attributes apply directly to nature as a whole; and nature does not indirectly express the purposes or personal will of either a God who transcends the world or a God who is immanent in the world. Impersonal nature has no values, pursues no goals, makes no judgments about good and evil or right and wrong, has no aims or intentions, does not care what happens, takes no attitudes towards anything, whether favorable or unfavorable, thinks no thoughts, knows not what it does, has no awareness or consciousness of its own, and does not consciously and purposefully try to do what it does or try to achieve anything at all. All personal, psychological, or "anthropomorphic" attributes must be excluded from our thinking about nature as a whole, no matter how appropriate these categories are for thinking about local earthly organisms within nature like animals and human beings, and no matter how impressive and powerful the creative natural forces are that bring living things into being. Cosmic-level teleology has no reality.

C. Nature as Infinite, Eternal, and Necessary

If we ask whence came matter, we say it has existed always. If we be asked whence came motion in matter, we answer that, for the same reason, it must have been in motion from all eternity.... These elements...are sufficient to explain the formation of all the beings that we see. Mirabaud[19]

The law of the conservation of energy includes in its operation an unceasing transformation of one form of energy to another, so that the basic energy, but none of its individual manifestations, is eternal. Corliss Lamont[20]

The ultimate elements of the body, as the Law of the Conservation of Mass implies, have always existed in some form or other and will go on existing forever. The indestructible matter that makes up our physical organism was part of the universe five billion years ago and will still be part of it five billion years hence. The infinite past comes to a focus in our intricately structured bodies; and from them there radiates the infinite future. Corliss Lamont[21]

Nature stands on its own feet and explains itself. Roy W. Sellars[22]

Why assume an absolute beginning for reality? If change is an event in nature, may not both change and nature always have been?....Neither science nor philosophy, then, assume any absolute beginning for reality. Roy W. Sellars[23]

Nature is ontologically ultimate and self-sufficient. Roy W. Sellers[24]

Religious humanists regard the universe as self-existing and not created. Humanist Manifesto I[25]

Undoubtedly, our knowledge of the universe is meager, given the vast infinity of space and events. Paul Kurtz[26]

Naturalists believe that "The most general attributes of nature like time, space, and the basic physical stuff within them exist infinitely, eternally, and necessarily." The fundamentals of the natural order of things, whatever they are, are uncreated, everlasting, indestructible, and self-sufficient in their being. Space and time are fundamentals that every naturalistic philosophy affirms, but considerable room must be allowed for Naturalists to disagree about the constitution of the basic stuff of the world. Thales, the first philosopher, thought that it is water and that all things in the universe are composed of transformations of water. The most sophisticated Naturalistic metaphysics developed in the ancient world was that of the Greek and Roman atomists, Democritus and Leucippus, according to whom nature is composed entirely and only of uncreated and indestructible atoms swimming everlastingly in an uncreated infinite void of empty space. This atomistic view is no longer tenable because modern nuclear physics finds no uncreated and indestructible atoms or sub-atomic particles. Contemporary views of the basic stuff of nature identity it with energy as such.

Naturalists ascribe the same metaphysical attributes to nature as a whole that traditional Theists ascribe to God. They argue that the rational ideal of simplicity requires that we think of nature, not God, as infinite, eternal or everlasting, necessary, self-sufficient, and uncreated reality. Theists think that it is appropriate to ask: Who created the world? but it is not appropriate to ask: Who created God? Divinity, in its very conception, is the uncreated creator of all else. Metaphysically, Naturalists move to pinpoint necessary being before getting to Divinity. Nature, not God, is the infinite, everlasting, necessary, self-existing, uncreated creator of all. So who created nature? No one, for nature exists necessarily; nature itself is everlasting, self-sufficient, and uncreated; and nothing that transcends nature is required to explain it.

D. Nature Causes Everything

> *Natural processes (including those of human living) do not imply anything beyond themselves and do not require for their explanation any grounds but the further stretches of natural processes, which we observe or inductively infer to be their context.* William R. Dennes[27]

> *The occurrence of all qualities or events depends on the organization of a material system in space-time, and...their emergence, development and disappearance are determined by changes in such organization.* Sidney Hook[28]

> *And naturalism is the metaphysical theory which maintains that everything that exists comes into being, endures for a time, and then passes away because of the interactions of the things and forces of the natural world.* Sterling Lamprecht[29]

> *The universe as a whole has no cause, since by definition, there is no thing outside it that could be its cause.* Hans Reichenbach[30]

> *[Secular humanists] consider the universe to be a dynamic scene of natural forces that are most effectively understood by scientific inquiry.* Paul Kurtz[31]

Naturalists claim that "All events have natural causes; there are no supernatural causes." With this metaphysical causal principle, Naturalists rule out the creation of the universe *ex nihilo* by God, for that makes God the supernatural cause of all creation. All miracles and direct acts of God upon the world are also excluded because miracles, by definition, are temporary suspensions of the laws and causal processes of nature, accompanied by causal interventions by a Divine Being who transcends nature.

The naturalistic assumption that "All events have natural causes" should not be confused with the more general causal principle that "All events have causes." About the second, Naturalists and Theists are in complete agreement. Since supernatural causes are indeed causes, Theists wholeheartedly agree that all events have causes, including the creation of the universe, miracles, and acts of God. Naturalists insist that the general causal principle is insufficient without the qualification "natural." "All events have natural causes" means that every happening in spacetime is brought into being by some other happening or set of conditions or happenings within our spacetime system, which as a whole has no cause beyond itself. Nothing comes to be through spatiotemporally transcendent causes. No acts of God or any other realities transcend our system of spacetime.

Descartes and many later rationalists regarded the metaphysical claim that "All events have causes" as one of many self-evident truths of reason. This truth needs no empirical confirmation, though it is confirmed in every experience. To understand it is to be certain that it is true. Most philosophers today think that Descartes confused logical certainty with psychological certitude; they reject all rationalistic synthetic or substantive axioms of knowledge. But Naturalists need not be epistemological rationalists; they can and do advance their principle of universal natural causation as a broad generalization from experience.

Supposedly, Cartesian self-evident truths are composed entirely of clear and distinct ideas; but the idea of causation is anything but clear and distinct, whether we are Cartesian rationalists or not. The concept of "cause" greatly needs clarification. To clarify we must distinguish between conditions that are necessary and those that are sufficient for bringing about events. Necessary conditions are those in the absence of which events cannot occur, and sufficient conditions are those in the presence of which specific events must occur. Causes are either necessary or sufficient conditions, or both.

Naturalists may or may not be determinists, may or may not believe in free will, despite their conviction that all events have natural causes. "Cause" in this formula can mean necessary conditions, or it can mean sufficient conditions, or both. Determinists construe the principle of universal causation to mean that antecedent conditions are completely necessary and sufficient to explain the occurrence of all events, including human choices. Since necessary and sufficient conditions exist for absolutely everything, they believe, only those events and choices that actually come to be could ever occur; and no other events and choices were or are really possible. Destiny, like some other things, just happens! What is actual is all that could have transpired. All events are rigidly determined or necessitated. Many determinists hold that the strongest desire or set of cooperating desires is sufficient to explain every choice. Naturalists may or may not be determinists, depending on whether they interpret the principle of universal causation in the strong sense just explained or give it the weaker interpretation explained next. Deterministic Naturalists may or may not be reductive materialists who think that stupid matter causes everything. Many non-reductive alternatives are open to Naturalists, and so is belief in free will, depending in part on how seriously they take their commitment to sense experience alone as integral to and admissible by glorious scientific method, which alone yields truth, they say. If they (inconsistently?) allow a place for non-sensory modes of experience like introspection, perhaps they can avoid psychological behaviorism and mechanistic determinism.

Naturalists may hold that universal causation means that only necessary but not sufficient conditions exist for at least some choices–those that are free. They (and non-naturalists, too, who affirm free will) may claim that active or effortful choices occur only when character and motivation are still being developed, only when no desires, inclinations, or habits clearly prevail. When some desires or disposing mental states are decisively strongest, we simply act on them without making creative, effortful choices. Free choices function to create strong desires where none previously exist. Choices add the weight of effortful attention to selected alternatives when no conditions are sufficient for appetitive dominance. Of course, some conditions are necessary for every choice. We cannot consciously and effortfully choose an alternative unless we are aware of it and have some desire for it or attraction to it; but neither this desire nor its cooperating motivational determinants needs to be stronger than all others. Determinism may be avoided by subscribing to this weak interpretation of the principle that every event has a natural cause. Theists too may avoid determinism by holding that all free human choices have necessary but not sufficient causal conditions–usually natural.

The distinction between necessary and sufficient conditions will come up later in discussions of the Heisenberg Uncertainty Principle, according to which no absolute knowledge or predictability exists at the level of atomic or sub-atomic particles. At the level of quantum events, we know only necessary but not sufficient conditions for what occurs. At the atomic and sub-atomic levels, no ace predictor like God could ever know everything that might happen, because conditions in nature are not sufficiently definite to ground such knowledge. Even so, total chaos does not reign in the domains of elemental physical particles, or of free moral agents; and we are not absolutely ignorant of what is going on, or absolutely incapable of predicting the future within limits. Degrees of order and disorder are found both in human complexity and in sub-atomic simplicity; about this, Naturalists and Theists can agree.

E. Scientific Method Alone

> *Whatever knowledge is attainable, must be attained by scientific methods; and what science cannot discover, mankind cannot know.* Bertrand Russell[32]

> *Naturalism...wholeheartedly accepts scientific methods as the only reliable way of reaching truths about man, society, and nature....* Sidney Hook[33]

The mind of man is being habituated to a new method and ideal: There is but one sure road of access to truth–the road of patient, cooperative inquiry operating by means of observation, experiment, record and controlled reflection....There is but one method for ascertaining fact and truth–that conveyed by the word "scientific" in its most general and generous sense. John Dewey[34]

In short, Naturalism is the expression of the desire for explanation in terms of objects which can be handled and studied in accordance with scientific method. Roy W. Sellars[35]

Any account of nature should pass the tests of scientific evidence; in our judgment, the dogmas and myths of traditional religions do not do so. Humanist Manifesto II[36]

There seems nothing in Humanism or human life that is inaccessible in principle to scientific methods of inquiry, either in primitive or highly elaborated forms. Stuart C. Dodd[37]

Naturalism is committed to certain methodological principles, primarily scientific and empirical methods, as the most effective way to arrive at reliable knowledge. Thus, to be warranted, a descriptive belief (1) must be experimentally verified; (2) must be logically consistent, internally with itself and externally with our other beliefs; and (3) may be judged convenient in part by its role in inquiry and its relation to the situations in which it arises. Paul Kurtz[38]

If we want the best answers to what things there are, it is to science that we should turn....It is science which yields our most reliable knowledge. Kai Nielsen[39]

There is only one way of knowing, the empirical way that is the basis of science....From a naturalistic perspective, we should deny that there is any a priori knowledge. Michael Devitt[40]

Naturalists contend that "Scientific method is the only legitimate method for discovering truth." This methodological claim must be understood both for what it excludes and what it embraces. It affirms that many commonplace methods for fixing belief and bringing about social agreement are totally unreliable and unacceptable, including rationalistic appeals to a priori or self-evident truths, and religious appeals to faith and divine revelation. Competing philosophical beliefs seem self-evident to different persons in diverse times and places; and by blind faith or revelation, we can have either side of any issue that we happen to want. Reliance on a blind faith response to religious revelation alone leaves us completely vulnerable to any and every superstition that comes along. Religious persons who appeal only to faith and revelation have no methods for resolving profound and interminable disagreements among themselves. Only scientific methods acquaint us with reality, correct their own mistakes, and can in principle resolve disagreement rationally.

Naturalists identify rationality itself with scientific method; but what is scientific method? Some contemporary philosophers like Richard Rorty[41] and Paul Feyerabend[42] maintain that there really is no such thing as scientific method; in doing their work and learning about the world, scientists really do an incredible variety of unpredictable, unformalizable, non-algorithmic things. According to Feyerabend, careful scrutiny of what scientists at work actually do shows that in science "anything goes," and nothing should or could replace this methodological anarchy.

By contrast, Humanistic Naturalists are convinced that there is such a thing as scientific method, and it alone can give us truth. In their disputes with theologians, Naturalists employ an extremely narrow understanding of scientific method to "prove" that religious belief is unscientific and unfounded. When refuting theologians, scientific method is understood to consist primarily if not entirely in empirical (sensory) verification and/or falsification of descriptive statements, in making inductive inferences, and in advancing and testing empirical explanatory hypotheses. Logical Positivists also emphasize empirical verification and falsification; but they insist that all metaphysical beliefs are meaningless. Unlike them, Naturalists regard theological beliefs as meaningful but false and their own metaphysical beliefs as meaningful, verified, and true. Naturalists may at times succumb to the allure of Positivism.

In their most polemical anti-theistic moods, Naturalists can be quite insensitive to difficulties that plague the methods of the natural sciences, such as that: (1) "factual" beliefs are inescapably theory-laden and are often constituted in part by, and/or derived from, formal mathematical and logical systems rather than from experience;[43] (2) how appearances or sensory observations are connected to realities is very uncertain (as illustrated in Chapter Six by the realism/idealism controversy); (3) scientific progress depends upon creative insights, not just upon collecting facts; (4) scientific disputes are often resolved, not by observation, but by appeals to aesthetic criteria like simplicity, harmony, beauty, symmetry, elegance, and intuition; and (5) scientific language, not just religious language, is often inescapably metaphorical.[44] Let us indulge Naturalists for the moment and try to understand their contention that the methods of science exclude religious beliefs.

Naturalists contend that beliefs in God, Heaven, Hell, Angels, and other-worldly entities are groundless because scientific methods do not and cannot directly disclose or indirectly and inductively reason to their existence, and all happenings supposedly explained by other-worldly causes can be better explained by appeal to this-worldly causes. Scientific explanations are closer to experience, simpler, and more elegant than religious explanations. Religious affirmations about other-worldly entities or occurrences cannot be verified directly, cannot be inferred as inductive generalizations or inferences from experience, and cannot be justified as explanatory hypotheses, Naturalists insist.

To Naturalists, belief in God is like belief in Santa Claus. We can actually go to the North Pole and search for Santa. When we do, we find neither the residence, the workshops, the toiling elves, the flying reindeer, nor the persons of Mr. and Mrs. Claus. Likewise, we can search the world over without finding God or any other-worldly entities. Observable worldly entities have properties that are incompatible with and exclude the properties of other-worldly entities, so we cannot generalize inductively from this world to the next. Thus, we have no more reason for believing in God than for believing in Santa, invisible elves, or angels dancing on the heads of pins.

Furthermore, the phenomena that the Santa Claus hypothesis is supposed to explain–the presents under the tree, the noises in the night, the missing milk and cookies on Christmas morning, and so on, are much more plausibly, simply, and verifiably explained: Daddy and Mommy do it. Likewise, whatever happens in the world always has a more plausible, simpler and (in principle) verifiable explanation: Natural causes do it (*ad infinitum*). Neither Santa Claus nor God is required to account for anything that we ever experience.

When Naturalists claim that scientific method cannot be used to establish religious truths, they usually mean that objects of religious interest and belief like God and Heaven can be neither directly perceived, which is obviously true, nor inferred inductively, since induction gives us only more of the same and thus cannot justify belief in transcendent realities. They usually do not ask whether the hypothetico-deductive method, so essential to theoretical science, has any sound religious uses. This aspect of scientific method is very complex. It involves (1) creatively constructing hypotheses and theories, (2) making deductions and predictions from them about what might be observed to verify or falsify them, (3) performing experiments and making additional observations that actually confirm or disconfirm them, (4) excluding alternative hypotheses, (5) appealing to abstract rational/aesthetic criteria like consistency with other scientific beliefs, simplicity, elegance, fruitfulness for further research, practical usefulness, and coherence or conceptual interconnectedness with other scientific concepts and theories, (6) modifying hypotheses to take care of anomalies, (7) occasionally abandoning generally accepted paradigms when the anomalies are too overwhelming, and (8) making creative gestalt switches, revisions, and revolutions. Scientific methodology is not value free but involves being committed (often quite passionately) to the values of truth (empirical alone?), knowledge, honesty, scientific subject-matter, scientific methodology, and objectivity–which is not uninterestedness but is disinterested willingness to play fairly with ideas, to follow out the logic of a position, and to change our minds when warranted.

F. Humanism

> *We need to extend the uses of scientific method, not renounce them, to fuse reason with compassion in order to build constructive social and moral values. Confronted by many possible futures, we must decide which to pursue. The ultimate goal should be the fulfillment of the potential for growth in each human personality–not for the favored few, but for all of humankind.* Humanist Manifesto II[45]

> *What these modern ways of life have in common is a devotion to the this-worldly welfare of men. The most enlightened of them, such as Humanism, Materialism and Naturalism set up the happiness, freedom and progress of all humanity as the supreme goal. This ultimate loyalty to the ultimate interests of all mankind, including one's own finest possibilities, is, I would suggest, a thing high enough and broad enough for any man to integrate his life around.* Corliss Lamont[46]

> *Most contemporary humanists have a commitment to some form of the greatest-happiness-for-the-greatest-number principle; they consider that the highest moral obligation is to humanity as a whole. This involves the view that since all men are members of the same human family, it is our obligation to further the welfare of mankind.* Paul Kurtz[47]

Most Naturalists believe that "A humanistic ethics and philosophy of man or humanity are adequate." This book is primarily concerned with metaphysical and methodological issues, but the usual association of naturalistic metaphysics with scientific method alone and with humanistic ethics and anthropology cannot be ignored.

Not all Naturalists are Humanists. Naturalistic metaphysics and scientific methodology have no obvious logical connection with humanistic ethics. Frederick Nietzsche combined naturalistic metaphysics with a non-humanistic might-makes-right ethics, according to which strong, fit, master-race supermen are fully justified in exploiting and destroying weak human beings. The Nazis, who greatly admired Nietzsche, showed us how this thoroughly naturalistic but anti-humanistic ideal works out in practice. Nothing in naturalistic metaphysics or methodology generates or logically implies Humanism. Logic permits Naturalists to be racists, nationalists, egoists, or anything but Humanists.

As a matter of brute fact, most Naturalists have been Humanists who affirm universal human rights and human equality, and who sponsor moral beliefs, attitudes, and behaviors designed to bring about a better and happier life for humankind in this world. There is no other life, they contend; belief in another world merely siphons off energies that might otherwise be devoted to human welfare here and now–the only place where moral effort ever counts.

Humanists tend to believe in a technological fix for almost everything,[48] but they are not irrevocably wedded to this. They do insist that human values, moral principles, and virtues are completely human in origin and neither have nor need a Divine origin or sponsorship. Theistic religion is not essential for moral motivation; many religious persons are very immoral, and many atheists are extremely virtuous.

Naturalists also accept a Naturalistic/Humanistic anthropology according to which human beings are only products of nature, not creatures of God (as if the two were mutually exclusive). Evolution accounts entirely for our existence, they contend, and human life is continuous with all of life. They reject all radical dualisms of body and mind, flesh and spirit, human and non-human.

Yet, only human beings count in humanistic ethical theories. Humanism does not recognize moral duties directly to non-human animals, other living things, or the environment. We have duties involving these things because they are beneficial to humanity, but we have no duties to them directly since they are not human.

2. How Scientific Is Humanistic Naturalism?

Naturalists claim to derive all their beliefs, the only legitimate beliefs, from scientific method alone. That they actually do this is very doubtful. Their failure, most conspicuous at the level of humanistic values, extends as well to their metaphysical commitments.

A. Humanistic Values and Scientific Method

The trouble is, Naturalists cannot derive their humanistic values and moral principles from either their metaphysics or from their methodology. Scientific method only describes the world, they concede. If so, it cannot prescribe; it cannot derive a very powerful "ought" from an "is" or readily bridge the "fact-value gap." Humanistic Naturalists endeavor to give the impression that "Science is on our side," and that their own metaphysical and moral beliefs are derived from the only rational methodology that they accept as legitimate–scientific method; but this is not so.

The recent scientific emphasis on evolutionary psychology attempts to generate some weak prescriptive statements by repudiating the absolute gap between "is" and "ought." Evolutionary ethicists like Michael Ruse[49] and Neil O. Weiner[50] propound a few "natural" norms and virtues. Traditional Humanistic Naturalists did not have access to evolutionary psychology, but they still endorsed very strong prescriptive statements about universal human rights and human welfare that cannot be derived from purely descriptive statements or from any elemental natural norms disclosed by evolutionary psychology.

Patricia A. Williams argues effectively that although the Christian ethical ideal of loving neighbors as self demands far more of us than our natural capacity for kin favoritism and reciprocal altruism, something within us nevertheless recognizes a more demanding saintliness and heroism as a higher ethical ideal. Williams indicates that both philosophical and religious ethical universalism extend the scope of moral obligation far beyond the provincial prescriptions of evolutionary ethics.[51] We seem to have evolved to act morally toward "insiders," immorally toward "outsiders," but not to respect or love all human beings as such. Originally, outsiders were members of different hunter-gatherer clans; and this perfectly natural, powerful, clannish, anti-humanistic propensity still manifests itself today in racism, sexism, nationalism, and every other distinction that divides us and sets us at enmity with one another.

Humanists actually admit that they cannot derive their own rich universalistic humanistic ethics solely from the methods of empirical science. Paul Kurtz, for example, espouses a "naturalistic ethics" according to which "Ethical judgments are empirical or may be supported by scientific knowledge."[52] Yet, he concedes, ethical Naturalism "cannot hope to derive universal values or principles that are objectively verified in the same way as descriptive hypotheses and theories are";[53] and he admits that scientific method merely helps us to collect the relevant facts, including facts about causation that pertain to choosing effective means to ends. But scientific method does not give us value-ends themselves.

Naturalistic Humanists have a much broader understanding of what passes for reasonable belief when universalistic humanistic ethical norms and value-ends are at stake. Kai Nielsen explicitly repudiates the claim that empirical scientific method alone yields all legitimate knowledge, mainly because it yields no ethics.[54] To scientific methodology he deliberately adds the axiological method of "wide reflective equilibrium" for rationally establishing and applying ethical, political, and social norms.[55] Nielsen extends this extra-scientific methodology to all of philosophy, science, and every rational quest for human knowledge. He calls it

> a coherentist method of explanation and justification...[that] starts with a society's or cluster of similar societies', most firmly held considered judgments (convictions), principally their considered moral judgments or convictions, and seeks to forge them into a consistent and coherent whole that squares with the other relevantly related things that are reasonably believed and generally and uncontroversially accepted in the society, or cluster of similar societies, in question.[56]

Nielsen, a Naturalist, clearly repudiates some prominent family traits of Naturalism, particularly its commitment to empirical scientific method alone as the sole source of human knowledge; and we can find much that is congenial and illuminating in his writings. I am substantially in agreement with most of Nielsen's philosophical outlook, including his fallibilism, the method of wide reflective equilibrium for questions of value, and even his denial of the existence and intelligibility of God as classically or standardly conceived. We disagree primarily in our assessment of Process Theology, to which Nielsen gives very little attention, as a plausible alternative to Naturalism.

B. Naturalistic Metaphysics and Scientific Method

Most important for present purposes, scientific methodology as usually conceived cannot justify the metaphysical beliefs of Naturalism. Naturalists insist that scientific procedures fail to justify the metaphysics of transcendent Theism; but they do not readily realize or acknowledge that natural science cannot justify their own naturalistic metaphysics.

Many well respected beliefs in science are not validated by scientific method when construed so narrowly as to exclude all theistic belief, as the concluding chapters of this book will argue. Put more positively, the beliefs of transcendent Theism are better justified by empirical methods, broadly construed, than those of Naturalism, especially with respect to many hypotheses about the origin of the universe offered by today's naturalistic cosmologists in the name of science. Scientists may not be interested in the kinds of empirical evidence that support Theism; but such evidence is abundantly available, as we will see.

Naturalists often intimate that their metaphysical beliefs are high order empirical generalizations, well supported by scientific investigation. Theistic beliefs are metaphysical, they decree; but naturalistic beliefs are scientific discoveries, conclusions proved by science. Yet, this is not so. If only beliefs *verified* by scientific methods are known to be true, naturalistic metaphysics cannot pass for scientific knowledge. Neither can the most theoretical parts of natural science!

Both Naturalism and Positivism emphasize verification, but Naturalism should not be confused with Positivism. Unlike Positivists, Naturalists do not deny the meaningfulness of the metaphysical and theological beliefs that they reject. They only deny their truth. Yet, Naturalists often feel the allure of Positivism and occasionally lapse into it. To understand the difference, let us consider the widely accepted positivistic Principle of Verification as a criterion for distinguishing between science and metaphysics. This principle asserts that meaningful, and thus "scientific," beliefs are those that are (or that someone believes to be), confirmed or at least confirmable (or falsifiable)–at least in principle–by sensory observation, experience, or empirical investigation. By contrast, metaphysical beliefs supposedly are matters of pure speculation, totally devoid of sensory observational import; (or at least someone believes this about them).

Karl Popper's slight variation on Positivism affirms that scientific propositions are falsifiable, at least in principle; some conceivable experience could count against them; whereas this is not so for metaphysical propositions. Verifiable or falsifiable "in principle" means that someone can imagine human sensory experiences or experiments that would count for or against them.

Meaningfulness, so understood, depends entirely on someone's powers of imagination, but whose? The class of meaningful or verifiable empirical or sensory beliefs is much more inclusive than the class of true or verified ones. What we can know to be true empirically or scientifically must be accessible directly or indirectly to human observations. Positivism claims that only empirically confirmable or falsifiable propositions are scientifically meaningful; but, Naturalists would agree, only those actually confirmed are known to be true. Many scientists and philosophers accept some variant of Positivism. They think that scientists ask questions that have empirically verifiable or falsifiable answers, whereas the questions of metaphysicians and moralists have no empirically verifiable or falsifiable answers. Scientists observe and prove; philosophers merely guess, speculate, or emote. Scientists discover hard facts, but philosophers only concoct ethereal theories that are empirically empty.

Though attracted to Positivism, Naturalists usually distinguish metaphysical from scientific propositions on the basis of their generality or particularity rather than in positivistic terms. In truth, the line between science and unverified metaphysics is not always very sharp. The history of science shows that beliefs that appear to have no empirical reference or support at one point in history are found to have such import at another. Einstein's general theory of relativity is a good example. The theory had no empirical reference or confirmation whatsoever in 1915 when Einstein first conceived of the curvature of space. Most nonscientists and scientists at the time were not able to imagine a way to confirm it, but Einstein soon showed them. The theory of the curvature of space implies that light moving through a gravitational field will be bent by that field; and this was confirmed during an eclipse of the sun on 29 May 1919 when stars in the background of the immediate border of the sun were observed to shift outward slightly from their usual positions as they were about to pass behind the darkened sun. Since then, the theory has received innumerable confirmations. By Positivistic standards, Einstein's general theory of relativity was a meaningless metaphysical theory between 1915 and 1919; but it became a meaningful and verified scientific theory thereafter.

Again, in 1929, when Paul Dirac deduced from quantum theory that antimatter exists, no one could imagine experiences or experiments to verify (or falsify) the proposition. At the time no cyclotrons were powerful enough to produce antimatter; but in 1932, Carl Anderson determined that it might be possible, by using a cloud chamber, to detect cosmic rays bearing antimatter particles.[57] Was the concept of antimatter meaningless between 1930 and 1932? Surely it was by positivistic standards, but not by broader naturalistic standards.

Naturalists usually regard as unknown or false what Positivists typically brand as meaningless.

The initially attractive but excessively simplistic positivistic distinction between science and metaphysics is fraught with difficulties. For one thing, the Verification Principle is itself meaningless when applied to itself, but it can be defended against this charge by indicating that it is a methodological rule and not the sort of descriptive statement to which it is intended to apply. More seriously, observations are usually if not always theory-laden; so we tend to see what our theories tell us we should see, and within limits we are disposed not to see what does not fit our preconceptions.

The meaning of the Principle of Verification is very ambiguous. When and by whom must relevant observations be made?[58] Must they be made in the present moment, or do remembered past and anticipated future observations count? If only present observations count, the very notions of past and future are meaningless; and Positivism yields a solipsism of the present moment, as it once did for Ludwig Wittgenstein. Must the observations be directly accessible to human beings; or do animals, extraterrestrials, or God count as observers? Who must the observer be to determine the meaningfulness of "The dinosaurs had halitosis," or "Adam did not have a navel"? If only present human observations count, all statements about early humans and pre-humans are meaningless. If only *direct* human observations count, our ideas about the Big Bang, inflation, and the activities of dinosaurs and other creatures who lived before human beings evolved are meaningless. No human observers saw the dinosaurs at play or prey, or smelled their breath. Only a transcendent God could directly observe the Big Bang, inflation, and either the infinite duration or finitistic origin of our universe. Being directly observable "in principle" by human beings is just equivalent to being directly observable by God, an ideal omniobserver.

Beliefs and theories can be supported by scientific method through: i. Direct Observation, ii. Inductive Inference, and/or iii. Hypothesis Formation and Testing; and such support comes in many degrees. Direct observation immediately perceives some object or process. For natural science, only sensory observation counts; moral and mathematical intuitions, religious experience, and introspective awareness do not count. Inductive inference reasons from perceptually observed samples to more of the same, which is how natural laws are established and justified. The hypothetico-deductive method postulates unobserved conditions or abstractions to account for observed realities, presumably better than alternative postulates. Science, philosophy, and religion all appeal to the unobserved to explain things observed. The best postulate is one that (1) predicts new phenomena not predicted by other postulates that experience actually confirms; or it is superior to other hypothetical explanations in (2) simplicity, (3) comprehensiveness, (4) consistency with itself or other established beliefs, (5) coherence or interconnectedness with other established concepts and beliefs, (6) elegance or beauty, and/or (7) fruitfulness for future research. These criteria are not exhaustive. Are the metaphysical beliefs of Naturalism supported by scientific methods in any of these ways?

i. Direct Observation

Science, especially astronomy, relies more heavily on sight than on other senses like touch, smell, taste, or hearing, though these senses too may be employed in scientific investigations. At the level of direct observation, seeing is believing. Many empirical beliefs like "The sun rises," "The moon shines," and "The stars are out tonight" are confirmed by direct observation. Even these elemental beliefs are theory laden, however; for us moderns, they presuppose the Copernican heliocentric theory, theories of optics, theories about the speed of light, theories about the correspondence of perceptions with realities, neurological theories, theories of mind/body relations, and theories embodied in the conceptual categories of common sense.[59]

Let us agree for the sake of the discussion that if something can be seen or otherwise consciously sensed while our senses are functioning normally, then our beliefs about that something are supported by direct observation. Let us also charitably extend the scope of "direct observation" to perceptions available through sense-extending apparatuses like microscopes, telescopes, and radioscopes. Stars and galaxies that are observed directly are, in a sense, perceived only indirectly through the light, sounds, or signals that they emit and the media that transmit them to human consciousness. These signals can be amplified by optical or radio telescopes, fall upon receptive human sensory modalities, and ultimately register in the brain and consciousness of human observers. Both direct and extended perceptions are mediated; yet, for simplicity, let us agree to say that heavenly bodies are directly observed even when amplified by complex scientific instruments and modified by our own sensory modalities.

Now, to get down to business, are the four metaphysical theses of Naturalism really high order empirical facts confirmed by direct human observations, broadly construed? Obviously, they are not. No *direct* human experience or scientific apparatus shows them to be true, or to be empirically meaningful in positivistic terms. This is most obviously so because no human being, scientist or not, and no living creature on earth, has ever directly observed the whole of reality–the subject matter of all metaphysics, including the naturalistic variety.

A. The claim that nature is the whole of reality, meaning that only nature (our system of spacetime) exists, cannot be a truth confirmed by direct human observation, for we have not directly observed all of our spacetime system, to say nothing of all of reality. God may do so, but that would be contrary to the basic suppositions of Naturalism.

B. The claim that nature as a whole is non-purposive and non-personal cannot be a truth of direct observation for a variety of reasons. Obviously, no human being ever has or ever will observe nature as a whole directly. Cosmological reasoning about nature or reality as a whole is inferential, not direct. It must be analogical or inductive; it does not and cannot simply report direct observations.

How can we tell if anything, including the universe, is or is not purposive? We know ourselves to be directly purposive mainly through immediate self-awareness, not through sensory observation. To capture all we know, the notion of "observation" must be extended beyond the sensory (and thus beyond the scientific) to include first person introspective experiences of our own conscious interests, desires, purposes, valuations, thoughts, intentions, volitions, attendings, and so on. We know (with less than certainty) that other people and animals are directly purposive because their bodies are structured like our own, because they behave vocally, linguistically, and otherwise in seemingly purposive ways, and because linguistic criteria indirectly indicate the presence (or absence) of inner conscious processes, activities, and values in other persons or animals. We infer that comatose persons are no longer directly purposive because they no longer engage in purposive behavior, and because medical scanners determine that their upper brains are not working sufficiently, or at all.

Some things are only indirectly purposive, if purposive at all. Rocks and rivers are not directly purposive or personal beings because they lack brains, sense organs, and self-originated goal-directed behaviors. Yet, rocks and rivers can be indirectly or instrumentally purposive and personal if they serve the interests, values, thoughts, desires, or choices of directly personal and purposive beings like humans, animals, and God. Artifacts like houses, bridges, and airplanes are only indirectly purposive. Without minds of their own, they are made to serve our purposes.

Is the universe as a whole purposive? Naturalists insist that it is not directly purposive because it has no mind of its own; and we have no good reason to believe that it is indirectly purposive, that it expresses the purposes of a transcendent divine World-Designer. How does the Naturalist know this? Our limited human acquaintance with the universe reveals that most of spacetime is not organized or structured as brains or sensory organs, but Theist may agree completely that the Universe is not directly purposive. Theists claim that the universe is only indirectly purposive and personal, that it indirectly expresses the purposes, values, wisdom, benevolence, and other personal attributes of a God who transcends nature while interacting immanently with and within it. Naturalists deny that nature is either directly or indirectly purposive.

Whether Theists or Naturalists are right about the purposiveness of nature will be explored in greater depth in later chapters. Both positions arrive at their conclusions by analogical or inductive reasoning, or by hypothesis, but not by direct observations. For Naturalists, the universe resembles a rock; for Theists, it resembles an artifact, a clock perhaps, made by an intelligent and benevolent manufacturer. We are never outside the universe as a whole to observe it directly; we are always inside the rock or the clock trying to figure out what makes it click or tick; we can only reason to the whole by analogy with familiar parts that are known to click or tick.

C. Do scientist observe directly that the most general attributes of nature like time, space, and the basic stuff within them exist infinitely, eternally, and necessarily? Again, obviously not. Infinite duration can never be observed directly by finite beings like us. No nuclear physicists, astronomers, other scientists, or ordinary persons have ever experienced anything directly that exists necessarily, infinitely, and eternally. All experience is against this claim of naturalistic metaphysics, not for it.

D. Finally, does anyone directly observe that all events have natural causes, and that there are no supernatural causes? Again the answer must be negative, partly because no human observer has ever directly experienced all events, partly because some observations actually count heavily against this family trait of Naturalism. That the Big Bang, for example, has a natural cause is doubtful. If produced by another world, that world is a supernatural cause, as Jastrow indicated. Later chapters will scrutinize naturalistic efforts to show that our universe was caused by natural processes–by other worlds regarded as natural causes; but no one claims that we human beings can observe these processes directly. Precise causal explanation breaks down at the level of singularities and/or quantum events, according to the Heisenberg Uncertainty Principle; and some cosmologists submit that the origin of our universe had no cause at all.

Lastly, the world religions report many direct observations of unusual events that they regard as miracles, which, by definition, lack natural causes. A naturalistic explanation probably works for most of them; but at least a few alleged miracles are not readily explained solely in terms of antecedent spatio-temporal conditions and are highly resistant to naturalistic explanation. The essential point is that some human experiences count, even if not decisively, against the metaphysical dogma that all events have natural causes. Naturalists just refuse dogmatically to allow them to count. Even the most highly unyielding instances of alleged miracles can be explained naturally, they insist, precisely because all events have natural causes.

The circularity of this naturalistic reasoning is obvious, as is its metaphysical nature. Naturalists "prove" that all events have natural causes by appealing to the premise that all events have natural causes! Propositions are metaphysical, and thus meaningless by positivistic standards, say many Positivists, if they are held in such a way that no experiences are allowed to count against them. By reasoning in a circle, Naturalists rule out all anomalies *a priori*, while claiming that no knowledge is *a priori*!

Plainly, none of the metaphysical theses of Naturalism are truths of direct observation; but they may be observational truths of a more indirect sort. Perhaps they are truths of inductive inference, or perhaps they are just the best available hypotheses for understanding our universe. Perhaps not!

ii. Inductive Inference

Inductive logic permits inferences from the observed to the unobserved, from parts to wholes, from particular samples to broad classes, from given instances to more of the same. If we reach into a barrel and take out three rotten apples in succession, inductive logic permits us to conclude that most, perhaps all, of the apples in the barrel are rotten. Inductive reasoning introduces inescapable elements of uncertainty, so scientists do not claim absolute certainty for their inductively supported conclusions. Probabilities increase as more and more rotten apples are extracted. Yet, some apples in the barrel might be sound, and we might have good reasons for thinking so. Realizing that life is chancy, we are generally content to live with the uncertainties of inductive inference. Inferring more of the same from observed samples is exactly what it means to think rationally and scientifically about the unobserved parts of the world of nature. If all perceived manifestations of energy conform to Einstein's $E = mc^2$, scientists predict that this formula fits all the energy in the entire universe. If energy is dissipating in all observed closed systems, scientists inductively infer that it is dissipating in all closed systems, whether observed or not, at least in the present expansion phase of the universe.

Can the metaphysical theses of Naturalism be transformed into scientific truths by showing that they are products of inductive reasoning from observed samples to more of the same? This is extremely doubtful.

A. Can we reason inductively that all realities belong to our objectively existing system of spacetime because all observed realities do so? This is very doubtful. Anti-realistic philosophers would question the basic premise. That all observed entities belong to an objectively existing spacetime system is philosophically controversial and depends on hotly contested theoretical or philosophical commitments. Immanuel Kant, for example, thought that space and time are mere forms of appearance, and that only appearances but no realities or things in themselves can ever exist or be observed to exist in spacetime outside our minds. We will see in Chapter Six that some versions of quantum physics are antirealistic and heavily stress the role of observers in structuring what is observed. Let us be generous again, however, and assume with realists that we perceive realities, not just appearances, and that these are located in objectively existing spacetime. Can we now infer that all realities are located in this mind-independent system of spacetime, given that all observed realities are so located? This strong presumption may be honestly doubted, if for no other reason than that the cause of the Big Bang, if it had a cause, was clearly located outside our spacetime system.

In making inductive inferences, we must be cautious not to generalize too hastily. Some apples in our barrel may be sound because they belong to a rot-resistant variety. Also, energy is not really dissipating everywhere–not for example in or near black holes where gravity is strong enough to concentrate energy. Perhaps energy can be created in a quantum vacuum in which "virtual" particles become ephemeral actual particles. To account for the Big Bang, some naturalistic cosmologies explored in later chapters postulate the existence of other worlds that are unobserved by and unobservable to us. On the basis of some (extremely inconclusive) evidence, many cosmologists now believe in the existence of many other universes beyond ours, perhaps an infinite number of them. This, we will see, is very problematic, but if true, then some realities are not located in our own system of spacetime or nature. Objectively existing nature was created by another objectively existing reality beyond and before itself. The contemporary version of the Cosmological Argument for the existence of God, as developed in our concluding chapter, will give good reasons for doubting that our universe exhausts reality. If all created things have causes, Big Bang Cosmology itself provides powerful evidence against the claim that our system of nature is all that really exists. Our system of nature was brought into being around fifteen billion years ago, presumably by something–perhaps God, perhaps another universe, perhaps quantum-fizzy Superspacetime–outside of itself. The naturalistic claim that nature, our system of spacetime, is infinitely old and causally uncreated and self-sufficient has been falsified decisively by modern astrophysics; but Naturalists do not concede this without a fight!

B. Is the claim that "Nature as a whole is nonpurposive and nonpersonal" derived from inductive generalization? It is definitely not an inductive inference from human acquaintance with many sample universes. We observe only one universe from the inside, and only parts of that. Naturalism itself is vulnerable to David Hume's inductive objection to the Teleological Argument for the existence of God. We cannot infer that order in our universe is created by a purposive and intelligent Divinity, said Hume, because we have not seen many worlds created before our eyes and thus cannot compare the order of this world with that of other worlds known to be made by a purposive and intelligent World Designer.

Naturalists are in exactly the same position; they have not seen many purposeless worlds created by other pre-existing godless systems of spacetime. Thus, they cannot establish inductively that the order (or disorder) of our world is similar to that of other purposeless worlds known not to have been made by Divinity. We cannot reason inductively to more of the same when an entity is one of a kind; and, as far as human experience goes, our universe is one of a kind. Empirically, Charles Sanders Peirce was right: "Universes are not as plentiful as blackberries."

Naturalists may contend that the orderliness of our natural world fails to express purposive intelligence because nature as a *whole* compares favorably with certain *parts* of nature like rocks and cabbages that are known not to have, or to express, purposive intelligence. Like Theists, Naturalists must reason analogically from parts of nature to the whole; their reasoning is definitely not an inductive generalization from several observed non-purposive universes to more of the same. Naturalists contend that nature as a whole is purposeless and dumb because it resembles certain parts of itself like purposeless and dumb rocks and electrons. Theists, by the same analogical logic, infer that nature as a whole is purposive and intelligent because it resembles human beings or animals who are directly purposive and intelligent, or, more appropriately, because of its similarity to houses, bridges, watches, and the like, that are purposefully produced by intelligent beings. The outcome of this debate turns finally upon which analogy between parts and wholes actually holds; but the conclusion is not derived by simplistic inductive inference from same-kind samples to more of the same. We will return to this topic in later discussions of the Anthropic and Biopic Principles; but we should note in advance that contemporary Anthropic Cosmologists find the universe to be exquisitely fine-tuned for the purpose of producing and sustaining life. This is decisively at odds with Naturalism's anti-teleology. Chapters Eight and Nine will examine this issue more carefully.

C. Can we know inductively that the most general attributes of nature like time, space, and the basic stuff within them exist infinitely, eternally, everlastingly, and necessarily? Surely not! Induction just gives more of the same, but all observed natural processes and objects are spatially finite, limited or transient in duration, and contingent or dependent in mode of existence. We never observe any natural objects or processes that are spatially or temporally infinite, everlasting, necessary, self-sufficient, self-caused, and uncreated. Inductive reasoning can give us additional spatiotemporal finitude and contingency, but it cannot give us their opposites! No inductive evidence whatsoever supports the conclusion that nature as a whole exists infinitely, everlastingly, or necessarily. Nature just might be infinite, everlasting, and necessary; but we cannot know this by inductive inference.

Naturalistic critics of the traditional arguments for the existence of God deny the legitimacy of reasoning from a finite observed world to an infinite God precisely on the grounds that we cannot reason inductively from the finite to the infinite. If this is so, Naturalists themselves are in the same boat. They too cannot reason inductively from observed finite portions of the world to the world's infinity in time and/or space. Inductively, from finitude we can only infer more finitude, from contingency only more contingency. Naturalistic logic returns eventually to devastate its own metaphysics!

D. Lastly, can the causal principle, "All events have natural causes," be known inductively? Initially, this seems to be a scientific inductive generalization from experience. Granted, we have not observed all events; still, Naturalists contend, all events have natural causes because all observed events have natural causes. Unfortunately, solid evidence against this argument is provided by the Big Bang, if not also by rare experiences of miracles, which cannot be ruled out *a priori* without a vicious circularity of reasoning.

Is the principle of universal natural causation a necessary presupposition of scientific inquiry? Wouldn't it be impossible for scientists to do their work without believing that all events have natural causes? Not at all! To do their work, scientists need only the imperative: "Look for natural causes!" They do not need an *a priori* metaphysical guarantee that they will always find that for which they are looking.

Naturalists are on no firmer ground in trying to derive their metaphysics from induction than they are in trying to derive it from direct observation. Can the empirical truth of their metaphysical claims be salvaged by demonstrating that they are rationally justified postulates or explanatory hypotheses?

iii. Hypothesis Formation and Testing

Scientists do more than merely switch on their eyes, ears, and other senses, and more than just reason from given samples to more of the same. They also explain things by appealing to often unobserved hypothetical or theoretical constructs. They always try to find the best available explanations, but how do we know which explanations are best?

Scientists theorize about what accounts for what, what in turn explains this, what finally explains that, and so on. Scientists creatively construct and defend hypothetical unobserved and unobservable entities, processes, and principles; they postulate the reality of innumerable things unseen (like initial singularities, inflation, or laws of nature) to explain things seen. Some initially hypothetical entities might be observed later; but some, like those just mentioned, will never be humanly observed. Some postulates predict outcomes that can be observed eventually; others make predictions that are falsified by later observations. Usually, several hypotheses can account for the same data, and the right one, the best explanation, must be identified, sought, and defended. Much scientific inquiry is just a quest for the best explanation using what Charles S. Peirce called "abductive reasoning" or what others call "the hypothetico-deductive method."[60]

Entities and processes that initially are nothing more than theoretical constructs may be later confirmed observationally, as was Einstein's curvature of space, and Dirac's antimatter. Hypothetical entities may be postulated because scientists hope to observe them in the future, walking by hope and faith and not by sight; but their hopes are not always fulfilled. Theoretical realities postulated by scientists like natural laws, singularities, inflation, and the Big Bang itself, are likely to remain totally unobserved, even with the aid of the most powerful instruments; still, scientists affirm them, even when more than one purely hypothetical construct explains same the facts. Why?

What distinguishes rationally justified from unjustified hypothetical explanatory constructs? Explanatory constructs tend to proliferate, so how can we tell which ones are true or best? Scientists (and philosophers) often choose between explanatory hypotheses on purely aesthetic grounds like simplicity, symmetry, harmony, elegance, and beauty.[61] Aesthetics is ultimately integrated into rationality itself; truth merges with beauty and goodness; the beautiful and the rational are ultimately identical. In these respects, no sharp distinction exists between theoretical science and speculative metaphysics; and plenty of room remains for honest disagreement about whether an explanatory hypothesis is rationally justified.

Still, some hypothetical explanations are clearly better than others. How so? Some hypotheses are much more powerful than others in pulling more loose threads together and unifying otherwise unordered disarray. Some are more fruitful than others, more suggestive of agendas for future search. Some theoretical constructs actually make predictions that come true, or that can be falsified. Some hypotheses can be tested in crucial experiments by searching for consequences that they alone, but none of the alternatives, foretell. When explanatory entities cannot be observed directly and no crucial experiments exclude alternative explanations, scientific theories shade off into sheer guesswork. But some guesses are more educated, more beautiful, more elegant, than others.

In both natural science and philosophy, the best justified hypothetical constructs are simple, symmetrical, harmonious, elegant, beautiful, powerful, fruitful, and testable in crucial experiments that exclude competing hypotheses. These criteria can be met in varying degrees, but the best explanatory hypotheses (1) make predictions not made by other hypotheses that are confirmed, not disconfirmed, by experience, and (2) are superior to other hypotheses in simplicity, comprehensiveness, coherence, consistency, elegance, beauty, and fruitfulness for further research. Some illustrations might help.

a. The Hypotheses of "Creation Science"

"Creation science," advocated by many religious conservatives, affirms two central theses: (1) our universe was created by God out of nothing at some time in the finite past, and (2) the theory of evolution is false because all earthly species of living things were created directly by God within six days after God produced the universe itself. If Big Bang Cosmology is correct, as it seems to be, "creation scientists" are right about (1), although the "by God" and "out of nothing" parts remain to be established, and the relevant "finite past" is closer to fifteen billion years than to the six thousand years favored by religious fundamentalists. But "creation scientists" are clearly mistaken about (2) for many good reasons, such as that the universe existed for over ten billion years before the simplest living things began to exist and to evolve into more complex beings on our earth. Without exploring the evolution controversy in depth, let us consider how the hypothetical-deductive method applies to the fossilized remains of innumerable extinct plants, animals, and microscopic life forms.

Both "creation science" and "real science" must explain the same observed facts. Extinct fossilized animals, plants, protozoa, and bacteria exist in geological strata. Real science hypothesizes that the fossils are best explained by evolutionary processes that transpired over at least four and a half billion years. The full course of evolution itself cannot be observed directly or repeated under laboratory conditions, but this explanatory hypothesis has testable implications. Although Naturalists are evolutionists, so are the Theists who do not insist upon a literal interpretation of the word "day" in the first chapter of *Genesis*. Evolution is incompatible with biblical literalism, but not with belief in God.

"Creation science" hypothesizes that either (a) God created the earth with all the fossils intact, but no such creatures ever actually lived, or (b) all the fossilized creatures once co-existed but were killed by Noah's flood, which deposited their bodies in existing geological strata as it receded. Hypothesis (a) is so egregiously *ad hoc* that few if any advocates of "creation science" take it seriously. It implies no additional observations by which it can be tested.

The most viable "creation science" hypothesis is (b). Predictions made from the deluge postulate are: (1) fossilized remains could be distributed randomly in areas of great aquatic turbulence; but in calmer areas the bones, shells, and other parts of the largest and heaviest animals will be in the lowest geological strata, and the remains of the lightest and smallest animals will be in the highest strata; and (2) carbon 14 and other dating techniques will show that all fossilized creatures lived at the same time, around six thousand years ago. Unfortunately, innumerable observations decisively disconfirm these predictions. Besides, if Noah carried breeding pairs of all living animals to safety on the ark, he presumably did this for all species of dinosaurs. So where are all the dinosaurs today?

"Creation scientists" claim to be doing science, but scientific methodology shows that they are wrong. The predictions about the fossils made by real science do not suffer this fate. The evolutionary hypothesis predicts that (1) fossilized bodies will be generally arranged so that the simplest (often the lightest weight) organisms will appear in the lower geological strata, and the fossils of more complex organisms will be in higher strata, and (2) carbon 14 dating techniques will show that many of them lived millions or billions of years apart, not all at the same time. These predictions are well confirmed by innumerable observations and tests. We can observe directly neither the lengthy evolutionary processes that created the fossils nor a direct divine creation of the dinosaurs and other extinct creatures; but the evolutionary hypothesis is supported by, and "creation science" is refuted by, its testable predictions.

b. Hypothetical Cosmological Entities and Processes

Turning now to cosmological hypotheses, most astronomers now believe in the existence of black holes which, almost by definition, cannot be observed directly. They did not originally generalize inductively from a few observed black holes to many others. Rather, unperceived black holes were postulated to explain other things that were visible in the heavens. Astronomers observed that stars were being pulled in certain directions by unseen celestial objects, and that various kinds of radiation were emitted by unidentified sources. The eccentric orbital movements of the star Cignus X-1 cannot be explained by the presence of any visible star, and powerful X-ray emissions discovered in the Large Magellanic Cloud of stars and elsewhere come from no visible source. Astrophysicists accounted for these phenomena by reifying hypothetical black holes that suck in almost all surrounding matter/energy. Some, like Stephen Hawking, conjecture that black holes emit some radiant energy, so not everything in their vicinity gets inhaled if he is right.

Until quite recently, black holes were purely hypothetical or theoretical constructs. The Russian astrophysicist I. L. Rozental wrote as late as 1988 that black holes were "created by the fantasy of the theoreticians," and that "Black holes are unrivaled in their popularity as theoretical objects yet to be reliably observed."[62] Rozental wrote before the definitive identification of a black hole at the center of the M87 galaxy in May 1994 by astronomers using the repaired Hubble Space Telescope. In one sense, this black hole can be seen, but in another, it cannot. The core, the hole itself, presumably a singularity, emits no light, and in that technical sense it cannot be seen; but some spectacular fireworks in its close vicinity are very visible. Since 1994, the Hubble Space Telescope has located many black holes.

Black holes were thought originally to be so dense, so completely dominated by gravitational attraction, that neither light nor any other forms of energy could escape from them or their vicinity. Supposedly, their awesome density and gravitational effects would devour all nearby objects. Stephen Hawking's more recent black hole theory hypothesizes that X-rays and gamma rays escape from the outer edges of black holes, and that black holes themselves gradually decay and radiate their energy back into the observed universe. Given enough time, all black holes will eventually evaporate completely.[63] As Hawking puts it, "Black holes are not completely black."[64] They do "have hair" after all. Radiation emitted by black holes is now called "Hawking radiation," even by those who do not accept its existence. If Hawking is correct, black holes are not gateways to other universes; though not verified, given enough time, they just dump their stuff back into our universe. Hawking also suggests, again without verification, that black holes occasionally just explode and release their contents back into our universe.[65] But no known laws of physics permit or predict these explosions.

Singularity cores of black holes cannot be seen because they allow no light to escape; but light definitely does escape from their close environment. If Hawking is right, black holes emit detectable X-rays, gamma rays, and flourescent electrons. Black holes themselves are hypothetical, invisible, imperceptible entities; but their existence predicts and explains certain visible, perceptible effects for which we have no other plausible explanations. At the center of the M87 galaxy, astronomers actually see a vortex of swirling gases, the innermost portions of which are disappearing into an unseen black something at the center of the vortex. Optically, black is the absence of light, so we are not "seeing" anything optically when we see black. Scientists calculate that the mass of the black hole in M87 equals two to three billion suns or solar masses, all of which it has already consumed.

Great uncertainty is involved in abductive reasoning because other hypothetical constructs might also explain observed phenomena. Scientists aspire to construct hypotheses with extensively testable consequences that are not predicted by other hypotheses; but crucial experiments that decisively rule out other hypothetical explanations are rare in the natural sciences (and in philosophy). Presently, only the black hole hypothesis explains what astronomers have found at the center of the M87 galaxy and in many other places like the center of our own Milky Way and perhaps of every other galaxy.

Scientists often postulate the existence of things that are not themselves visible or directly perceptible. Like metaphysicians, they frequently appeal to the unobservable to explain things observed. Purely hypothetical or theoretical and unobserved constructs abound in theoretical physics and cosmology–like singularities, inflation, cosmological constants, magnetic monopoles, dark matter, antecedent universes, or tiny vibrating strings within primitive particles. Consider one other example.

Big Bang Cosmologists think that matter and antimatter particles were initially created together, and that we live in a material world because, in the earliest fractions of a second of the universe, in some mysterious way matter came to prevail over antimatter. How did it happen? We can only theorize, not observe. "Baryonic Asymmetry" means that baryons (like neutrons and protons) came to prevail over their antimatter counterparts in the very early universe. The most widely accepted explanation of matter dominance, developed originally by the Russian physicist Andrei Sakharov, postulates purely hypothetical explanatory entities and processes. I will not give all the details;[66] but they involve the existence in the earliest universe of never-observed, massively large, primitive particles called "X bosons." X bosons supposedly decayed into matter and antimatter particles at an irregular rate through processes that have never been observed, with matter ultimately prevailing. The observed fact that we live in a material world is explained by constructs that are purely theoretical and uninspected. But why weren't all X bosons destroyed by their own anti-particles?

One serious problem with this is that a very different hypothesis can explain the same facts. The universe just might have been divinely created with a massive matter/antimatter imbalance. Matter dominance could be an initial condition of existence as we know it, not a product of inaccessible natural processes. All existing phenomena would be explained by the God hypothesis, which scientists don't like because it appeals to unobservables! Similar considerations apply to the equally unobservable inflation postulate. Perhaps God just created the world with similar or isomorphic generic properties throughout; perhaps nearly flat space (if that is indeed what we have), and few if any magnetic monopoles were just initial conditions of creation. Is selecting the most reasonable or best-justified explanation just a matter of aesthetics–or of prejudice–where the most reasonable hypothesis is the most beautiful, or the most scientifically conventional, or the most atheistically naturalistic? At this level of abstraction, scientific or philosophical truth/beauty may be in the eye of the beholder, as later explained.

So, what has become of the simplistic but appealing dogma that scientific beliefs are truths of observation? What remains of the naturalistic claim that beliefs about God, like those about Santa, are false because they cannot be confirmed by observation? Theoretical science regularly postulates ultimate unobservable explanatory constructs like black holes, X boson decay, singularities, inflation, antecedent universes, Superspacetime, and infinitely many universes; but these are no more empirical than God! Physical science gradually shades off into metaphysics. Is purely theoretical science just bad science, as Eric Lerner maintains in the next chapter? Even if X bosons and inflation have predictable and testable consequences, alternative but equally invisible conceptual constructs can explain the same observed effects. The most abstract scientific theories are shot through with unobservables and uncertainty, and modesty becomes their endorsement. At some point, we know not just where, theoretical physics becomes raw conjecture, indistinguishable from speculative metaphysics at its worst; but most scientists do not openly admit it. Some spectacular examples will be discussed in coming chapters.

c. Naturalistic Metaphysical Hypotheses

Now that we understand the hypothetico-deductive method, let us ask the really crucial question. Are the four metaphysical truths of Naturalism justified explanatory hypotheses that adequately account for the world we observe?

A. "Only our system of spacetime exists" explains all that we see, but so does "In the beginning, God created our system of spacetime." These metaphysical hypotheses may be debated on their own merits, but the naturalistic hypothesis yields no testable predictions that differentiate it from the theistic alternative. The world, our system of spacetime, would be here and might look much the same as it does whether its transcendent creator is Divine or not. Claiming that our spacetime system is "in principle" all that exists, the whole of reality, adds nothing to what we can see or predict about our universe. No crucial experiments show that Naturalism can, but Theism cannot, explain any or all of the observable features of our world. Actually, cosmic teleology and contingency, considered in the concluding chapters of this book, show just the reverse.

Theism can adequately account for many observable features of our world that Naturalism cannot plausibly elucidate.

B. Perhaps the purposelessness, indifference, and stupidity of nature as a whole best explains the presence of evil and pervasive lifelessness in the world; but the purposefulness and intelligence of Deity may best explain the presence of life and goodness in the world. Whether Naturalism or Theism best explains both good and evil is an ancient and interminable debate, to be considered more carefully in Chapter Eleven. Final conclusions about the meaning of value and disvalue in the world are primarily philosophical, not scientific. With simple empirical claims, we can tell the difference between natural science and philosophy; but in the middle and at hypothetical extremes where scientific theories wax philosophical, it is often hard to tell. The presence or absence of ultimate cosmic teleology is not an obvious issue for purely descriptive science. The facts of life pose questions about both good and evil in the world, and about the pro-anthropic orderliness of the cosmos; but the answers belong more to philosophy than to empirical science. So, too, do Naturalism's metaphysical claims.

C. Naturalists contend that nature and its most basic ingredients–time, space, and energy–are infinite, eternal, and necessary. How does this fare as a scientific explanatory hypothesis? This and other Naturalistic metaphysical claims about the most fundamental features of the universe cannot be confirmed by direct observation, and they cannot be inferred inductively since we observe only finite, transient, and contingent parts of nature. Induction yields only more of the same, not diametrical opposites. As an explanatory hypothesis, the eternity and necessity of our spacetime system fares no better. Naturalistic claims about the infinity, eternity, and self-sufficient necessity of our universe are definitively falsified by modern science's massive empirical and theoretical evidence for the Big Bang.

Recall the seven converging lines of evidence that convince most scientists that our universe began in a cataclysmic explosion about fifteen billion or so years ago. In asserting that our universe is infinite, eternal, and necessary, Naturalism affirms just the opposite. The naturalistic hypothesis seems to predict that: (1) distant galaxies should exhibit no redshift, (2) the universe should not be expanding at a regular pace, (3) order and energy should not be dissipating with time, (4) Einstein's field equations should not be satisfied by an expanding universe that had a beginning in time, (5) the heavy elements should predominate over hydrogen and helium in the universe since they have had an infinite amount of time to be cooked up by stellar furnaces; indeed, there should be no hydrogen or radioactive elements left at all, (6) no relatively uniform cosmic microwave background should exist, and (7) the sky at night should be as bright as day! These empirical implications of naturalistic metaphysics have been falsified overwhelmingly.

D. That "All events have natural causes" may be hypothesized, but so may "Some events have supernatural causes." To do their work, natural scientists require only the imperative, "Look for natural causes." They do not require a metaphysical guarantee of success. Ancient and contemporary debates over miracles proceed at the fuzzy borderline of science and metaphysics. Input data and conclusions drawn from them often just presuppose rather than establish metaphysical convictions. Natural science does not settle all such disputes, but it has established the truth of creation. Something outside of and preceding our universe presumably created it. Just what this something was–whether God, a preceding universe, a system of Superspacetime, or what have you–and whether the universe actually had a cause at all will be addressed in following chapters. Clearly, if our universe had a cause, it was supernatural, transcendent, in relation to our system of spacetime.

In sum, natural science does not side with Naturalism, the fundamental beliefs of which are not verified by, and some of which are clearly falsified by, natural science, despite all the huffing and puffing of Naturalists themselves. Its metaphysical claims have no direct observational component; and we cannot inductively infer them as more of the same. Construed as explanatory hypothesis, the metaphysical "truths" of Naturalism make no verifiable predictions that Theism fails to make. Naturalistic reasoning is often viciously circular and "proves" itself only by presupposing itself. It is often defended so dogmatically and tenaciously that no experience is allowed to count against it; but much that modern scientists have discovered about our universe counts decisively against it. Naturalists can confess their faith, and insist that their beauty, or their prejudice, is everyone's truth; but they cannot appeal to empirical science to confirm their worldview. Naturalism is *a priori*, unconfirmed, and in some instances clearly falsified metaphysics, pure and simple. Its metaphysics is not derived from experience or "scientific method." Instead, Naturalists presuppose their metaphysical beliefs in interpreting all experience; and their ungrounded assumptions set limits to any beliefs, hypotheses, and theories that Naturalists are willing to take seriously. Some people are psychologically predisposed toward Naturalism, but no rational or scientific considerations require anyone to accept it.

Three

STEADY STATE AND PLASMA COSMOLOGIES

At least two twentieth-century cosmologies offered in the name of natural science would regard What caused the Big Bang? as a loaded question because it presupposes that the Big Bang really happened, and that it created our whole universe. According to Steady State Cosmology and Plasma Cosmology, there never was a cosmos-making Big Bang. As Eric Lerner, a Plasma Cosmologist, expressed it in the title of his 1991 book: *The Big Bang Never Happened*; but this is a small minority view among today's cosmologists. A few skeptics about the Big Bang still cling to Steady State Cosmology,[1] but this position is now mainly of historical interest, except to Fred Hoyle. Lerner's challenge to Big Bang Cosmology is serious, well developed, and deserves and receives the most attention in the following pages.

1. Steady State Cosmology

In 1948, Big Bang Cosmology did not dominate the field as it does today. The redshift of the galaxies was well confirmed by then, and Hubble's law was available to calculate their rate of expansion. George Gamow's associates, Ralph Alpher and Robert Herman, published a paper in 1948 in which they predicted the existence of background radiation as a lasting relic of the Big Bang. They calculated its temperature at five degrees Kelvin, two degrees higher than currently accepted; but for years no one paid much attention to their prediction or tried either to confirm or falsify it. The Big Bang gained its contemporary dominance when this microwave background was actually found by radio astronomers Arno Penzias and Robert Wilson in 1965, but this information was not available in 1948.

In 1948, alternative cosmologies had a chance, especially the Steady State theory developed and published that year by Fred Hoyle, Thomas Gold, and Hermann Bondi. This theory tried to explain how the universe could appear constant, orderly, and unchanging to all observers at all times despite the fact that it is also expanding and dissipating its order and mass/energy. Steady State Cosmology said that the observable universe of all observers throughout infinite time and space would contain essentially the same quantities of entropy, background radiation, redshifted galaxies, and stars or other configurations. The heavens would always be steady or constant, on average, despite the continuous disappearance of stars and galaxies over the horizon of visibility in accord with Hubble's law. How is this possible?

To reconcile sameness of observable mass with disappearance of observable mass, Steady State Cosmologists resorted to creation, but not the all-at-once creation of Christian or Big Bang Cosmology. Hoyle and Bondi posited an infinitely prolonged process of *continuous creation*. Throughout infinite space and infinite time, just enough hydrogen atoms are created to replenish the supply within every astronomer's observable universe. Given enough time, gravitational attraction forms new background radiation and new gaseous clouds from the newly created particles. Gravity concentrates these gaseous clouds into new stars and galaxies to replace those lost to view through the Hubble expansion process. Thus, throughout infinite time and space, the universe always looks basically the same, in a steady state, to all observers everywhere.[2]

Continuous creation is not wrought by God or any transcendent reality, according to Hoyle and Bondi. Their metaphysics was thoroughly naturalistic and this-worldly. Hoyle asked: "Where does the created material come from?" and he answered that matter is responsible for it. "Matter that already exists causes new matter to appear. Matter chases its own tail."[3] Matter does not create new matter out of old matter or out of pre-existent energy. If it did, nothing would be left after a finite interval to feed the Hubble expansion, and the universe would be empty and dead. Like God, matter creates matter out of nothing. Just how, Steady State Cosmologists did not say.

According to Hoyle, the creation of matter by matter *ex nihilo* is a slow process, but in an infinite amount of time it really adds up. "The average rate of appearance of matter," he wrote, "amounts to no more than the creation of one atom in the course of about a year in a volume equal to that of a skyscraper."[4] This is enough, nevertheless, to generate the microwave background, to cause the redshift, and to drive the Hubble expansion. "The new material produces a pressure that leads to the steady expansion,"[5] Hoyle proclaimed.

2. Critique of Steady State Cosmology

Most cosmologists today reject Steady State Cosmology, though Hoyle and a few others tenaciously defended it. What considerations make it implausible?

A. No Observational Evidence

Most seriously, no observational evidence supports the theory of continuous creation. Hoyle almost saw the difficulty when he conceded that "It would be quite impossible to detect such a rate of creation by direct experiment."[6] Bondi also acknowledged that "It is utterly impossible to observe directly such a rate of creation."[7] This is certainly true. However, the problem is more serious than detecting the *rate* of the creation of material particles out of absolutely nothing. Namely, *no instance whatsoever of such creation* has ever been observed! Even if hydrogen atoms were being created *ex nihilo* at the rate hypothesized, no one could find them, no one has, and no one will. In particle accelerators, a great variety of physical particles can be created out of *pre-existing* particles and radiant energy; but after the initial creation of the universe itself, no matter has ever been created out of absolutely nothing. No empirical evidence at all supports the theory of continuous creation *ex nihilo*. Bondi, who saw this very clearly, confessed that the continuous creation of matter by matter out of nothing "is not directly observable."[8] He argued for the theory nonetheless on the grounds that it is the simplest assumption compatible with the observable facts.[9] Once the microwave background was discovered in 1965, this was no longer true. A few more recent Quantum Cosmologists surmise that new *and enduring* particles may arise from quantum fluctuations in the vacuum of empty space; but this, too, has not been verified, and there is ample room for doubt. New particles that endure are created only when actualized energy is injected from outside the vacuum.[10] The vacuum of empty space is not pure *nihilo*!

B. The Dark Sky, Microwave Background, Redshift, and Hubble Expansion

Without continuous creation of hydrogen atoms from nothing, Steady State Cosmology can not account for the dark sky at night, the microwave background, the redshift, or the Hubble expansion of the universe. Olbers' Paradox, according to which there should be no dark sky at night in an infinite universe, could be resolved, Bondi believed, by the supposition that redshifted light loses energy with distance–enough energy to darken most of the night sky.[11] "High-entropy energy (in the form of radiation)," he wrote, "is constantly being lost through the operation of the Doppler shift in the expanding universe, while low-entropy energy is being supplied in the form of matter."[12]

Unfortunately for the theory, low entropy energy is *not* being supplied in the form of new matter. No fresh hydrogen atoms come into being from nothing to replace what is lost to view, to drive the redshift and Hubble expansion, to resupply the ebbing microwave background, or to explain why the sky is dark at night. The omnipresent background radiation, the almost ubiquitous redshift, Hubble's law of uniform expansion, and the dark night sky count decisively against Steady State Cosmology. *No* observational evidence exists that *any* matter is being created continuously out of nothing, much less that *just the right amount of it exists* to replenish what is being lost to view by a relentless Hubble expansion.

C. The First Law of Thermodynamics

Critics of Steady State Cosmology protested that continuous creation violates the First Law of Thermodynamics, which prohibits the creation and destruction of mass/energy. Hoyle replied that the theory is actually necessary to account for conservation only in the *observable* universe. Energy is conserved because exactly enough matter is being created to counterbalance its loss over the horizon due to the Hubble expansion.[13] Bondi claimed that continuous creation "prevents the approach of the heat death, the state of thermodynamic equilibrium in which no evolution can take place and in which the passage of time has no significance."[14] These responses are inadequate.

Hoyle displaced the "Energy *is* constant" of the First Law of Thermodynamics, with "Energy *that can be observed at any given time* is constant." His position is that the total *existing* amount of mass/energy in the universe is constantly increasing. *Real* energy is being created continuously out of nothing to preserve the *appearance* of constancy. Both Steady State and some Big Bang Cosmologies postulate the creation of mass/energy out of nothing, either gradually, or all at once. If energy conservation is construed as a metaphysical principle that guarantees the eternal and necessary existence of all existing mass/energy, Steady State Cosmology violates this incrementally, but in infinite quantities over infinite time. Big Bang Cosmology violates it in finite quantities during a single resplendent burst. However, as explained in Chapter One, the law of conservation is not a metaphysical principle, despite the claim of the Oscillation Cosmologies introduced in the next chapter that the energy of our universe is derived from an infinite number of prior universes. Understood scientifically, all natural laws, including the laws of thermodynamics, come into being together with the universe and do not antedate it. Steady State Cosmology violates the First Law of Thermodynamics because it posits creation of mass out of nothing within our existing universe; but Standard Big Bang Cosmology does not violate it because "prior to" the universe-creating Big Bang, nothing existed to which the law could have applied, so no such law existed.

D. Antimatter

In particle accelerators, matter and antimatter particles are always produced in pairs; and particle physicists believe that symmetrical particle/antiparticle production is a universal law of nature.[15] In the thermonuclear furnace of the Big Bang, conditions somehow permitted enough matter to prevail over antimatter to produce the universe as we know it. Within our universe, when particle/antiparticle pairs are produced, they immediately annihilate one another in detectable explosions that leave traces of gamma rays.

The law of symmetrical production obviates Steady State theory's continuous creation of hydrogen atoms. If the proton of a hydrogen atom is created, it will be accompanied by an antiproton; and if the electron of a hydrogen atom is created, a positron will accompany it. New particle/antiparticle pairs always immediately annihilate one another. Thus, endless hydrogen atom production in "empty space" is impossible. Continuous creation fails to provide for the renewal of matter and for the elimination of antimatter. If matter/antimatter obliteration were transpiring throughout space, the resulting ubiquitous explosions and the ensuing omnipresent gamma ray radiation would be readily detectable; but astronomers detect no such all-pervasive occurrences. According to Virginia Trimble, "a few things don't belong anywhere...[like] antimatter, at least not anywhere in the observable universe, or we would see gamma rays where it meets matter."[16] We just don't find what the theory predicts.

E. Verifying Infinity

Philosophically, Steady State Cosmology presupposes a naturalistic metaphysics that is totally unverified and unverifiable. Hoyle explicitly affirmed the infinity of both time and space, writing that "Theory requires the galaxies to go on forever, even though we cannot see them....The galaxies are expanding out into an infinite space. There is no end to it all....The same thing applies to time."[17] Bondi said that "There is no point origin and no initial catastrophe in this theory."[18] Unfortunately, there neither is, nor can there be, a direct or an indirect inductive confirmation of the propositions that time extends infinitely into the past, that it will extend infinitely into the future, and that space extends endlessly in all directions. Logically, inductively, we cannot infer the infinite from the finite, which is all that we ever observe.

Is the infinity of spacetime plausible as an explanatory hypothesis? Since no empirical evidence favors it, why would anyone want to adopt this postulate? Cosmologists like Sandra Faber[19] and Roger Penrose[20] confess being attracted formerly to Steady State Cosmology on purely aesthetic grounds, because they found the ideas of endless time and space to be beautiful and those of an initial singularity and an absolute beginning to be ugly. In adopting a general worldview, aesthetic considerations may be inescapable in the *final* analysis; but most rational persons are skeptical of attempts to ground metaphysical worldviews primarily or prematurely on aesthetic preferences. Atheistic philosophers and astrophysicists definitely would not allow theologians to get away with believing in God merely because God is beautiful; so why should they get away with affirming an infinite Godless universe just because they find the thought of it beautiful?

In 1989, Fred Hoyle still defended Steady State Cosmology in an article[21] and in an interview conducted in August of that year. In the interview about his earlier work, he commented: "I don't really work in terms of belief. I didn't go beyond saying that the steady theory is a *possibility*."[22] However, his books and articles do not disclose that Hoyle cared *only* about possibilities. Yes, anyone who wants to know the actual world must explore possibilities, but mere possibilities are very cheap. They require no empirical knowledge at all, no experiments, and no verification whatsoever (all dear to the heart of Hoyle). As Leibniz proposed, being possible means nothing more than that the concept of something is free from logical self-contradiction. Matter that is actually antimatter is not possible; but matter that really is matter is possible. Pure possibilities give us no information at all about our actual world, except that it too is a possible world. Philosophers play interminable games with abstruse possibilities; but we expect more from scientific cosmologists. Usually they expect more of themselves.

To summarize, Steady State Cosmology was developed as an alternative to the somewhat primitive versions of Big Bang theory that were available in the late 1940s. Steady Staters were convinced that Big Bangers did not know what they were talking about, and *vice versa.* According to Hoyle and his associates, our universe is infinite in space and time; it has always been here; so no Big Bang is required to explain it. Their confidence that on a large scale the universe has always looked, on average, very much as it does today had to be reconciled somehow with the Hubble expansion. How can the observable universe always contain the same amount of intergalactic mass/energy if and when galaxies are constantly disappearing over the horizon of visibility due to the Hubble expansion process? Hoyle and Bondi resorted to continuous creation of matter out of nothing by pre-existing matter to account for the eternal renewal of mass/energy in the visible universe.

For many reasons, Steady State Cosmology is not plausible. No empirical evidence supports the continuous creation of matter out of nothing by pre-existing matter. Without continuous creation, Steady State Cosmology cannot account for the background radiation, the redshift of the galaxies, the Hubble expansion, and the dark sky at night. Continuous creation of matter out of nothing violates the First Law of Thermodynamics, according to which total mass/energy in the entire universe, not just the observable universe, is constant. Particles of matter and antimatter are always created together and then annihilate one another immediately; but no observations confirm the ubiquitous and continuous annihilation that should accompany ongoing matter/antimatter creation; and if it happened, we and our material world would not be here. Finally, the boundlessness of space and time are totally unverified and unverifiable, infinitely beyond the limits of scientific methods and knowledge. These metaphysical beliefs cannot be counted as scientific knowledge, and no convincing philosophical arguments support them. Crucial evidence for Steady State Cosmology is lacking. The evidence against it is overwhelming.

Perhaps another cosmology that affirms the boundlessness of spacetime is more credible. We turn now to Plasma Cosmology. Can it succeed where Steady State Cosmology failed?

3. Plasma Cosmology and Eric Lerner's Critique of the Big Bang

Today, occasional minor anomalies may not quite fit, but Big Bang Cosmology is massively supported and far from discredited. Big Bang Cosmology indisputably dominates all others; but it is not without its critics. A few astrophysicists emphasize transient discrepancies between the theory and available data, but these discrepancies are constantly being resolved. In 1991 Eric Lerner published *The Big Bang Never Happened.* Lerner's primary objections to the Big Bang, the central physical and metaphysical features of his Plasma Cosmology, and the basic flaws in his position will be discussed next. Interested readers should peruse his informative and challenging book for themselves.

A few cosmologists announced during the 1980s that they had found serious difficulties with the Big Bang theory of cosmic origins. Eric Lerner builds on their work and argues that discrepancies between theory and observation are significant enough to falsify Big Bang Cosmology. As he summarizes his main criticisms,

> The test of scientific theory is the correspondence of predictions and observations, and the Big Bang has flunked. It predicts that there should be no objects in the universe older than twenty billion years and larger than 150 million light-years across. There are. It predicts that the universe, on such a large scale, should be smooth and homogeneous. The universe isn't. The theory predicts that, to produce the galaxies we see around us from the tiny fluctuations evident in the microwave background, there must be a hundred times as much dark matter as visible matter. There's no evidence that there's *any* dark matter at all. And if there is no dark matter, the theory predicts, no galaxies will form. Yet there they are, scattered across the sky. We live in one.[23]

The following doubts arise about Big Bang Cosmology, according to Lerner.

A. Large-Scale Structures

Is the universe too immense for Big Bang Cosmology? Astronomers have now mapped and measured large sectors of the heavens. They found galaxies clustered together in vast strings, sheets, webs, and tapestries, separated by huge voids of seemingly empty space. These clusters of galaxies are themselves clustered into superclusters, separated by immensities of near emptiness. The superclusters are grouped into patterned megaclusters, divided by vast expanses of barrenness, almost without end. Age is correlated with size and velocity in an expanding spacetime system, and the age and magnitude of space in the observable universe are unimaginably vast.

Both the age and the size of the universe are far too great for Big Bang Cosmology, according to Eric Lerner. Big Bang Cosmology can allow for no objects older than twenty billion years at most, and no expanses greater than one hundred fifty million light years across; but the universe accessible to the best modern telescopes is radically different from what the Big Bang theory predicts. Lerner claims that some supergalaxies are between one hundred to one hundred and fifty billion years old, at least five times older than Big Bang theory can accommodate.[24] If some galaxies formed from an earlier state in which matter was spread smoothly through space, this matter must have moved for at least eighty billion years to arrive at its visible distribution; but the Big Bang says that the entire universe is at most only twenty billion years of age.[25] This theory cannot be saved by moving back the time of the initial explosion, Lerner contends, for that would be incompatible with the measured pace of the Hubble expansion.[26]

B. Dark Matter and the Galaxies

Does enough matter or mass/energy exist in the universe to explain how the galaxies were formed? This question is closely related to others, Lerner thinks. Is our universe closed rather than open? Does enough mass/energy exist to enable gravity to halt and reverse the universe's expansion? Enough mass to form the galaxies would equal that required to close the universe, according to Lerner.[27] Other cosmologists doubt that the relation of mass to forming galaxies and closing the universe is so exact. Stephen Hawking, who concedes uncertainty about it, suggests that "A tenth of the critical density would be enough matter for galaxies and stars to form."[28]

In Big Bang Cosmology, galaxies form as a result of tiny initial irregularities or fluctuational flukes in the early universe that spread over time to detach large irregular chunks of radiant energy and gases from one another. Later, many of these gaseous expanses solidified as gravity concentrated them into supergalaxies, galaxies, their stars, halos, and smaller clouds of gas and dust. Lerner insists that not enough physical mass/gravity exists in the galaxies or elsewhere to do the job. Big Bang Cosmologists assume that gravitational energy is the dominant force in the universe and that it caused the galaxies to form. Many believe that it will ultimately reverse the expansion of the universe. Without sufficient mass, there is not enough gravity for either purpose; and, Lerner contends, the *observable* mass definitely is not there.

The *visible* mass in the universe is only two percent at most of what is required to close the universe and form the galaxies, Lerner claims. Other cosmologists readily acknowledge that the luminous mass in the universe is barely one to two percent of the critical mass necessary to close the universe.[29] Ninety eight percent or so of the mass required to close the universe and form the galaxies is missing, says Lerner.[30] If Big Bang Cosmology is right, there should be no galaxies; but galaxies exist; so Big Bang Cosmology is wrong, Lerner concludes.

To save the theory, Lerner suggests, Big Bang Cosmologists conjure up massive quantities of *cold dark matter* and claim that it comprises ninety-eight percent or so of the physical universe. The trouble is, no one can find it! Astronomers can detect only the hot luminous matter that appears in the optical, ultraviolet, X-ray, gamma-ray, infrared, and radio regions of the spectrum; but this luminous matter falls significantly short of what Big Bang Cosmology requires. Rather than give up the Big Bang theory, its advocates concoct imaginary stuff that is not there. Big Bang Cosmology can be defended, Lerner charges, only by arbitrarily postulating the existence of purely fictional entities, a process akin to adding epicycles to defend the Ptolemaic geocentric theory of the solar system.

The cold dark matter of Big Bang Cosmology is merely deduced from theory, Lerner says, without a trace of observational support; and other prominent astrophysicists agree.[31] This missing mass, theorists concede, is not conventional matter like protons, neutrons, and electrons; if it were, it would be luminous. Radically different unobserved kinds of matter *must* exist in massive quantities to close the universe; so theorists concoct exotic particles like heavy neutrinos, axions and WIMPs (Weakly Interacting Massive Particles) which, by their own admission, no one has ever found. Astronomer Vera Rubin mentions "neutrinos, gravitinos, photinos, sneutrinos, axions, magnetic monopoles, and dozens more."[32] *If* they exist, they could easily supply the missing mass required to form the galaxies and close the universe. "Their only drawback," Lerner observes, is "that as in the case of cosmic strings, there is no evidence that they exist."[33] This drawback, if sustained, is quite serious indeed!

Actually, we possess significant empirical evidence for the existence of *some* cold dark matter. It is obtained by first measuring the velocities of stars in galaxies, and of galaxies in clusters. Then the gravitational force, and thus the mass, required to hold them in orbit or cause their observed movements is calculated. Finally, the mass of detectable luminous matter is subtracted from the required mass. The results indicate that observable matter is only a small fraction of what is necessary to prevent orbiting stars and galaxies from flying apart. According to Vera Rubin, at least 90 to 99 percent of the matter in the universe is dark matter that is "detected by its gravitational attraction on the matter which we can see."[34] Whether enough dark matter really exists to close the universe is a matter of heated controversy among contemporary cosmologists. As William Fowler put it, "According to the Caltech religion, the universe is open, and according to the Princeton superstition, it's closed."[35]

Lerner attempts to explain away all empirical evidence for dark matter. Citing research done by other astronomers, Lerner argues that dark matter is accepted by many astronomers because they greatly overestimate the masses of measured galaxies and clusters in two ways. These astronomers either count "interlopers," on this side or the other of observed galaxies as if they actually belong to the galaxies they are weighing,[36] or else they count small galaxies completely beyond the gravitational field of larger galaxies as if they belong to them. Either way, the mass and gravity of a measured galaxy will be greatly overestimated.[37] Lerner concludes that "These two errors would account for *all* of the "missing mass"; in pairs of galaxies, groups of galaxies, and clusters there is *no* dark matter."[38] Both of Lerner's explanations seem irrelevant, however. Whether or not interloper or nearby galaxies are beyond, before, or outside the gravitational fields of focal galaxies, their mass is *there* in the universe nonetheless.

Very few astronomers are persuaded that all cold dark matter can be explained away so easily. Much more evidence is available for the existence of cold dark matter than Lerner's exposition suggests. As Vera Rubin points out, the velocities of the outermost stars in spiral galaxies are not significantly less than the velocities of the innermost stars, which indicates the presence of dark matter because visible matter cannot explain this motion. Spiral galaxies are very different from our solar system, where planetary velocities decrease with increasing distance from the Sun, in accord with Newton's law that gravitational attraction decreases as distance increases.[39] The luminous centers of spiral galaxies seem to contain the most matter, so we would expect the velocities of their outermost stars to decrease, like the outermost planets of our solar system; but this does not happen.[40] Why? According to Rubin, "The conclusion is inescapable: matter, unlike luminosity, is not concentrated near the center of spiral galaxies. In short, the distribution of light in a galaxy is not at all a guide to the distribution of matter."[41] Thus, much dark matter must be there within the galaxy and/or its surrounding "dark halo."

Lerner concludes that even if some dark matter exists, contemporary measurements of celestial mass disclose "far too little to 'close the universe' and solve the various problems confronting the Big Bang theory."[42] Many contemporary cosmologists agree. Lerner probably loses the battle over the non-existence of dark matter, but he wins the war over whether enough of it exists to close the universe. Vera Rubin's studies of galactic dynamics disclose some missing mass; but, she admits, when all this dark matter is added in, the universe still has a critical density of .2 at best,[43] still far short of the critical density of 1. James Trefel remarked in 1988 that after including all detectable dark matter, we are left with only thirty percent of the critical value required for closing the universe.[44] Similarly, in 1990 H. Reeves wrote,

> The best estimates of the total (baryonic and nonbaryonic) cosmic density, from dynamic effects on galactic motions, yield values around ten percent of the closure density....There is no sound proof of the existence of a nonbaryonic matter contributing in a major way to the total density of the universe.[45]

In late 1995, Joshua Roth and Joel R. Primack affirmed that "counting the universe's luminous inhabitants–galaxies...only adds up to at most about one percent of the critical density," and that "galaxy halos typically contain enough dark matter to contribute at least 13 percent of the critical density, with preferred values exceeding 30 percent."[46] Many astronomers estimate the total dark and luminous mass to be around thirty to forty percent of critical mass.[47] Sixty percent or more of the mass required to close the universe can be provided only by postulating the existence of an odd attracting, not an Einsteinian repelling, Cosmological Constant, or by assigning significant mass to neutrinos. Neutrinos were found in 1998 to have a tiny mass,[48] but they add only another tenth of critical mass at most; and an attracting Cosmological Constant is nothing more than an *ad hoc* hypothesis unsupported by empirical evidence.[49] Any implausible theory can be saved if we are willing to posit enough epicycles and to violate egregiously the scientific principle of parsimony.

As far as we can tell, the density of our universe is only a small fraction of one, Omega, or critical density. Some non-luminous matter may be tied up in black holes, brown dwarfs–as confirmed by the Hubble Telescope in 1995, dim stars, and interstellar gas and dust; but astronomers were stunned in 1994 by the failure of the Hubble Telescope to find such things where most expected.

Some dark mass may be totally different in kind from any matter with which we are familiar; but when all the dark matter we can find is added to luminous matter, the total mass is at best only four percent or so of critical mass, according to an extensive review of all available evidence by Peter Coles and George Ellis in 1994.[50] Recent attempts by more than one research team to "weigh" the universe by examining the redshift of distant supernovae also favor a low-density universe.[51] Estimates of the grand total of dark and luminous matter obviously differ somewhat, but astrophysicists agree substantially that far too little exists to close the universe. Thus, even if some dark matter exists, this does not help very much. Emphasis on insufficient mass is not merely a Lerner eccentricity; many prominent astrophysicists agree that too little mass/energy exists to reverse the cosmic expansion process and close the universe.[52] As Coles and Ellis indicate, "On the balance of the evidence, an open Universe should be preferred."[53]

Conclusive evidence that we live in an open universe now seems to be available. Astronomers were stunned again in early 1998 when two research teams arrived independently at the unexpected result that the *rate* of cosmic expansion is actually increasing, not decreasing, as almost everyone since Hubble assumed. Gravitation/mass is not even powerful enough to *slow down* the pace of cosmic expansion, as universally assumed prior to 1998, much less to *stop* it. Supernovae in far distant galaxies were discovered to be much dimmer and further away (by ten to fifteen percent) than predictions based on a slowing rate of cosmic expansion could explain. The best explanation of their having traveled so much further than expected is that the rate of Hubble expansion is increasing, being driven by the pervasive energy of a repelling Cosmological Constant.[54] These initial findings were confirmed many times during and since 1998; and by the end of that year, the prestigious journal *Science* declared the increasing rate of cosmic expansion to be the "Breakthrough of the Year"![55]

Most astrophysicists, including Stephen Hawking who at first resisted,[56] are now convinced that the issue is definitively resolved: we live in an open universe, one that will never end in a Big Crunch. Thus, all cosmologies that presume a closed universe are utterly implausible. Some astronomers caution that the dimmer-than-expected light from these Supernovae might be best explained by the presence of cosmic dust, or by their taking longer than usual to achieve their maximum brightness. So far, the increasing rate of Hubble expansion has withstood every challenge;[57] and even if the Hubble expansion rate is not increasing, too little mass exists in the universe to close it. High precision observations of the microwave background made in 1999-2000 seem most compatible with a flat universe having an Omega of 1,[58] but neither flat nor open universes ever collapse, and both kinds are absolutely incompatible with oscillationism.

We probably live in an open universe with a finite past that will expand forever, but Lerner contends that an open universe would falsify all forms of Big Bang Cosmology. Readers will soon see that and why this is not so.

C. Cosmic Heterogeneity

Is the universe as homogeneous and isotropic as required by Big Bang Cosmology, or is it too clumpy and structured? Without dark matter and its gravitational effects, Big Bang theory cannot account for the heterogeneity of the universe, Lerner contends; but Plasma Cosmology can readily explain the formation of stars, solar systems, galaxies, supergalaxies, and cosmic heterogeneity. Big Bang Cosmology assumes that among the four basic forces of nature, only gravity is available to consolidate gaseous regions of mass/energy into heavenly bodies or into galactic and supergalactic structures.[59] Relying on the theories and research of Hannes Alfvén[60] and his associates in Sweden, Lerner proposes that electromagnetism can diversify the cosmos, where gravity alone cannot;[61] once we comprehend how electromagnetism operates on plasmas, we will know how the universe came to be structured.

Plasmas are "hot, electrically conducting gases;" and "Over 99 percent of the matter [in the universe] is plasma," Lerner writes.[62] Plasma Cosmology says that flowing electromagnetic currents pervade our plasma universe; the entire universe is a gigantic electrical power grid: "Plasma cosmologists envision a universe crisscrossed by vast electrical currents and powerful magnetic fields, ordered by the cosmic counterpoint of electromagnetism and gravity."[63]

Big Bang Cosmology ignores electromagnetism and relies on gravity alone to explain the lumpiness of the universe. This does not work, Lerner insists; but when both electricity and magnetism are considered, the heterogeneity of the universe is adequately explained. Gravity is not the only physical force that draws things together; electromagnetism does also. On a cosmic scale, it and gravity together structure the universe. Lerner's hero, Hannes Alfvén, contends that "Plasma becomes inhomogeneous naturally," that "Plasma, electrical currents, and magnetic fields work to concentrate matter and energy, to make the universe the complex, dynamic, and uneven place that it is."[64] Plasma pervaded by electromagnetic currents naturally produces swirling filaments, whirlwinds, and vortices through a "pinch effect" that pulls currents together when they are flowing in the same direction.[65] On a cosmic scale, these processes produce solar systems, stars, vast strings of galaxies, and enormous filaments of super-galaxies.[66] Furthermore, "Magnetic fields and currents can concentrate matter and energy far faster and more effectively than can gravity."[67] Indeed, "Plasma interactions can, given a few hundred billion years, form the super-cluster complexes." This is far too long for a Big Bang universe with only twenty billion years at most to spare, but for Plasma Cosmology "time is no problem."[68] A few hundred billion years hardly matter in an infinitely old universe.

D. Homogeneity and the Microwave Background

Does the microwave background really originate with the very earliest universe? The discovery in 1965 of the microwave background by Penzias and Wilson was crucial in persuading most cosmologists to accept the Big Bang theory of origins. Both the remarkably homogeneous black body radiation coming from everywhere in the universe, and the uniform distribution of matter on a large enough cosmic scale, seem to confirm Big Bang Cosmology's prediction of homogeneity and isotropy in the universe. Big Bangers construe the microwave background to be an enduring remnant of the age of radiation that began only minutes after the eruption of the primordial fireball. Lerner challenges the presumption that this is the only plausible explanation and offers an alternative hypothesis as much more credible.

Lerner's theory is simple. Electrons flowing through magnetic fields emit radio waves and microwaves. The microwave background originates in intergalactic magnetic fields that first absorb and then reemit microwave radiation.[69] The microwave background appears to come from everywhere rather than from specific intergalactic locales because, after a number of reabsorptions and reemissions, microwaves are scattered in all directions, and their radiation is "smoothed out."[70] Thus, the microwave background originates in intergalactic space and requires no Big Bang for its explanation.

Another problem, says Lerner, is that the microwave background is too smooth. Big Bang theory proposes that fluctuations and aberrations very early in the universe eventually created supergalaxies, galaxies, and stars. If this actually happened, some inhomogeneities in the microwave background should show up, but they do not, Lerner declares. Preliminary reports from the Cosmic Background Explorer (COBE) Satellite launched by NASA in 1989 indicated that "the Microwave spectrum is 'too perfect'," and this "rules out any way of forming the large scale structure of the universe from the Big Bang."[71] This challenge to Big Bang cosmology is very serious, but since Lerner wrote, the tables have turned decisively against him.

E. The Hubble Expansion and the Infinite Universe

Could the universe expand in accord with Hubble's law without a Big Bang? If so, how so? Lerner's Plasma Cosmology is not easily reconciled with the Hubble expansion of the universe and the redshift from distant galaxies. Like Steady State Cosmology, Plasma Cosmology incorporates a non-empirical metaphysics of infinite space and endless time. Steady State Cosmology tried to reconcile the Hubble expansion with an infinite and eternal universe by postulating everlasting continuous creation. How can Plasma Cosmology repudiate both the Big Bang and continuous creation, yet affirm the Hubble expansion? Lerner admits that this is not easy!

The Hubble expansion cannot be explained away, though Lerner would like to do so. Most cosmologists identify the redshift of the galaxies with the Doppler effect, and Lerner agrees after examining alternative proposals. In an Appendix,[72] Lerner discusses and rejects two alternative explanations for the redshift, first that light simply gets tired or loses energy as it travels through long distances (as Bondi believed), next that some unknown physical law causes the scale of everything to expand with time. Lerner repudiates the first because it requires a much greater density of matter than is available and because there is no evidence that anything absorbs energy from traveling photons. The second view is unacceptable because it cannot be confirmed and because it involves new, unverifiable, and implausible laws of physics. Lerner concedes that since light arriving from a source moving away from an observer shifts toward the red end of the spectrum, the galaxies must be moving away from us as the universe expands.

To explain the Hubble expansion, Lerner prefers a proposal developed by Alfvén, according to whom it results from what I will call a "Mini-Bang," though this is not Lerner's terminology. Lerner thinks that a Mini-Bang (perhaps more than one) occurred when a limited region of the infinite universe was blown apart by a colossal matter/antimatter collision to form our observable universe.

Alfvén and Lerner are convinced that substantial quantities of antimatter exist in the infinite vastness of space. It is normally separated from our material corner of the universe by electromagnetic vortices, but occasionally matter and antimatter collide. Billions of years ago, in our small corner of the infinite cosmos, matter contracted gravitationally, not to a singularity, but to a hundred million light-years across, a tenth of its present size.[73] Massive quantities of matter and antimatter just happened to be in the same vicinity, so one or more matter/antimatter explosions occurred, producing our observable expanding universe or "metagalaxy." If the Big Bang never happened, Lerner must still answer the question: "What caused the Mini-Bang?" He answers that ten to twenty billion years ago it resulted from a gigantic collision of matter with antimatter in our little corner of infinite space. Residual kinetic energy from that explosion accounts for the Hubble expansion,[74] which was "in no way a Big Bang that created matter, space, and time. It was just a big bang, an explosion in one part of the universe."[75]

Lerner's Plasma Cosmology thus resorts to a Bang after all, but only a metagalaxy-producing Mini-Bang. Not surprisingly, he abhors the outcome that he embraces.[76] He concludes, "The question of the Hubble relation remains unanswered," and "Far more theoretical and observational work is needed."[77] But "Why the Hubble expansion?" remains unanswered only if Big Bang Cosmology is rejected! Big Bang theory provides a very plausible answer!

Plasma Cosmology invites a number of serious questions. Is *the entire* infinite universe expanding like the *observable* universe? Is this a material world through and through? Plasma and Big Bang Cosmologies give very different answers.

What would Plasma Cosmologists expect to find if we could see very deeply into infinite spacetime? What are those parts of the universe like that were *not* affected by our localized Mini-Bang? Lerner does not answer, but we can make some educated guesses and predictions. Since most of the allegedly infinite universe was not affected by our Mini-Bang, it would not be involved in *our* Hubble expansion. Other metagalaxies might express their own expansions or contractions. Some of them might be headed directly toward us. Some might be composed of antimatter or, heaven forbid, even dark matter; and they could exhibit radically different natural laws.

How could our own metagalaxy belong to the "same universe" with innumerable causally unrelated metagalaxies? Lerner neither asks nor answers, but the question deserves some serious consideration. Could any two metagalaxies belong to a single universe without any causal connections or lawful spatiotemporal continuities, at least at their edges? These edges might be so far away from local astronomers that they could not observe any metagalactic interactions. *Totally independent universes* might exist without causal contact and spatiotemporal continuity; but if metagalaxies belong to *the same universe,* surely they must affect one another, be spatiotemporally continuous, and share many if not all natural laws. Actually, Plasma Cosmologists do not have the slightest trace of scientific evidence that the universe is infinite, that other metagalaxies exist (like Lerner's alleged matter and antimatter worlds that collided to produce our world), or that they are distantly continuous with our own. Even if these presumptions were true, we could never know it. Very few, if any, metagalaxies in Lerner's infinite universe ever affect our observable world, so how could they belong to our universe?

Big Bang cosmologists, by contrast, do not identify the whole expanding universe with the *observed or observable* universe, but they think that it is finite and in principle observable and continuous with our own causally, spatially, temporally, and in other ways. They expect to investigate more and more of it as better instruments like the Hubble Space Telescope are deployed, repaired, and upgraded. They predict that the most general features of the as yet unseen far distant universe will significantly resemble what we have seen already. This has always been the case as better scientific instruments have allowed us to probe deeper and deeper into the unknown universe, and it should not change. Big Bang Cosmologists expect the now unknown universe out there to be composed mostly of matter, not antimatter, and to obey familiar physical laws, including Hubble's law of uniform expansion. Results obtained from the Hubble Space Telescope consistently confirm these expectations.

The further out we look in space, the further back we see in time; even at the speed of light, considerable time, billions of years in some instances, lapses before photons and radiant energy reach us from far distant objects. Big Bang Cosmologists expect the most distant objects in the universe to be moving away from us and from one another at speeds roughly proportional to their distance, with minor variances due to local gravitational fields. These speeds, they anticipate, will gradually approach the speed of light. They expect that the most distant objects in the heavens will be observed in their youth, as they were billions of years ago when their light and radiant energy now arriving here departed from there. So far, with every improvement in technology, their predictions pan out.

Some of the most distant objects that astronomers have discovered thus far are the quasars–starlike objects that contain as much mass/energy as entire galaxies. In late 1998, quasars that are thirteen billion light years away were detected.[78] Quasars, seen in their youth, are some of the most distant and thus the oldest celestial objects known to us, but they obey familiar physical laws. In January 1993, radio astronomers, using the facilities at Kitt Peak Observatory in Arizona, discovered giant clouds of gas twelve billion years old in which galaxies are being formed. These clouds contain as much mass as entire galaxies, but no stars had formed in them twelve billion years ago. Since 1996, many additional observations with the Hubble Space Telescope disclosed galaxies as they existed nearly twelve billion years ago, within three billion years of the Big Bang, and found them to be smaller but much closer together and more numerous than those nearer to us in space and time; the small early ones probably merged over time to form the large later galaxies.[79]

The Hubble Space Telescope now provides us with a vision of embryonic protostars and new stars being formed in and ejected from dense clouds of space dust.[80] Big Bang theory predicts that galaxies go through gaseous phases before they develop into stars and galaxies; and this prediction is now confirmed. In 1994, in a closer galaxy, astronomers identified for the first time a planet in orbit around a distant sun in our Milky Way. By the end of 1995 at least four were recognized, with new ones being added almost every month thereafter. By the end of 2000, nearly fifty or so were known, some earth-size, most Jupiter-size.[81] As we learn more and more about our universe, we find no far distant metagalaxies or alternate antimatter universes. We just find more and more of what Big Bang Cosmology predicts.

Nothing can be accelerated through existing space faster than the speed of light and cross the speed of light barrier, says the theory of relativity; but the Hubble expansion of space itself can separate celestial objects at faster-than-light speeds, thereby making it impossible for these objects ever to communicate causally. This is the "horizon problem" discussed earlier. According to Hubble's Law, each time the distance doubles, the speed doubles. Without violating relativity, some galaxies are moving away from local observers at speeds that exceed 186,000 miles per second, the approximate speed of light. Objects moving away from us faster than light are forever beyond our horizon of visibility. Their light will never reach us. This is one good reason why we cannot identify the observable universe with the total universe that originated with the Big Bang. Though his estimate is admittedly only approximate, Alan H. Guth surmises that "...The entire universe is expected to be at least 10^{23} times larger than the observed universe;...the observed universe is only a minute speck in a universe that is many orders of magnitude larger."[82] This largeness should not be confused with infinity, however.

Astronomers look back in time as they see further away in space, but Big Bang Cosmologists do not expect to see the Big Bang itself, or anything beyond and before the Big Bang. No optical telescope will ever see through the fog of cosmic radiation that lasted for the first 700,000 years. Since the microwave background dates back to the onset of this age of radiation around 300,000 years after the Bang, radioscopes already see at least that far back in time. Milton Munitz speculates that if we could develop telescopic instruments capable of detecting neutrinos, gravitons, magnetic monopoles, or free relic quarks, we could access information coming from the earliest fraction of a second after the Bang.[83] Unfortunately, at present we are far from having such instruments. Commenting on the Ligo and Lisa observatories now being constructed to detect gravity waves, Gary H. Sanders and David Beckett observe,

> As early as 10^{-43} second after the Big Bang, space and time became differentiated and gravitational radiation was able to stream freely in all directions. In principle, LIGO could detect these primordial waves, enabling us to hear at last the hush of the universe's birth.[84]

From this also we are infinitely far removed in practice. We do not know much about our universe, but everything that we do know favors Big Bang Cosmology.

To summarize, Eric Lerner avidly repudiates the Big Bang theory of cosmic origins. He contends that the Big Bang's ten to twenty billion year age for the universe is far too small to accommodate recently mapped supergalaxies that are a hundred and fifty billion years old. Without massive quantities of dark matter, there is too little gravity for the Big Bang cosmos to form stars, galaxies, and supergalaxies; and no good evidence indicates that any cold dark matter exists. Big Bang theory is oblivious to the power of electromagnetism to form heterogeneous astronomical structures. Electromagnetism accounts for the microwave background without a Big Bang, and the COBE satellite found no evidence of galaxy-producing fluctuations in the cosmic background radiation.

Defenders of the Big Bang can offer intelligent responses to these doubts.

4. Critique of Plasma Cosmology

Plasma Cosmology has its own grave defects, some concerning its physics, and some its metaphysics.

Lerner's contention that the universe is infinite in space and time sounds remarkably like old fashioned naturalistic metaphysics. Is he a Naturalist? Many Theists, he notes, also believe that the universe is unbounded in space and time, and they regard an infinite world as an expression of God's infinite creative power. Merely believing in infinite spacetime does not make one a Naturalist. Other Theists believe that the universe is finite in time and space, even though a finite universe implies no more than finite Divine power and creativity.[85]

Lerner avows that Plasma Cosmology is neutral with respect to atheism and theism, that it "does not demand a creative God and is perfectly compatible without one. But nor does it preclude a creative deity."[86] This cunning evasion does not successfully conceal Lerner's commitment to a Humanistic Naturalistic worldview. He contends that scientific method alone yields truth, and he rejects all "mythological" explanations.[87] He develops an optimistic, futuristic, and humanistic understanding of mankind and human values.[88] He makes no appeal to God, and twice he expresses a decided preference for a naturalistic theory of origins.[89] He vehemently defends the principle that all events have (natural) causes against what he regards as irrationalism, occultism, mythology, and mysticism. He even finds antiscientific perspectives, which he opposes, in fashionable but fanciful metaphysical interpretations of quantum physics.[90]

But Lerner's naturalistic Plasma Cosmology has its own serious problems.

A. A Universe Infinite in Space and Time

Unlike Big Bang Cosmology, Plasma Cosmology affirms that the universe is infinite in both space and time. Is it reasonable, is it scientific, to adopt these features of Lerner's Naturalism? If scientific method alone yields truth, what scientific evidence proves that we live in an infinite universe? None at all, even if we do! Early in his book, Lerner proclaims that "Philosophers such as Nicholas of Cusa and Giordano Bruno had advocated the idea of a universe unlimited in time and space, eternal and without beginning. But no scientist had justified these notions with hard data."[91] These words tacitly promise to provide hard data later to prove that space and time are infinite; but Lerner never fulfills this promise. Later he reaffirms that the finitude or infinitude of the universe "is a scientific question that must be answered by observation;" but almost immediately he concedes in a footnote that "Strictly speaking, it is not possible to *prove* scientifically that the universe is infinite. But it is quite possible to claim that we have no observational evidence that it is finite."[92] Thus, he argues, lack of evidence for finitude counts as evidence for infinity. Are his claims plausible?

First, we observe only finitude, so all our observations count only for finitude. More importantly, we really cannot show that the universe is infinite merely because we have no evidence for its finitude, or against its infinitude. All appeals to lack of evidence are fallacious arguments from ignorance, and they establish nothing.

The naturalist Sidney Hook wrote,

> The existence of God, immortality, disembodied souls or spirits, cosmic purpose or design, as these have been interpreted by the great institutional religions, are denied by naturalists for the same generic reasons that they deny the existence of fairies, elves, leprechauns, and an invisible satellite revolving between the earth and the moon.[93]

So, why should we disbelieve in the existence of such fanciful entities? We could argue that God, disembodied souls, cosmic purpose, pixies, elves, and invisible satellites between us and the moon really exist just because there is no observational evidence against them. No one with any knowledge of logic would accept such an argument from ignorance. Yet, Lerner's central argument for the infinity of nature is no better. Neither logic nor scientific method entitle us to believe something simply because no evidence disproves it. We would surely have to accept the reality of every occult entity imaginable if having no observational evidence *against* their existence counts decisively as evidence in their favor. Yet, this is all that Lerner or any other naturalist can say for their infinite and self-sufficient universe–there is no evidence against it (which ignores the Big Bang). Lerner's Naturalistic metaphysics hangs on an *argumentum ad ignorantium*.

We really should disbelieve in pixies, elves, and invisible satellites because, after a thorough search, we find no observational evidence *for* their existence. Their existence is *excluded by the presence of other things* that we know to exist. Thorough searches yield knowledge, not ignorance, and justify many negative conclusions. Negative propositions like "There is no butter in the refrigerator," "Unicorns don't exist," and "No stars are composed of antimatter" can be confirmed by observation to a very high degree of probability. If we are in no position to make a thorough search, we may not know some things for certain; but we can still know many negative truths within limits of high probability. Still, complete lack of evidence for one theory does not establish its opposite.

So what does the available evidence actually show about our universe? Our most thorough searches disclose only finitude. Empirically or scientifically we can only *know* finitude, even if the universe is actually infinite. Further, all the *positive empirical evidence* for the Big Bang (given in Chapter One) counts decisively against the infinity of space and time. Many allegedly scientific cosmologies covered in later chapters accept the Big Bang; but they postulate other universes in ways that leave all accessible evidence and all natural science far behind. For them and for Plasma Cosmology this is a great weakness.

Another subtle but unsound argument for infinite space and time runs through Lerner's book, one from historical association. He says that historical figures like Anaxagoras, Nicholas of Cusa, and Giordano Bruno, who captured the spirit of science and scientific method also believed in an infinite universe. He insinuates that these historical facts somehow imply that the universe really is infinite; but this does not follow. How did Lerner's historical heroes know that the world is infinite? What empirical evidence for it did they adduce? None!

The truth about how Naturalists like Lerner arrive at an infinite universe surfaces when he writes, "Plasma cosmology assumes that, because we now see an evolving, changing universe, the universe has always existed and always evolved, and will exist and evolve for an infinite time to come."[94] The key word here is "assumes"; but his opponents can just as easily *assume* the contrary, if that is all that there is to it. Lerner's Infinite-world Metaphysics is merely an unjustified assumption, merely an expression of groundless metaphysical faith. No logic warrants reasoning from finite observational premises to conclusions that affirm infinity, and all the positive evidence for the Big Bang weighs heavily against Lerner's position.

B. Hydrogen and An Infinitely Old Universe

Plasma Cosmology confronts scientific as well as metaphysical obstacles. Our universe is around seventy five percent hydrogen, as spectrographic scans of the heavens repeatedly confirm. If the universe were infinitely old, as Plasma Cosmology maintains, no hydrogen would exist in it today, given the laws of physics as we know them. Hydrogen is constantly being synthesized into helium and heavier elements in the stars; *and this hydrogen cycle is irreversible.* If nucleosynthesis of hydrogen into helium and the heavy elements has occurred throughout infinite space for an infinite amount of time, every hydrogen atom in the infinite universe would have passed through an infinite number of exploding supernovas, each of which would have converted some hydrogen into heavier elements. If this has been transpiring for an infinite amount of time, no hydrogen would remain in the universe today. Yet, the hydrogen is there. It makes up nearly seventy five percent of the universe as we know it, and its existence is overwhelming evidence against an infinitely old universe.

Fred Hoyle raised the foregoing objection against his own Steady State Cosmology.[95] To avoid the difficulty, he resorted to the implausible continuous creation of hydrogen atoms. Can Plasma Cosmology provide an alternative hydrogen-renewing mechanism? Electromagnetism cannot save the theory, for it does not reverse the process of nucleosynthesis. Could Lerner's Mini-Bang solve the problem for Plasma Cosmology? Can hydrogen be renewed periodically in the infinite universe by matter/antimatter annihilations that create infinitely many Hubble-expanding metagalaxies like our own? Lerner has not worked out the physics for this; and the physics that we know is decidedly against it. Matter/antimatter explosions always produce gamma rays, never hydrogen or helium atoms, the dominant elements in our universe.

Besides, no convincing evidence indicates that antimatter exists in sufficient quantities to produce metagalactic Mini-Bangs.[96] Lerner concedes that evidence for the existence of massive quantities of antimatter is inadequate, and he pleads for "more observation."[97] Plasma Cosmology replaces the open universe's enigma of the cold dark matter that is not there with that of the antimatter that is not there!

Observation actually shows that very little antimatter exists in the universe, as far as we can tell. If some far away galaxies are composed of antimatter, it might seem difficult at first to know this, according to I. L. Rozental, because "Antimatter emits photons in absolutely the same way as matter does."[98] To us, light from antimatter galaxies would look exactly like light from galaxies composed of matter. However, Rozental adds, galaxies emit particles as well as photons; and antimatter galaxies would emit antiprotons and positrons. If vast quantities of each exist, then somewhere in space antiparticles would be colliding constantly with material protons and electrons. Significant and detectable matter/antimatter annihilations would result; but none have been discovered after extensive searches; so the existence of antimatter galaxies is extremely doubtful.[99]

Lerner might reply that although no discernible antimatter exists now, it might still have existed at the time of the Mini-Bang. According to the Big Bang theory itself, the initial universe-creating thermonuclear explosion was driven primarily by stupendous matter/antimatter collisions; so it too must reconcile the past existence of antimatter with its virtual non-existence at present.

Which theory offers the most plausible account of the origin and presence of the antimatter that fueled its Bang? In Big Bang theory, either a surplus of matter prevailed from the outset as a given initial condition of creation, or else both antimatter and matter emerged from X bosons (derived, in turn, from pure energy during the first few pre-matter minutes of creation) that decayed irregularly into a material world. How does Plasma Cosmology explain the origin of the massive quantity of antimatter involved in its Mini-Bang? How was it separated from matter? How did it survive annihilation by matter prior to its collision with a material world undergoing metagalactic gravitational collapse?

Lerner recognizes that antimatter is produced on earth whenever matter is derived from energy, so it would be an immense puzzle, he suggests, if this did not happen throughout the universe.[100] Yet, whenever this happens on earth, matter and antimatter immediately annihilate one another, so it would also be a great enigma if this too does not happen throughout the universe; but it doesn't. Lerner claims that antimatter can be separated from matter if both pass through a magnetic field containing a current that squirts matter out in one direction and antimatter out the opposite way. Clouds of matter and antimatter would then form and be pushed apart by the Hubble expansion. Multiple metagalactic gravitational collapses might later bring them together again to produce a Mini-Bang.[101]

This explanation cannot be verified directly, Lerner admits.[102] Neither can the standard Big Bang explanation of why we live in a material world, we must concede. The true test of a scientific theory, Lerner contends, is the correspondence of its predictions with observation. Apparently, he regards his cosmology as an explanatory hypothesis that is justified by its fruits; but Plasma Cosmology's explanation of why we do not live in an antimatter world flunks this test. Big Bang Cosmology, with all the supporting evidence presented in Chapter One, passes.

C. The Vastness of the Universe

Lerner proclaims that the visible universe is too vast to have been produced within Big Bang's paltry ten to twenty billion year-old universe. Recall his claim that astronomers have now mapped supergalaxies that are at least one hundred to one hundred and fifty billion years old. This would make them at least five times older than the Big Bang's entire universe. Something has to give!

In astronomy, ages are related to velocities, distances, magnitudes, and masses; and our estimates of age depend on what we know about such things. Lerner's predicament may appear at first to arise because estimates of galactic and supergalactic ages, velocities, distances, sizes, and masses are notoriously inexact. Methods available to astronomers do not yield very precise measurements; but improved methods, enhanced by the power of the Hubble Space Telescope and the Hipparcos Satellite have already given us a much more accurate perspective on the age and composition of the stars and of the cosmos. Astronomical distances are judged by such methods as the parallax, which indicates the displacement of celestial objects when observed at the same moment by two different observers in two different places, by the intrinsic brightness of certain stars like the Cepheid variables that function as standard candles, and by spectral analyses of redshifts.[103] The Hubble Space Telescope now views galaxies so far removed from us that Cepheids are not detectable; so today's astronomers are using the intrinsic brightness of exploding supernovae as standard candles for measuring vast cosmic distances.

Hubble's law of uniform expansion affirms that celestial distances increase uniformly in proportion to velocity, and the Hubble constant sets the scale of the universe as the ratio of distances to velocities. Much more detailed and exact measurements of cosmic distances and ages made by the repaired and upgraded Hubble Telescope and by the Hipparcos Satellite[104] have already produced more evident and precise estimates of the age of the stars and of the universe; and it will continue to do so.

Still, the inexactness of measurement on an astronomical scale does not procure more than an eight to twenty billion year variance for the age of the universe. It fails to supply the hundreds of billions of years required by Lerner's "ancient structures." Lerner does not say where he gets his figures. He seems to do his own calculations, but most astronomers simply do not accept them. Discrepancies between dates and data may be dismissed as only a "temporary difficulty," but not a "permanent breakdown of the Big Bang itself."[105] In 1993, George Smoot concluded from his research team's momentous study of the cosmic background radiation by the COBE satellite that doubters about the Big Bang are now proved wrong because

> The existence of the wrinkles in time as we see them tell us that big bang theory, incorporating the effect of gravity, can explain not only the early formation of galaxies but also the aggregation within 15 billion years of the massive structures we know to be present in today's universe. This is a triumph for theory and observation.[106]

Today, most astronomers would agree, not without justification, that Lerner and others of like mind plainly miscalculate the age and the vastness of the most ancient structures in the visible universe. The best available numbers clearly support the "Standard Big Bang" model of the origin of the universe.[107]

D. Mini-Bangs and the Age of the Universe

If the visible universe is really too vast to have been produced in a mere ten to twenty billion years, as Lerner maintains, then neither Big Bang nor Plasma Cosmology can accommodate the discrepancy. Plasma Cosmology resorts to one or more Mini-Bangs to explain the Hubble expansion, so it has exactly the same problem. If the Big Bang cannot account for the age and expanse of the visible universe, how could Plasma Cosmology's Mini-Bang do any better? In describing what followed the Mini-Bang, or the series of them, that allegedly produced our visible world, Lerner fumbles to avoid this difficulty.

> Trapped in magnetic fields, these particles drove the plasma apart over hundreds of millions of years. The explosions were small enough not to disrupt previously formed filaments of plasma, so these far more ancient objects still exist today, in expanded form–just as designs printed on a balloon persist while it is inflated.[108]

Exactly what Lerner is getting at here is obscure. Are his "ancient objects" within the visible universe, or do they exist only in some theoretically constructed but unexperienced universe? Assuming the first, Lerner may have something like this in mind: The most distant parts of the visible universe contain galaxies and supergalaxies that are a hundred to a hundred and fifty billion years old, he claims. These did not participate in and were not affected by the Mini-Bangs that created the Hubble expansion. Thus, one hundred and fifty billion-year-old structures can exist within a metagalaxy that is only ten to twenty billion years old. The force of the Mini-Bangs that created our visible universe bypassed these structures, but somehow they were swept into our visible universe. How did this happen? Why did the Mini-Bangs not destroy them?

Because vast cosmic structures that antedate our Mini-Bang(s) (which other astronomers cannot find) would defy Hubble's law of uniform expansion, Lerner's explanation does not work. Except for gravity-bound galaxies like the Andromeda Galaxy and our own Milky Way, *all* structures in the visible universe, no matter how vast, participate *uniformly* in the Hubble expansion on a cosmic scale. *None* proceed through the heavens at a pace that drastically defies Hubble's law. Beyond gravitationally bound local groups, all galactic and supergalactic structures take part uniformly in the Hubble process. According to *Sky and Telescope*, "that hundreds of galaxies, including our own, are collectively swarming toward a super-massive entity dubbed the Great Attractor...describe(s) a subtle distortion in the universe's otherwise stately expansion, not its wholesale reversal."[109] Only a single Bang with cosmic-wide effects can account for such ubiquitous cosmic homogeneity. No vast visible plasma filaments exist that are exempt from the effects of a single initial blast.

Lerner concedes that his "ancient objects" exist today "in expanded form"; but what caused their expansion if they were not brought into being, affected by, or disrupted by his own initial Mini-Bang? A matter-antimatter explosion powerful enough to initiate our expanding observable universe would either destroy all nearby pre-existing structures or push them forever beyond our horizon of visibility. Lerner himself cannot account for structures that are hundreds of billions of years old within a metagalaxy he concedes to be only ten to twenty billion years of age.

E. Electromagnetism and the Age of the Universe

Gravity is too weak, Lerner insists, to concentrate mass/energy into galaxies and supergalaxies in only ten to twenty billion years. However, if, as Lerner maintains, electromagnetic forces can concentrate matter and energy much more effectively and rapidly than gravity, and *both gravity and electromagnetism* are available to concentrate mass, then his own plasma physics should help to explain how the stars, nebulae, and large scale structures of the universe came into being within Big Bang time. Plasma physics itself seems to make Plasma Cosmology's infinite and eternal universe unnecessary and comes to the support of the Big Bang. Calculations involving causal processes that can concentrate mass into galactic structures and help close the universe must encompass both electromagnetic and gravitational attraction, if Lerner is right. Perhaps electromagnetism and gravity conjointly determine large-scale cosmic structures; and together they determine whether our universe is open–to expand forever, or closed–to slip some day into reverse. Astrophysicists must do the computations; but the recent discovery that the rate of Hubble expansion is increasing, not decreasing, clearly indicates that the combined force of gravity and electromagnetic pinching is insufficient to reverse the course of cosmic enlargement.

F. Background Radiation

Lerner holds that the microwave background can be explained adequately without *any* appeal to the Big Bang. If he is right, his own metagalaxy-creating Mini-Bang would produce background radiation equal to that produced by a cosmos-creating Big Bang. Yet, all observable background radiation, he maintains, was produced by intergalactic electromagnetic fields. He takes no notice of all this extra background radiation. How much background radiation should astronomers expect to find if Plasma Cosmology is correct? If the radiation Lerner believes to be produced by intergalactic electromagnetic fields is added to the radiation residue of his Mini-Bang, the total would far exceed all the measurable background radiation in the universe. One Bang, whether Mini or Maxi, accounts quite sufficiently for all detected cosmic background radiation. Lerner's Mini-Bang is incompatible with his electrodynamic solution to the background radiation problem. If intergalactic electromagnetic fields could generate all observable background radiation, and if his Mini-bang would do exactly the same, there should be twice as much background radiation as there actually is. But there isn't.

G. The Smoothness of the Universe

With new instruments of exploration like the Cosmic Background Explorer (COBE) Satellite, astronomers can now "see" back to within 300,000 years after the Big Bang. In the preliminary COBE report cited by Lerner, the universe appeared to be too smooth to support the Big Bang account of the origin of heterogeneous structures in the universe like supergalaxies, galaxies, and stars. Originally, as Lerner suggests, COBE investigators found no evidence of fluctuations in the microwave background radiation.[110] After he published *The Big Bang Never Happened* in 1991, another more definitive report on COBE data based on much more detailed analysis was presented to the American Physical Society meeting in Washington, D. C. on 23 April 1992 by George Smoot, who headed the team of COBE researchers.[111] Members of the Society were electrified by the announcement that "ripples" were indeed detected in the cosmic background radiation.

Variations in temperature thirty millionths of a degree warmer or cooler than average were discovered on a scale of five hundred million light years across. These were originally very tiny, probably no greater than quantum level fluctuations; but over time minute differences become vastly expanded. The astrophysicist Michael Turner remarked that "The Holy Grail has been found. It's that important. If this evidence holds up to scrutiny, it is what we've been looking for 20 years. It confirms our ideas of how structures form."[112] Stephen Hawking, exaggerating a bit, called the COBE findings "the discovery of the century, if not of all time."[113] COBE Satellite data also provided evidence for the existence of dark matter, though still far from enough to close the universe. Thus, a primary source of information about the cosmos cited by Lerner to support his position now counts decisively against it.

To summarize, like Steady State Cosmology, Plasma Cosmology affirms the infinity of space and time. Naturalistic atheists like Lerner assume that an infinite universe has always been here, is totally self-sufficient, and requires no Big Bang and no God for its creation. Yet, empirical knowledge reaches its limits long before we arrive at spatiotemporal infinity, so Lerner's avowedly scientific metaphysics is scientifically baseless.

Lerner tries to refute Big Bang Cosmology and offers Plasma Cosmology in its place. He emphasizes the dominance of gaseous plasmas in cosmic structures and the enormous cosmic effects of electromagnetism; but Plasma Cosmology comes to grief over the redshift and the Hubble expansion. Lerner reluctantly postulates a Mini-Bang resulting from the gravitational collapse and collision of antecedently existing regions of matter and antimatter to explain the creation of our observable expanding metagalaxy. He affirms that the universe beyond our metagalaxy is infinite and implies that it is not involved in our Hubble expansion. He cannot confirm this, and we should doubt it for many good reasons. If the universe is infinitely old, no hydrogen would remain in it today; but it is almost seventy-five percent hydrogen. Plasma Cosmology offers no mechanism to reverse the process of nucleosynthesis of hydrogen into heavier elements in the stars. No matter-antimatter Mini-Bang could reverse the hydrogen cycle.

Lerner's visible universe was itself supposedly caused by a Bang ten to twenty billion years ago, so his own theory has the same (unsolved?) problems of scale that he attributes to the Big Bang. If some structures in the visible universe really are a hundred and fifty billion years old but nevertheless participate uniformly in the Hubble expansion, Lerner's own twenty billion-year-old visible universe cannot accommodate them. Most cosmologists do not accept the exaggerated age and size that he assigns to anomalous cosmic structures. If electromagnetism concentrates energy more rapidly than or in addition to gravity, this mechanism actually supports the contention that all visible structures in the universe came into being within Big Bang time.

Lerner's Mini-Bang would also produce a microwave background, so there would be far too much cosmic radiation if, as he contends, intergalactic electromagnetism generates all discernible background radiation. The latest results from the COBE Satellite actually confirm the reality of diversifying structural fluctuations in the microwave background. They also confirm that our universe is exceptionally fine-tuned for life. Calculations by Max Tegmark and Martin J. Rees "show that if the CMB's (Cosmic Microwave Background's) temperature variations were as little as ten times greater or smaller, life as we know it would not exist today."[114]

Lerner cannot establish an infinite spatiotemporal metaphysical background for his Mini-Bang. On the whole, his case against the Big Bang is very weak. The Big Bang theory of cosmic origins is still without a serious rival. But how did the Big Bang itself originate? What caused the Big Bang? At this point, we still have no plausible answer; but other possibilities remain to be explored.

Four

ANTECEDENT UNIVERSE COSMOLOGIES

The Standard Model of the Big Bang assumes that the universe was created all at once *out of nothing*, and the question of *what caused* its creation is left unanswered. It does not affirm that the universe was created *by* nothing, only that it was created *out of* nothing or *from* a singularity, which empirically is nothing. In Steady State Cosmology, the universe as a whole is uncreated, everlasting, and spatially infinite; but individual hydrogen atoms are constantly being created out of absolutely nothing by pre-existing matter through some totally mysterious and unverified process. In Plasma Cosmology also the universe as a whole is uncreated and infinite in both time and space, but the Hubble expansion of our visible universe or metagalaxy was caused by the explosion of massive quantities of pre-existing matter and antimatter that were brought together by chance in a limited region of infinite spacetime. Assuming that the Big Bang really happened, many causal explanations of it are available in contemporary scientific cosmology.

Antecedent Universe Cosmologies affirm that the Big Bang was a rebound from the collapse (Big Crunch) of an antecedently existing universe; it was caused by an influx of energy from a preexisting world. This naturalistic account of origins was given by George Gamow's Infinite Contraction/Squeeze/Bang/Rebound Cosmology which allows for only one contraction, one terminal squeeze, one Bang, and one rebound, and by Oscillation Cosmologies that postulate up to an infinite number of successive crunch/bang/rebounds. These two theories agree that our universe was brought into being, not by God, but by the collapse of an antecedently existing universe. In the next chapter we will explore a theory which claims that our universe was caused by indeterministic spontaneity within a pre-existing infinite Superspacetime.

Antecedent Universe theorists usually try to avoid the question of absolute origins by presupposing a naturalistic metaphysics that affirms the eternity and self-sufficiency of a system of nature that requires no God for its being, structure, or explanation. This metaphysics is not science, even if it is disguised as science. Without pretending to do natural science, throughout this chapter serious philosophical problems of intelligibility about prominent Antecedent Universe Cosmologies will be raised. If good philosophical or physical reasons are available for rejecting Antecedent Universe Metaphysics, one of many major obstacles to theistic belief has been removed.

Most Antecedent Universe Cosmologists intensely dislike creation *ex nihilo* and wish to avoid it because it seems to call for a Divine creation of the universe. Alan M. MacRobert, a regular contributor to *Sky and Telescope,* suggested in 1983 that "The idea of an oscillating universe, in which the Big Bang resulted from the recollapse of a previous phase of the universe, gained currency merely because it avoided the issue of creation–not because there was the slightest evidence in favor of it."[1] As I. L. Rozental said, "The only way to account for all these facts in the framework of physical concepts, without invoking a metaphysical power, is to admit the existence of an (infinitely large?) number of universes."[2] It looks as if, having made a prior decision that there is no God, Oscillationists just postulate an infinite series of antecedent universes as a way to avoid metaphysical appeals to Divinity. Unfortunately, Antecedent Universe Cosmologies are just as metaphysical, non-empirical, unsimple, unsupported, and unscientific as Theism, if not more so. It really is more so, as discussions to follow will show. Atheistic cosmologists seem to assume that the world's existence without God is totally unproblematic, and that only God's existence is problematic. Not so!

1. Gamow's Infinite Squeeze/Bang/Rebound Universe

George Gamow thought that prior to the Big Bang an antecedent universe had been collapsing from an infinitely expanded state throughout an infinite past; our universe came into being as a rebound from the minimal-size contraction state of this antecedent universe. In 1947, Gamow wrote,

> The universe is now expanding because in some previous period of its history (of which, of course, no record has been left), it contracted from infinity into a very dense state and then rebounded, as it were, propelled by the strong elastic forces inherent in compressed matter.[3]

Gamow never explained adequately what caused the rebound, or how to convert his concept of "elastic forces" into the formulas of physics; but he was convinced that the quantity of matter in our universe is insufficient to permit gravity to close it and bring about a second Big Squeeze. He predicted that "The distances between the neighboring galaxies are bound to increase beyond any limit, and *there is no chance that the present expansion will ever stop or turn into a collapse.*"[4] Thus, his theory of origins makes a place for only one Bang, preceded by a single but infinitely prolonged collapse, and followed by an unprecedented but infinite expansion. The expansion phase of the universe in which we happen to live will proceed forever toward an infinite fizzle. It will end with a whimper, not a Bang. About that, Gamow was probably right.

2. Critique of Gamow's Cosmology

George Gamow's cosmology does not survive serious critical examination for many reasons, two in particular. A. It cannot reconcile scientific knowledge with its postulate of an infinite past, and B. it cannot explain why gravity and/or mass were so different before the Big Squeeze.

A. Science and an Infinite Past

The central problem of Infinite Contraction/Squeeze/Bang/Rebound Universe is common to all Antecedent Universe Cosmologies. It affirms, presumably on "scientific" grounds, that our universe existed in some form throughout an infinite past; but this claim cannot be confirmed scientifically. According to Gamow, "Our Universe has existed for an eternity of time."[5] He conjectured that the antecedent phase "contracted from infinity," but he hesitated to speculate about the preceding stage of the universe. He cautioned that "There is nothing that can be said about the pre-squeeze era of the universe,"[6] because "the maximum compression of the universe, which squeezed all matter into a uniform nuclear fluid, must have completely obliterated all the records of the earlier compressive stages."[7] But our universe, the most impressive record of all if an antecedent crunch does indeed explain it, was not obliterated.

Gamow actually said a great deal about the Squeeze Era, despite his misleading warning that absolutely nothing can be said about it. He claimed that it existed, that it was of infinite duration, that it underwent gravitational collapse, that it was composed of matter not antimatter, that this matter was gradually compressed, that in its maximal compaction state it was 30 times larger than our sun, that this state contained "elastic forces," that it rebounded or exploded, and that our phase of the universe was created from this rebound. That is quite a lot to know to be nothing!

How did Gamow know all of these things that he was not supposed to know? If *all* scientific or empirical evidence was destroyed, as he conceded, then nothing that he tells us about the antecedent universe, including its existence and infinite duration, can be regarded as scientific knowledge. Good natural science must be grounded in experience as well as theory; but we experience only finite time and *only one phase of the universe*. We have no *direct* experience of infinite time, other cosmic phases, or other universes. We cannot inductively infer infinite time from finite time or many universes from one that is experientially unique. We may postulate the existence of an infinitely prolonged antecedent physical state, but no human experience could ever confirm the hypothesis directly or inductively. Belief in an infinite past is uncorroborated naturalistic metaphysics, not natural science. Naturalists complain that Theists believe in things that science cannot verify, but Naturalism suffers from exactly the same defect.

All Antecedent Universe Cosmologies are unscientific metaphysical guesses because they are totally unconfirmable. This failure is disastrous for theories offered in the name of science. Antecedent Universe Cosmologies presuppose that some universe in some form has endured perpetually through an infinite past. Yet, no empirical evidence for the infinity of space or time is available to us. Even if space and time really are infinite, we can have no scientific or empirical knowledge of it. No set of finitely enduring human scientists, no matter how large, can directly observe infinite space or time; and inductive logic forbids us to infer infinity from finite samples, just as it forbids us to infer white from samples of black or gold from instances of dross. As explanatory hypotheses, Antecedent Universe Cosmologies imply no observations that would confirm an Infinite World Metaphysics or that would falsify the hypothesis that space and time are immense but finite and created *ex nihilo*. Naturalists are welcome to their faith, but they cannot call it scientific knowledge. In fact, as we shall see, much of what we know scientifically and philosophically counts heavily against an infinite chain of successive universes.

B. Gravity and Mass/Energy in the Squeeze Era

Why does our present infinitely expanding phase of the universe lack sufficient mass/energy for gravity to shut it down, as Gamow believed, if the antecedent Squeeze Era contained enough mass/energy to contract and close it? How could the Squeeze Era include enough mass/energy to be closed by gravity if our present era does not have enough to shut it down eventually? Two possibilities suggest themselves. First, perhaps there actually was enough mass/energy in the Squeeze Era to close it, but much of it was destroyed in the Big Squeeze through some gross but unexplained violation of the law of the Conservation of Energy. Second, perhaps gravity was a much more powerful force during the Squeeze Era than it is now.

Neither option is very attractive for much the same reason. Both require radical changes in fundamental constituents of the universe during the Big Squeeze, and Gamow did not explain how gravity or mass/energy could make such fundamental changes. He did not even recognize the problem, but if he had it is doubtful that he could have solved it. In discussing Oscillation Cosmology, we will soon see that if Big Squeezes result in singularities, there can be no spatiotemporal or causal continuity from era to era. If Big Squeezes do not result in singularities, there is no good reason to think that basic laws, forces, and masses change dramatically from era to era, which creates additional problems.

In sum, by its own logic, the Infinite Contraction/Squeeze/Bang/Rebound Universe of George Gamow lies beyond the limits of scientific knowledge. Throughout an infinite past, supposedly, an antecedent universe contracted to a point of maximal compression and minimal size thirty times larger than our sun, then exploded in a Big Bang to create our open and infinitely expanding universe; but, Gamow conceded, we can know absolutely nothing about the antecedent universe. Even if one existed, all evidence for and about it would have been destroyed in the primordial fireball. Gamow clearly affirmed scientific agnosticism but inconsistently practiced unscientific gnosticism. Science cannot establish that there was an antecedent universe, much less that it collapsed throughout an infinite past. Gamow conceded that many things cannot be known, then pretended to know them. He did not identify the physical forces and laws that initiated the rebound; and he did not explain why the present infinitely expanding phase of the universe lacks sufficient mass/energy for gravity to close it, despite the antecedent Squeeze Era's containing enough to shut it down. No cosmologists today defend a temporally infinite universe with only one Squeeze/Bang/Rebound; but other versions of Antecedent Universe Cosmology are vigorously affirmed.

3. Oscillation Cosmology

According to Oscillation Cosmology, an antecedent collapsing universe ending with a Big Squeeze or a Big Crunch preceded our Big Bang. Our Big Bang was caused by a massive influx of energy from this earlier era. Did only one universe precede our own, or did an infinite number of collapsing/exploding/expanding universes antedate us? Conceivably, only a few Bangs preceded the one that initiated our universe. If so, the question of absolute origin arises with respect to the *first* member of this finite series. What caused the *first* Bang? God, chance, or what? What selected just those few closed and oscillating universes for actualization out of an infinite number of different possible universes? Many cosmologists are convinced that some energy would be used up and lost with each oscillation, so in a finite number of rebounds, all mass/energy whatsoever would be lost. Nothing would exist today; yet, here we are!

An ancient myth explains what holds up the earth–it rests on the back of a turtle! And what holds up the turtle?–another turtle; and another; and another.... If prying minds persist, the final answer is, "The turtles go all the way down!" As does this myth, Oscillation Universe Cosmologists mistakenly believe that they can avoid the question of ultimate origins if our universe was preceded by universes that go all the way back (to infinity), each of which commenced with a Bang and ended with a Crunch. All these supposed antecedent universes start with Bangs, then expand, halt, contract, collapse, and finally explode to create new universes *ad infinitum*.

Oscillation Cosmologists are reluctant to calculate the duration of a complete cosmic cycle from start to finish; but a few make educated guesses. A. Karel Velan maintains that our universe is presently about 18 billion years old, that it will continue to expand for another 17.5 billion years, after which it will contract again to a point of maximal compression, bounce back, and start all over again. The whole process, he claims, takes 71 billion years from start to finish.[8] Mark Israelit and Nathan Rosen calculate that "The period of oscillation of the universe is $\sim 1.2 \times 10^{12}$ yr,"[9] (1.2 trillion years). The difference in these two estimates is quite remarkable!

If our own cosmos was generated by another cosmos that antedated it, and it by another, and so on endlessly into the past, what were they like? Laws of nature, physical constants, and initial conditions could have varied immensely, perhaps infinitely, in antecedent universes; but one commonality is certain. All of them had to be closed universes, no matter how different they were otherwise. All of them had to collapse to create succeeding universes. The Oscillationist requirement of universal closure permanently excludes an infinite number of possible laws of nature, physical constants, and initial conditions, specifically, all that would yield an open universe. Oscillationists must deny that all possible universes (open universes, for example) are actualized by and within infinity. Once they admit this, they are in deep trouble!

Let us call a single universe falling within an extended oscillating series of successive worlds a "cosmic epoch." Alfred North Whitehead used this terminology, though he was not very specific about its scope. Under the influence of early quantum theory in the 1920s,[10] Whitehead thought that our own cosmic epoch is dominated by electromagnetic energy existing only in discrete quanta, and he defined a "cosmic epoch" as "the widest society of actual entities whose immediate relevance to ourselves is traceable."[11] Our present cosmic epoch can be traced "to an aboriginal disorder, chaotic according to our ideals,"[12] Whitehead believed; but there are other cosmic epochs "far beyond our immediate cosmic epoch" that are ordered very differently from our own.[13] He did not know anything about Big Bang Cosmology, which was still in its infancy when these words appeared in *Process and Reality* in 1929; and he did not explain whether his "beyond" is to be construed spatially, temporally, or both. Mainstream Process Theology has interpreted Whitehead's wording temporally; but "widest" and "beyond" are actually spatial words, not temporal words; he did not say "oldest" or "before."

A. Singularities

Significantly, Oscillation Cosmologists disagree about whether a *singularity* begins and ends each cosmic epoch. This notion, very unfamiliar to most philosophers and to common sense, is very familiar to astrophysicists. A singularity is a physical state resulting from total gravitational collapse that is infinitely dense, infinitely compressed, infinitely hot, infinitely small, infinitely curved. Exactly how cosmologists interpret the reality of singularities depends upon the metaphysics they presuppose.

Would an initial singularity contain no energy at all, or would it contain a finite, or an infinite, amount of infinitely condensed energy? Exactly what cosmologists believe about the presence or absence of energy in an initial singularity varies according to whether they presuppose Oscillationism or Quantum Big Accident Cosmology (developed in Chapter Seven), which affirms that the universe arises spontaneously from absolutely nothing, not even a singularity. On the simplest interpretation of the Big Accident view, and in theistic creation *ex nihilo* (a world created out of nothing by God), an initial singularity is construed to be an infinitely small state of nothingness containing no mass, energy, space, or time at all. Many Oscillationists, by contrast, think that an initial singularity would contain all the mass/energy of an antecedent universe in an infinitely condensed state. On either view, a singularity is *empirical* nothingness. No one, not even God, could perceive or empirically verify claims about an entity that is infinitely small. Yet, many Oscillationists hold, a singularity is still a physically real something. Others submit that a singularity is just nothing at all, and that physical realities emerge only as a universe erupts from nothingness.

A complex view of the nature of original nothingness is defended by Quentin Smith in his debate with William L. Craig in their book titled *Theism, Atheism, and Big Bang Cosmology*, 1993. According to Craig, who is a creation *ex nihilo* Theist, the initial singularity out of which our universe erupted was just a state of physical nothingness; a singularity is an unreal and merely theoretical idealization or fiction.[14] The words "infinite density" suggest the presence of something physically real; but, says Craig, "a condition of 'infinite density' is precisely equivalent to 'nothing'. There can be no object in the real world that possesses infinite density, for if it had any size at all, it would not be *infinitely* dense."[15]

By contrast, Smith, a Big Accident Atheist, contends that "Big Bang cosmology represents the singularity as a unique sort of reality, a *physical reality*, but it is represented as real none the less."[16] The initial singularity somehow contained matter or mass, Smith insists, "not ordinary mass, three-dimensional mass, but *infinitely compressed mass*."[17] Smith does not succeed in giving empirical meaning to the notion of something that is physically real but totally devoid of physical, spatiotemporal properties.

Says Smith, the physical reality of an initial singularity means three things. First, if we extrapolate the physical quantities of our universe backwards in time, we reach a point "arbitrarily close" to the singularity in which these physical quantities "have arbitrarily high finite values".[18] Second, "When the singularity is reached the values become infinite."[19] Third, the initial singularity has the topology of a dimensionless point, but "It assumes, at a subsequent time, the topology of a finite three-dimensional space."[20]

Smith's first two points show only that the concept of "singularity" is equivalent to the notion of the ultimate limits of measurable physical quantities, but not that a singularity actually contains or consists of some mysterious, immeasurable, imperceptible, and non-extended physical mass. By definition, nothing could be a body, something physical, without being spatially or spatiotemporally extended. Smith never explains how his third state could be achieved, how an infinitely small singularity could be transformed into a finite spatiality, how a timeless state could *do or become* anything at a later time, or exactly what the difference is between a dimensionless point's becoming dimensioned and something's being created out of nothing. Oscillation Cosmology shares all of these problems.

In Oscillation Cosmology, cosmic epochs are worlds that fall between Bangs in an infinite temporal series. Does each cosmic epoch begin with a dimensionless singularity of infinitely condensed mass, then expand to a maximum at which gravity overcomes the residual kinetic energy of the initial explosion, and finally recontract into another infinitely condensed singularity? Astrophysical cosmologists frequently discuss singularities that supposedly separate successive universes; but finding cosmologists who actually believe in them is very difficult. The notion that oscillating universes could be separated by singularities was first introduced and discussed by Richard C. Tolman in 1934; but it is not at all clear that Tolman himself actually believed this.[21] John A. Wheeler once accepted a form of Oscillation Cosmology in which cosmic epochs are divided by intervening singularities,[22] and John Gribbin[23] still affirms this in combination with a wider quantum World-ensemble Metaphysics.[24] In 1970, Roger Penrose and Stephen W. Hawking published a definitive proof that the theory of relativity, applied to closed universes, necessitates that they begin and end with singularities.[25] Now Hawking[26] seems (to some interpreters) to champion oscillating universes that avoid intermediate singularities through quantum effects that reverse cosmic contractions before universes totally collapse into nothingness. His view, which has other interpretations, will be explained later in more detail.

B. Quantum Effects and Singularities

For many reasons, a defensible Oscillation Cosmology must circumvent singularities. The existence of an antecedent universe cannot be inferred from known laws of nature if these laws break down completely at or in singularities; nor can they explain what causes an initial singularity to explode into a Big Bang. Roger Penrose says that, "A space-time singularity is, almost by definition, 'a place where the known laws of physics break down.'"[27]

Can quantum theory rescue Oscillation Cosmology from singularities? Quantum theory embraces many oddities called "quantum effects." Two of them, quantum indefiniteness and quantum discreteness, may help Oscillationists evade initial singularities.[28]

Oscillationism can make a place for spatiotemporal and causal continuity between cosmic epochs if no singularities intervene, if earlier collapsing universes round off and undergo Big Bounces before they totally collapse. Quantum theory will support Oscillation Cosmologies if it can (1) eliminate initial singularities and (2) explain how all entropy or memory of preceding disorder can be lost between epochs. Two quantum effects, indefiniteness and discreteness, seem to be incompatible with initial singularities; but entropy, persisting from epoch to epoch, is a stubborn problem.

i. Quantum Indefiniteness

According to Werner Heisenberg's Uncertainty Principle, quantum-size atomic and sub-atomic entities do not have definite simultaneous positions and velocities. It is not simply that we cannot find them together; they just never exist together. Real uncertainty, indeterminateness, and unpredictability obtain at the quantum level of physical reality, but not merely as expressions of human ignorance or of the limits of experimental investigation. Einstein defended "hidden variables" that would reconcile quantum with classical physics; but most quantum physicists agree with Heisenberg's repudiation of absolutely inaccessible physical unknowns; and many experiments in quantum mechanics now count decisively against hidden variables. Sub-microscopic quantum realities are very different in many respects from the macroscopic realities of ordinary perceptual experience.

In quantum theory, sub-atomic particles are thought to be more like waves than like billiard balls. Their existence is spread out indefinitely over a fuzzy region of spacetime, and they lack what Alfred North Whitehead called "simple location." As Stephen Hawking put it, "Particles do not have precisely defined positions and velocities but are 'smeared out' over a small region by the uncertainty principle of quantum mechanics."[29] Existing sub-atomic particles are so peculiar that we really cannot envision them. But we can try!

Applied to the universe as a whole, quantum indefiniteness implies that when a cosmic epoch contracts to the size of a quantum particle or wavicle, it cannot be squeezed down any further into a singularity because this would require it to have a definiteness that entities of that magnitude cannot and do not have. If collapse could progress indefinitely with uninterrupted continuity toward a singularity, at some point no room would be left for indefiniteness. Yet, indefiniteness is inescapable in a quantum universe. Stephen Hawking has something like quantum indefiniteness in mind when he asks, "Does time really have a beginning and, possibly, an end, as predicted by classical general relativity, or are the singularities in the big bang and the Big Crunch smeared out in some ways by quantum effects?"[30]

In several ways, quantum effects might prevent a contracting cosmic epoch from coalescing into a singularity.

First, if a universe expands asymmetrically due to quantum fluctuations and gravitationally retarded expansions in the loci of galaxies and superclusters, it will also contract asymmetrically; if riddled with black holes, some might explode or radiate themselves back into expansion while others are still contracting. Not all parts of a non-synchronous universe could ever meet again at a singularity. Its parts would not contract synchronously; some parts would bypass others; some would go in one direction and some in another; some would contract while others expand; and some might contract at one rate and others at other rates. It is like, "You are never going to get it all back in there again!"

Technically, this is called "differential collapse."[31] Such an uneven collapse would make impossible the collapse of an entire universe to either a singularity or to Planck dimensions.

Second, even if all parts of a given universe do contract concurrently, they could never coalesce into a singularity because of quantum indefiniteness. If singularities are impossible or unintelligible, Oscillation Cosmologists must endorse Big Bounce universes that avoid singularities through quantum effects. A Big Bounce is a rebound from a state of minimal size and maximal compression that is larger than a singularity. Through Big Bounce strategies, Oscillation Cosmologists can avoid singularities and all their concomitant difficulties, but Oscillationism may still have other serious problems.

ii. Quantum Discreteness

Quantum theory affirms that mass/energy can exist and can be transferred only in discrete, discontinuous, minimal units or quanta. Quantum theory began in 1900 when Max Planck realized that blackbody objects, those saturated with radiant energy, release so little high frequency radiation only because energy is emitted in discrete atomistic packets or quanta, not in infinitely divisible continuous gradations. Actually existing mass/energy is not infinitely divisible into smaller and smaller real units, although it may be so divisible in our imaginations. In the twentieth century, quantum physicists learned many amazing but now well-established things about the domain of very small physical particles. Electrons can assume some orbits around atomic nuclei but not others because some potential orbits or orbital shells are forbidden by nature. The spin of quantum particles takes only certain discrete values but not others. The physical world is not an infinitely divisible continuum, except in our imaginations. Physical units are required to have definite, finite, minimal magnitudes in order to exist at all, and in order to be transferred and absorbed. Mother Nature does not permit intermediate degrees between discrete quantities.[32]

Quantum physics clearly excludes the actualization of all possibilities. Mother Nature tolerates only discrete quantities in multiples of Planck dimensions. Planck numbers are invariant physical constants. Many numbers that are logically possible are forbidden in physics and chemistry; others are privileged, not just in quantum theory, but presumably also within the underlying reality that the science of physics reflects.

Cosmologists who combine Oscillationism with quantum theory think that Planck spacetime (10^{-43} of a second in age and 10^{-33} centimeters in diameter) is the ultimate minimal unit for anything physical to be at all. (See endnote 74.) If they are right, no mass/energy, and no space or time can exist that is smaller or earlier than Planck dimensions. Quantum physicists are not perfectly consistent in eschewing dimensionless entities, for quarks, electrons, and other sub-atomic particles are sometimes said to be the size of Euclidean points, which is no size at all;[33] but this embarrassment is avoided if small physical particles are regarded as tiny blotches of mass/energy or as vibrating strings with definite Planck length and magnitude.[34]

Quantum theory discloses that spacetime and mass/energy exist only in atomized quanta and are not physically or metaphysically divisible into infinitesimally small real units, though they are thus divisible by mathematical imagination. If they were infinitely divisible in reality, something real, something intermediate, could exist between Planck dimensions and the zero dimensions of a singularity; and a collapsing quantum universe could gradually and continuously shrink to nothing. In a quantum universe, a physical continuum is forbidden; and gravitational collapse to a singularity existing below Planck dimensions is physically impossible.

Additionally, quantum gravity may prevent a contracting universe from collapsing to a singularity. According to some theories, gravity becomes a negative or repulsive force under certain pressure conditions; and if this happens in the final stages of gravitational collapse, a universe could shrink just so far but no further before negative gravity sets in and terminates its contraction. Negative gravity may (or may not) be operable at Planck dimensions, for the theory has not been tested and confirmed. Better established is that the degeneracy pressure of compacted electrons and neutrons will prevent a universe from shrinking to a singularity, but only if, unlike our own, that universe has less than one and a half times the mass of our sun.[35] Another possibility is that if the Big Crunch is a vortex spinning all the matter in the universe at unimaginably great speeds, the powerful centrifugal force of the spin itself may forestall an ultimate collapse into a singularity.[36]

Quantum theory applied to the origin of the universe says that nothing can exist earlier than Planck time or with a volume smaller than Planck space. Thus, the universe was 10^{-43} of a second old and 10^{-33} centimeters in diameter when it was anything physical at all. No laws of physics could apply antecedently since nothing could exist to which physical laws apply. Planck dimensions are called "Planck's Wall," the beginning and end of all scientific knowledge, in Quantum Cosmology.

An initial singularity is inferred in non-quantum Oscillationism by extrapolating from the observed universe back to $T = 0$ in accord with *non-quantum* laws of physics. From *quantum* laws we cannot extrapolate back to an initial singularity because these and all other physical laws become inapplicable below Planck dimensions. Some Quantum Cosmologists disagree and suggest that the laws of quantum physics do not breakdown at or below Planck dimensions, but if they do we obviously cannot use these laws to extrapolate back either to an initial singularity, or to an antecedent universe.

Stephen Hawking does not specify exactly which quantum effects are relevant, but he may have had quantum indefiniteness and discreteness in mind in 1988 when he announced that he had changed his mind about singularities and was trying to persuade other physicists that "There was in fact no singularity at the beginning of the universe," that "It can disappear once quantum effects are taken into account."[37] Hawking seems to think that, like the surface of the earth at the North Pole, our universe began in a rounded-off, smeared-out, quantum state that precluded an initial singularity.

So what lies at the bottom of a black hole? Not an infinitely compressed singularity, we would expect Hawking and other Quantum Cosmologists to say, merely a great mass compressed to finite Planck dimensions. But Hawking and others are not consistent about this.

The preceding arguments from quantum effects against singularities are exceptionally powerful, but cosmologists like Hawking who rely upon them really do not consistently believe them! Hawking frequently changes his mind on many issues and is very difficult to pin down. First, considering only relativity theory apart from quantum theory, he and Roger Penrose proved that closed universes must begin with initial singularities. Then in *A Brief History of Time*, Hawking repudiated initial singularities by appealing to quantum effects, presumably indefiniteness and discreteness, and to quantum gravity. In his very recent debate with Penrose in *The Nature of Space and Time*, 1996, Hawking switched again; expressly repudiating quantum discreteness, Hawking says that he sees "no reason to abandon the continuum theories that have been so successful;"[38] and he affirms again that the Big Bang emerged from a singularity.[39]

Hawking and Penrose both believe that black holes, collapsing neutron stars, and collapsing universes regularly coalesce into real singularities below Planck dimensions. They agree that "cosmic censorship" hides the singularities in black holes and neutron stars from our view because they are cloaked to outside observers by their surrounding event horizon;[40] but the singularity of the Big Bang is "naked" and exposed to us. In *The Nature of Space and Time*, Hawking says that "Cosmic censorship may shield us from black hole singularities but we see the big bang in full frontal nakedness."[41]

One of Hawking's hands does not seem to know what the other hand is doing! Hawking appears to accept arguments from quantum effects against singularities when convenient, and to reject them when convenient! But we must allow him to change his mind. Penrose, by contrast, has never ruled out an initial singularity or appealed to quantum effects like discreteness and indefiniteness to exclude it. Penrose persuaded Hawking that the distinctness of an initial Big Bang involves a different kind of quantum effect, a peculiar sort of spatial curvature, based ultimately on quantum gravity.[42]

As Penrose explains, there is a fundamental difference between an initial world-creating singularity and the terminal singularities that appear in black holes or a collapsed universe, but it does not involve such quantum effects as indefiniteness and discreteness. The difference is in the kind of spatial curvature that each requires and manifests. Astrophysics distinguishes two fundamental kinds of spatial curvature, first, that involved in the increasing volume of a small sphere being influenced by gravity (called "Ricci curvature"), next, that involved in tidal distortions of this sphere (called "Weyl curvature").[43] In the initial world-making singularity, "The Weyl tensor was exactly zero at the big bang itself, while the Ricci curvature diverged to infinity."[44] By contrast, in the singularity of a terminal black hole or collapsed universe, the reverse is true: Weyl curvature is infinite and dominates over Ricci curvature. This explains why entropy or disorder is low in an initial Big Bang singularity and high in a Big Crunch singularity, and why time and the Second Law of Thermodynamics would not be reversed in a closed collapsing universe.[45] An initial singularity is infinitely ordered, completely free from quantum distortions or irregularities in space-time geometry with respect to Weyl curvature; and a terminal singularity is infinitely disordered in this respect, if Penrose is right.

Both Penrose and Hawking now believe that ripples that would tear everything apart would pervade the contracting phase of a collapsing universe. This is how its entropy would continue to increase. Quantum effects like indefiniteness and discreteness begin only when an initial singularity expands to Planck dimensions.[46] Despite quantum effects, our universe, the only one we know to exist, really could have begun in an initial singularity of nothingness. If Penrose is right, an initial singularity is different, improbable, but not impossible; but it could not be the product of an antecedent collapsing universe with ever-increasing entropy. Singularities still have serious shortcomings.

Quantum effects like discreteness and indefiniteness together with Weyl curvature should rule out singularities in collapsed universes if not also in black holes and dimensionless particles. Hawking, who affirms them, assumes at times that they do not rule out singularities as such. The Hawking who thinks that quantum effects would rule out an initial singularity still affirms singularities in gravitationally collapsed black holes and collapsed neutron stars. In them, presumably, mass/energy could retain sufficient definiteness, continuity, and freedom from distortion to collapse to zero size, despite any and all quantum effects. In them, mass/energy could gradually shrink continuously below Planck dimensions. In them, singularities are physically attainable, so the laws of both classical and quantum physics could apply all the way back to T = 0.

Why, then, are initial singularities impossible? Large exceptions should make one suspicious. Singularities may be impossible only when some cosmologists do not want to find them! If black holes and gravitationally collapsing neutron stars can shrink smaller and smaller to singularities, then quantum effects like indefiniteness and discreteness do not rule out singularities as such. According to Penrose, the absence of Weyl curvature from an initial singularity is what makes it infinitely ordered and free from entropy; its presence in a terminal singularity makes it infinitely disordered and riddled with entropy. But this should rule out Oscillationism, for antecedent universes all end with overwhelming entropy.

Hawking's arbitrariness in rejecting initial singularities, at least for a time, may be seen in his best selling book, *A Brief History of Time*. In explaining his position, Hawking makes the following contradictory claims about singularities.

1. "There was in fact no singularity at the beginning of the universe."[47]

2. "Only if we could picture the universe in terms of imaginary time would there be no singularities."[48]

3. "When one goes back to the real time in which we live, however, there will still appear to be singularities. The poor astronaut who falls into a black hole will still come to a sticky end; only if he lived in imaginary time would he encounter no singularities."[49]

Since 1. is presumably about the real time in which we live, it contradicts 3. Hawking's discussions often fail to clarify the role of quantum mechanics in permitting or excluding singularities, whether initial or terminal. His occasional retreat into an instrumentalist position, according to which physical theories are merely mathematical models that exist only in our minds but tell us nothing about reality,[50] also contributes nothing to the clarity of his views. His renewed discussion of real versus imaginary time in his 1993 book *Black Holes and Baby Universes and Other Essays* did little to elucidate his position.[51]

A boundary condition is a state out of which another state arises, normally, a cause. Hawking's peculiar brand of Quantum Cosmology affirms that the ultimate boundary condition for the universe is that there is no ultimate boundary condition; but the meaning of this is extremely obscure.[52] Carl Sagan took it to mean that the universe has no cause, hence there is nothing for a Creator God to do.[53] Neither Sagan nor anyone else noticed, but this also means that there is nothing for an antecedent universe to do! In one place, writing of a universe "without boundaries or singularities,"[54] Hawking seems to regard the terms as synonymns. So regarded, a universe without boundaries is just a universe that does not begin with a singularity, which is perfectly compatible with that variety of quantum oscillationism that dispenses with singularities but not with antecedent universes. Thus, our universe could be bounded by a collapsing temporally antecedent universe from which it bounced, one that crunched only to finite quantum dimensions but not to infinite singularity dimensions, while not being bounded by a singularity. It could be singularity unbounded but temporally bounded–in an extended sense of time that transcends our cosmic epoch.

Robert Jastrow interprets Hawking's no boundary condition to be just another version of Oscillation Cosmology. Any quantum Big Bounce Oscillation Cosmology with an infinite number of cosmic epochs not separated by singularities lacks ultimate boundary conditions (singularities) while having ultimate boundary conditions (causally efficacious antecedent universes). According to Jastrow, Hawking's "universe without a boundary means a 'closed' universe–one which oscillates between expansion and contraction, instead of expanding forever."[55] Jastrow's interpretation of Hawking is not entirely baseless. In comparing the Big Bang to a black hole in his *Black Holes and Baby Universes and Other Essays*, Hawking says that "It may be that there was an earlier phase of the universe in which matter collapsed, to be re-created in the big bang;"[56] and many things that Hawking says in 1996 in his and Penrose's *The Nature of Space and Time* about the Big Crunch and a closed universe sound very much like Oscillationism.[57] An important analogy between black holes and the Big Bang is that black holes might sometimes explode! They don't always just radiate and slowly fade away.[58] However, as caution warns, in no instance has anyone ever observed the explosion of a singularity; and if no laws of physics apply to them, we cannot appeal to the physical laws we know to explain how singularities ever could explode.

In developing his own view that the universe has no ultimate boundary, Stephen Hawking compares spacetime to the spherical surface of the earth, but the analogy is not very illuminating. In this comparison, spacetime begins at the North Pole, expands to its maximum at the equator, and then shrinks toward the South Pole.[59] There, presumably, the process reverses and is repeated over and over again, though this is not perfectly clear. Hawking says that he agrees with St. Augustine that time begins with the creation of the universe, and that it makes no sense to ask what happened before that.[60]

Yet, if the North to South–South to North–phases repeat themselves, the whole of time cannot be contained entirely within a single North to South sweep, unless North represents the absolute boundary condition of the creation of the universe *ex nihilo*, as it did for St. Augustine. This would rule out not just singularities but also temporally antecedent universes. Without an absolute temporal boundary, a South to North sweep could come *before* a North to South sweep; and the indefinitely large South Pole of one sweep could be identical with the North Pole of its successor; if so, time as such cannot begin absolutely at any given North Pole, and Hawking is an Oscillationist, as Jastrow claims. No ultimate temporal boundary condition is real if time already existed before *our* time, our sweep, began. Thus, if there was no time before our time began–which Hawking sometimes affirms, there was a time before our time began–which he also sometimes affirms; and an ultimate boundary condition exists after all! Hawking's position is puzzling if not riddled with flip-flops and contradictions. Like Hawking, many cosmologists cannot make up their minds, equivocate on the meaning of basic concepts ("singularity" versus "temporality" as the meaning of "boundary condition"), often change their minds, and occasionally argue for incompatible positions. Given the complexity of their subject matter, this is easily understandable.

Hawking may or may not be a real Oscillationist. Perhaps his universe without boundary conditions does not oscillate because time is unreal. At times, Hawking seriously doubts the reality of time, something that a true Oscillationist cannot do. Perhaps he confuses the absence of a singularity boundary with the absence of a temporal boundary; but perhaps he wishes to deny the reality of time altogether. Imaginary time, favored by Hawking, is really a spatialization of time which, by a mathematical trick, converts time to a fourth spatial dimension. Just what this means is extremely obscure. Hawking explains that imaginary time is "at right angles" to real time; but it is difficult to see how something that is not spatial could be "at right angles" to anything, or how there can be a singularity at the beginning of real time but no singularity in the imaginary time that is also somehow real and at right angles to it. Also, a fourth *spatial* dimension gets lost in Hawking's analogy with an expanding and contracting globe because these expansions and their cosmic evolutions are *temporal*.

In purely spatial terms, the North Pole, no matter how inexact, definitely represents a boundary between the earth and its atmosphere; so Hawking's analogy does not adequately illustrate the notion of finite but unbounded time; perhaps no analogy could succeed. If time really can be spatialized, Hawking's initial temporal singularity is avoided by eliminating time altogether. Once time is abolished, the whole universe is timeless: and our experience of time is an illusion. But why isn't space also an illusion? A purely spatial but timeless Hawking universe is no longer *our* universe, no longer a quantum universe in which time and space are inseparable but not identical. Hawking pays a very high price for avoiding an initial singularity. His theory becomes irrelevant to our world. As Benjamin Franklin said, "Time is the stuff life is made of." The real world of nature is temporal through and through. Contemporary physics does not spatialize time; it temporalizes space, despite many claims to the contrary.[61]

4. Critique of Oscillation Cosmology

Oscillation Cosmologies have troublesome defects. Their most serious flaws must be identified and considered. The greatest difficulties for all Antecedent Universe Cosmologies, including Oscillation Cosmology, are that we do not and cannot directly experience antecedent worlds; we have no inductive empirical access to them because inductive logic is inapplicable where we know only one of a kind; and the laws used to reason back to antecedent worlds break down before they take us that far. Science cannot establish their existence or know their essence. Other difficulties for Oscillation Cosmologies center upon A. their affirmation that space is finite with sufficient mass/energy to open and close it an infinite number of times, B. their treatment of singularities, C. their affirmation of universal causation, D. their appeal to quantum effects, and E. their approach to thermodynamics.

A. An Infinite Number of Closed Universes

Oscillation Cosmologists do not agree with Plasma Cosmology that space is infinite, but they share its view that time is infinite. Only a finite quantity of spatially extended stuff expands and contracts forever in oscillating universes. If singularities are rejected, oscillating universes have a finite mass/energy density and both a minimum and a maximum magnitude that are finite. When maximum allowable magnitude is reached, expansion ceases; and a Big Squeeze commences. When the allowed minimum is reached, contraction ceases, and another Big Bang begins. And this goes on forever.

Unlike Gamow's One-Bang theory, Oscillation Cosmologies assume that enough mass/energy existed in an infinite number of spatially finite consecutive universes to close them all and prevent them from expanding forever, no matter how much they differ otherwise with respect to laws, physical constants, and initial conditions. The requirement that all antecedent universes be closed places severe limits on admissible laws, constants, and initial conditions for all oscillating universes. It excludes an immense number of possible laws, constants, and initial conditions–all those that would engender flatness or openness. If only one universe in an oscillating series is flat or open, the series breaks and absolutely terminates. Oscillationists cannot explain why an immense number of logically possible open or flat universes are metaphysically impossible, or why an infinite number of only closed universes can or must exist.

Open and flat universes never collapse and can have no Big Crunch successors. If a prior universe in our own presumed series had been open or flat, our world would have been impossible. For Oscillationism, necessary conditions for the origin and existence of our universe are fulfilled only if all earlier cosmic epochs were closed.

Consider this very forceful argument against an infinite series of oscillating universes. Many cosmologists are convinced that a Big Crunch will so diversify initial conditions for its successor that eventually, given enough time, an open universe or epoch having no successors will inevitably occur. John Barrow and M. P. Dabrowski think that they understand the mechanism by which some if not all series of oscillating universes will eventually end in openness: the maximum size of each epoch increases with increasing entropy; and eventually they become so large that oscillations cease and expansion continues forever.[62] If this is possible at all, then no oscillating set of universes could be eternal, especially if, as some hold, all possibilities are actual.

By definition, no open or flat universe ends with a Big Crunch; thus, no open or flat universe can have a successor universe that rebounds from its Big Crunch. Mechanisms may exist by which any and every infinite set of universes will eventually produce and thus terminate with an open or flat universe. This would definitely happen if every temporally ordered infinite set of worlds diversifies to actualize every possible universe, as Atheistic and Finalistic Anthropic Cosmologists propose.

If an open or flat universe will come along eventually by accident in an infinite set of cosmic epochs actualizing all possible worlds, then *every* world in such a set must be open, for *every* member has an infinite number of predecessors. The predecessors of every universe in an infinitely prolonged set collectively endure for an infinite amount of time, which is quite long enough to engender openness, given the presumption of infinite diversification within infinite time. Thus *every* infinite oscillating series of universes must end with an open or flat universe; indeed, every individual universe will be open or flat because in an infinite series each would have an infinite number of predecessors; so the very idea of oscillationism is nonsense!

Closed universes must be finite in space and mass/energy. Robert J. Russell suggests that an open universe is actually infinite in size and mass/energy,[63] but in an important sense this is not true. If spatially infinite when maximally expanded, an oscillating universe that contracts at a finite temporal rate would never totally collapse because finitude cannot use up infinity. Similarly, and for the same basic reason, a universe that is initially finite and expands at a finite rate could never achieve spatial infinity.

Both closed and open universes must be actually finite, but open universes have an indefinitely large potential for expansion. In this sense they are finite but unbounded; their actuality is finite, but they are unbounded with respect to their potentiality for expansion. If an open universe begins with finite space/energy and expands at a finite rate, it will still be finite after fifteen billion years–and after fifteen billion billion years; only its potential for future expansion is infinite, never its actuality.

Oscillation Cosmologists must insist that in an infinite number of actual universes, including our own, Omega (total density) is greater than 1 (perfect balance, or critical density). Obviously, this cannot be verified; and our universe with an increasing rate of Hubble expansion, and with total density of from .1 to .3 or .4 at most, appears to be open. Even a flat universe with an Omega density of exactly 1 is incompatible with Oscillationism because, like open universes, flat universes never collapse and never have successors.

Recall that the case against the existence of an enormous quantity of dark or hidden cosmic mass that cannot be located is very strong. Astronomers can find only ten to thirty percent or so of the mass/energy required to close the universe, despite their best efforts. After devoting much of his book on *The Dark Side of the Universe* to examining the case for and against dark matter, James Trefel concluded:

> It used to be customary in discussions of this sort to entertain the idea that the universe was cyclical–that the Big Bang would be followed by a collapse (the Big Crunch) and another expansion (the Big Bounce). But if our current ideas are true, this will not happen. The Universe has one shot at existence–one explosion followed by an expansion that slows down for an infinite length of time.[64]

Yes, some dramatic discovery tomorrow might reveal the existence of the missing mass; scientists are constantly stumbling upon the unexpected, especially with the aid of the Hubble Space Telescope. New sources of both luminous and cold dark matter are regularly identified; but will we ever locate seventy to ninety percent more? Will astronomers or particle physicist ever find enough ordinary matter or previously undetected cold dark matter to close the universe? Maybe so. Maybe not. The existence of enough mass to close the universe is merely a "Philosophical Maybe," an abstruse possibility lacking probability, given our vantage point in history. To every "Philosophical Maybe," corresponds a "Maybe Not." In the realm of pure possibilities, "possibly so" is always checkmated by "possibly not." We should not be intimidated by mere possibilities, and we should never confuse possibilities with probabilities.

Given the present state of human knowledge, it is reasonable to think that enough mass/energy probably does not exist to close the universe even once, much less an infinite number of previous times. From seventy to ninety percent is missing; the odds against finding enough are just too great; and the expanding rate of cosmic expansion seems to rule it out definitively.

As explained in Chapter Three, the recent discovery that the rate of cosmic expansion is increasing, not slowing, as earlier assumed (for example, by Trefel above), counts overwhelmingly against a closed universe. Too little mass exists in the universe to slow down its expansion rate, much less to stop it. Yet, oscillating universes *must* be closed and reopened an infinite number of times, no matter how different other epochs are from our own. For exceptionally good reasons, we should reject the oscillation hypothesis. We really do not know that an antecedent Big Crunch happened even once, much less an infinite number of times. What we do know suggests that it never happened at all. We will return later to the problem of the missing mass.

B. Singularities vs. Finite Size Maximal Compression States

A few Oscillation Cosmologists like a younger John A. Wheeler and a contemporary John Gribbin maintain that each cosmic epoch begins and ends with a singularity, a state of infinite compression, infinite density, infinite heat, infinitesimal size, and infinite curvature. Other Oscillation Cosmologists like Mark Israelit and Nathan Rosen[65] repudiate infinitesimally small initial singularities and contend that the maximal compression state of the antecedent universe was finite in size or volume. Let us further considers rebounds with and without singularities.

i. Rebounds from Singularities?

The reality of an initial singularity from which an oscillating universe could rebound may be doubted for a variety of reasons. For one thing, there are serious problems about its empirical status. I. L. Rozental says that "The singularity situation in cosmology is in contradiction to all accumulated physical experience."[66] According to Rozental, singularities cannot be created in the laboratory because "under terrestrial conditions" a phase transition always occurs as pressure approaches infinity, and "the singular state is not achieved."[67] The idea that all sensory experience counts against singularities can be further expanded. It is absolutely impossible for our physical senses to perceive directly anything that has no size or spatial magnitude whatsoever. Conceptually constructed Euclidean points and spatiotemporal singularities are so small that they are indeed *empirically nothing*, and the claim that our universe was created out of an initial singularity is empirically indistinguishable from the claim that it was created *ex nihilo*.

An initial singularity for our universe could not have been perceived directly if we had been there, but confidence in an initial singularity may be inductively grounded in some other way. Ignoring quantum effects for the moment, perhaps an initial singularity can be extrapolated mathematically from observable processes such as increasing entropy, the redshift, and the Hubble expansion. If these processes are reversed and traced far enough into the past, in ten to twenty billion years, the results equal 0. This assumes that the laws and basic structures of nature do not break down before zero diameter or magnitude is reached, that abstract calculations can accurately reflect physical processes that contract all the way down to zero and not just up to the edge of it, and that quantum physics does not block the descent to nothingness.

If, as most quantum physicists believe, the laws of nature do not break down *at* Planck spacetime but nevertheless become inapplicable *below* Planck dimensions, it would definitely not be rational to believe that our universe began in a singularity at T = 0. Quantum Cosmology indicates that the laws and the most basic concepts that we apply to nature *do* become inapplicable to spacetime below Planck's Wall because nothing physical whatsoever can exist on its other side. Scientifically, we can trace origins back no further than when the universe was 10^{-43} ($1/10^{43}$) of a second old and its size was 10^{-33} ($1/10^{33}$) of a centimeter in diameter. Quantum Cosmologists like I. L. Rozental and Stephen Hawking (at times) are convinced that quantum effects make collapses to initial singularities impossible (despite the fact that Hawking still believes in black holes with singularities at their core). We will soon see that Quantum Cosmologies have serious problems of their own.

Because they suggest creation *ex nihilo*, singularities create an interesting philosophical problem for Oscillationism and all other naturalistic Antecedent Universe Cosmologies. Theistic creation *ex nihilo* affirms the known existence of only one universe and is not vulnerable to this difficulty. Its proponents would be quite happy if modern science discovers that the universe was created out of nothing in the finite past!

The philosophical difficulty is this. Naturalistic metaphysics maintains that *our* universe has existed throughout an infinite past; but the notion of "our universe" is not easily stretched to infinity, despite Oscillation Cosmology. The problem of personal identity through time has as its counterpart the problem of cosmic identity through time. Our cosmic epoch, which began around fifteen billion years ago, could not belong to the same universe as its predecessors if it is separated from them by even one singularity, much less by an infinite number of them. Conceptually, singularities involve the total eradication of space, time, physical causation, continuity, and all laws of nature; so no universe derived from a singularity, including our own, could have any temporal predecessors or physical causal progenitors.

If antecedent Cosmic Epoch A and subsequent Cosmic Epoch B are spatiotemporally and causally linked through a singularity, that is, by causal and spatiotemporal nothingness, then they are not linked at all. Two regions belong to the *same universe* only if they belong to the same spacetime system, that is, if it is at least theoretically possible to get from one to the other by traveling through space or time; all members of the spacetime system have physical causal relations with–are either the physical cause or effect of–some other members of that system; and all members function statistically in accord with an all-pervasive set of physical laws. "Sameness of universe" thus involves continuity of (1) space, (2) time, (3) physical causation, and (4) natural laws. Our universe is the totality of the spacetime system to which we belong, including all the laws, mass/energy, and cones of physical causation within it. To belong to the same spacetime system, any two things must be bound together lawfully by spatiotemporal continuities, contiguities, and causal processes. Entities belong to *our* system of spacetime only if spatiotemporal continuities and physical causal bonds link them lawfully to us directly or indirectly.

No lawful spatiotemporal and causal continuities whatsoever obtain between universes separated by singularities, where all of the above break down and disappear. Without something physical (spatiotemporal) there is no physical causation. Since physical causation and the laws of nature break down in a singularity, no universe separated from another by a singularity could be its physical cause or project its mass/energy and its laws into its successor. An antecedent universe that crunched to a singularity could not be the cause of our Big Bang.

In fact, it could not even be antecedent! Our world could not be part of an eternal universe, as Naturalists hold, if it began fifteen billion years ago and is separated from its antecedents by singularities. Singularities eliminate temporal as well as causal continuity and succession, so speaking of antecedent universes separated by singularities makes no sense. No temporal relations can exist between two cosmic epochs separated by utter nontemporality.

Nothing can be earlier than the first moment of time. If our universe begins and ends with a singularity, nothing could precede it or follow it. The Big Bang would be the beginning of time itself; nothing could antedate it. One universe can precede another only if they belong to *the same* spatiotemporal series. If singularities separate them, the spatiotemporal series is unconditionally interrupted; and the notion of "antecedent universes" becomes inapplicable and unintelligible. Antecedent Universe Cosmologies that affirm singularities are logically incoherent, and Oscillation Cosmologists *must* avoid them.

A naturalistic metaphysics that accepts singularities between cosmic epochs is itself incoherent. Our universe cannot be separated from its predecessor by a singularity because applying the concept of "predecessor" to such a relationship is unintelligible. Big Bang Cosmology shows that *our* universe is not eternal, contrary to the assumptions of Naturalism. If *some* universes are eternal, they can have no spatiotemporal or causal continuity with our own through singularities; and we can never know of their existence on lawful, inductive, or scientific grounds.

On non-scientific grounds, many theologians postulate Heaven and Hell as completely independent spacetime systems, that is, as Other Worlds, that have no direct spatiotemporal or causal relations with our own. Naturalists who accept singularities incoherently embrace disconnected, transcendent, Other-Worldly spacetime systems–the very things that they repudiate so vehement when quarreling with theologians about Heaven and Hell! Our world is not eternal and could not be connected to an everlasting series of worlds if it began fifteen billion years ago in a singularity of nothingness.

If and when a singularity forms at the end of a "Big Crunch," why doesn't it just stay there forever? We actually have neither empirical nor theoretical evidence that they ever explode, and this is absolutely devastating to any Oscillationism that separates universes by singularities or derives them from singularities. In no instance has a singularity ever been observed to explode. Even theoretically, no one knows what would cause a singularity to explode since no known laws of physics apply to them. According to Alan H. Guth, Oscillationism is now very unpopular with most scientific-minded cosmologists for this very reason. Guth indicates that what we know of gravity in relativity theory "does not allow a crunching universe to bounce into a big bang," but, he insists,

> This is not a fatal objection, however, since general relativity presumably breaks down at the extraordinarily high densities encountered in a big crunch. Nonetheless, since there is no reliable theory that describes how a universe might bounce, the basis of the oscillating universe theory relies solely on speculation.[68]

Since all natural laws break down in singularities, no known or knowable laws of nature could permit them to explode. Something very similar is true with respect to big bounces from finite compaction states. Given our ignorance with respect to high density physics, no *known* laws of physics could permit them to explode. As Martin Rees indicates, "Physical conditions in the 'bounce' would transcend the physics we understand, so that nothing could be said about the possibility of a rebound into a new cycle–still less about what memory would be preserved of what had gone before."[69] With or without singularities, quantum fluctuations really could not do the job, for we really know almost nothing about quantum fluctuations in high density physics. Inflation Cosmology does not derive its many worlds from exploding singularities or immensely compressed antecedent universes. Inflation, which does not appeal to explosions at all, requires just the right kind of diluted quantum-foamy "empty space"; and crunched-up antecedent universes just aren't the right stuff!

ii. Big Bounces Without Singularities

Singularities are not easily circumvented, but Oscillation Cosmology can be developed without them. If antecedent cosmic epochs are not infinitely compacted into singularities as they crunch to an end, if they rebound from *finite* size maximal compression states, some of the foregoing difficulties can be avoided, but at a price. If the laws of nature as we know them apply to high density situations, corridors of no more than Planck dimensions might link universes causally, spatially, and temporally and provide for the continuation of natural laws from epoch to epoch.

Some Antecedent Universe Cosmologists explicitly reject singularities and conjecture that antecedent universes rebound while still finite in diameter, curvature, density, and temperature, thus avoiding the embarrassing infinities and breakdowns of singularities. These minimal-size-maximal-compression states may be quite large. Georges Lemaître initiated Oscillation Cosmology with his image of the Phoenix that dies a fiery death, then rises again from its own ashes. His "primeval atom" was two hundred million miles in diameter when it exploded.[70] George Gamow speculated that when the previous universe crunched to its maximal pre-expansion density, all of the matter within the reach of a 200-inch telescope "must have occupied a sphere only thirty times as large as the sun."[71] Eric Lerner's Mini-Bang occurred when an antecedent metagalaxy collapsed to a hundred million light years across.[72] A. Karel Velan thinks that before a Big Bang explosion, a collapsed universe has a radius of 1.17×10^{14} cm,[73] that is, 1.17 trillion centimeters. Most Antecedent Universe Cosmologists actually prefer sub-microscopic but still finite dimensions as minimal-size, maximal-compression states for collapsing/erupting universes, partly because they think that at lesser volumes and densities, thermonuclear reactions would not be sufficiently hot and compact to destroy heavy elements and replenish the universe's supplies of hydrogen and helium. Defenders of large finite maximal-compression rebounds may find it difficult to explain the physics of hydrogen renewal.

Tiny Planck spacetime is the most popular minimal-size-maximal-compression dimension for contracting/rebounding universes, or for any physical realities at all.[74] Contracting universes may proceed almost to zero, but not quite, before a Big Bounce erupts to recreate another world.

Oscillationists must answer this question: What initiates the explosion? What makes a contracting universe bounce? Quantum fluctuations might make Big Bounces inevitable, but only if the laws of quantum physics obtain in all antecedent universes and at the junctures between universes, and only if all logically possible non-quantum universes are inexplicably non-existent. With Big Bounces, Planck size threads could connect cosmic epochs; but these are so minute that one still wonders if anything is left of the notion of *same universe*. All ordinary sameness is left far behind long before we arrive at Planck's Wall. From a commonsense standpoint, a Big Bang universe would be a distinct universe whether or not it issues from a singularity or a Planck size quantum corridor. Yet, some Oscillationists submit, the connecting link might not be implausibly thin.

C. Singularities and Universal Physical Causation

Perhaps singularities are ruled out because they are incompatible with Naturalism's metaphysical causal principle: "All events have natural causes." In 1965, the year that the omnipresent background radiation was discovered, Milton Munitz rejected an initial singularity for our universe for this reason; but his arguments confused scientific methodology with Naturalism's principle of physical causation.

Accounts of origins *must* transcend finite limits like those set by an initial singularity, according to Munitz, because scientific method presupposes that *all* spatiotemporal events are caused by other spatiotemporal events. An *original* event proceeding from the nothingness of a singularity would violate this methodological rule, (which is really a metaphysical rule in disguise), so science itself is incompatible with an initial singularity. According to Munitz, it is always possible to find "some more refined theory, in which inferences would be made to events even earlier than the one identified as "the beginning" in the theory of coarser grain."[75] As he expressed this argument in *The Mystery of Existence*,

> Science is grounded in the use of the Principle of Sufficient Reason and, therefore, always leaves open the possibility of finding the explanation of *any* event. To say there is some unique event, marking the beginning of the universe for which no explanation *can* be given, is to say something contrary to the method of science. It is for this reason, I should argue, that any conception of the beginning of the universe, when defended under the aegis of some supposedly scientific cosmology, is an indefensible notion.[76]

Here Munitz clearly identifies the Principle of Sufficient Reason with the naturalistic metaphysical principle that "All natural events have natural causes;" otherwise, the door would be open for the Theist who thinks that God's original act of creation is the sufficient reason for the existence of the world. A natural cause belongs to the same system of spacetime as its natural effect; but God presumably transcends the spacetime of our world; and so do all antecedent universes. Munitz has natural causes in mind when he says that "*All* events investigated by science are ones for which it is relevant to inquire into their causal conditions" and that "On scientific grounds, one could never hope to establish that the universe had an absolute origin, or came into existence."[77] Here Munitz assumes that a scientific inquiry into natural causes will always find them–*ad infinitum*; for this reason, he assumes, science could never discover absolute origination.

Munitz is mistaken in thinking that science could never discover an absolute origin for the universe merely because it must always inquire about the natural causal conditions for everything that it investigates. Methodology is confused with metaphysics. Scientific methodology always *inquires* about natural causes, but only a non-empirical Naturalistic metaphysics *guarantees* that inquirers will always find that for which they are looking. Scientific methodology needs only the imperative: "Look for natural causes!" It does not require an *a priori* metaphysical guarantee of success.

When divorced from naturalistic metaphysics, science can conclude that scientific method reaches its limits with an initial cosmic singularity, if this is indeed where time, space, natural laws, and physical causation commense, the point beyond which they can be traced and applied no further, the end of scientific explanation.

As Robert Jastrow indicated, "Science has its own religion. That religion is founded on faith in natural law and in cause and effect. Science has succeeded in posing, on its own terms, questions not answerable within the domain of science."[78] Jastrow could be wrong in thinking that scientific knowledge ends when it stumbles upon an initial singularity; but he correctly judges that any confusion of methodology with metaphysics, like Munitz's, is religiously grounded in a blind faith in absolute natural causation, a faith that empirical scientific methodology alone cannot substantiate. Even quantum physics calls universal causal absolutism into question, as we will see. That an initial singularity would be incompatible with scientific method and universal physical causation is only a minor obstacle for Oscillation Cosmology, but other difficulties are more serious.

D. Quantum Effects Near Singularities

We should be very cautious about accepting Quantum Cosmology's supposition that when the whole universe is squeezed down to the size of sub-atomic quantum objects, it will begin to obey the laws of quantum physics and manifest quantum effects like indefiniteness, discreteness, and spontaneity–if for no other reason than that Quantum Cosmologists seem to believe this only when it is convenient. Stephen Hawking suggests that size is the relevant consideration for appealing to quantum effects when he writes that "There must have been a time in the very early universe when the universe was so small, that one could no longer ignore the small-scale effects of quantum mechanics."[79] Other interpreters also treat size as the only relevant consideration in applying quantum physics to the universe as a whole.[80]

The real difficulty is that quantum demeanor may depend on more than or something other than just *size*. It may depend heavily on degree and kind of curvature, compression, density, temperature, or quantity of energy involved. An entire universe compressed to the size of an electron or quark may be too curved, compressed, dense, or hot to obey quantum laws; or the quantity of energy may be too great. Roger Penrose, who doubts that size and distance mark the boundary between quantum level and classical level events, suggests that all quantum level events involve "very tiny differences in energy."[81] If so, an entire universe of energy concentrated to the size of a photon just might be too much for quantum effects!

Without agreeing with them completely, Heinz R. Pagels notes that according to cautious critics, "Theorists exploring, on paper, the very early universe have gone too far. Extrapolating from theories that work in the relatively low-energy domain examined by terrestrial accelerators to such ultrahigh energies is a dubious enterprise."[82] Martin Rees recognizes that "...Physics at ultrahigh energies...is almost completely unknown."[83] John Gribbin and Rees warn that "Because the physics of the ultracompressed, high-density stages is speculative, we have no firm understanding of exactly where the fluctuations come from."[84] This is precisely the difficulty. The physics of ultracompressed high-density states is largely untested and unknown. As Mark Israelit and Nathan Rosen affirm, "We lack any knowledge whatsoever of the constitution of matter under such extreme conditions";[85] yet they make many assumptions about it. Jonathan J. Halliwell recognizes that "Quantum mechanics was developed to describe atomic-scale phenomena," and he recognizes the enormous contentiousness of "the most extravagant extrapolation possible: that quantum mechanics applies to the entire universe at all times and to everything in it."[86]

None of the sub-atomic particles that we know to be affected by quantum indefiniteness, discreteness, and spontaneity are experienced and tested at anything close to what the curvature, compression, density, temperature, and energy quantity of the universe as a whole would be if compacted to the volume of an atom, electron, photon, quark, or something even smaller. The differences in proportions are truly astronomical, and these enormous differences could easily distort or even completely negate quantum effects. In the absence of experimental confirmation, there is room for serious doubt that the whole universe obeys the laws of quantum physics when compressed to Planck (or smaller) dimensions. Every law of nature that takes us back to "the beginning" may be inapplicable sufficiently close to T = 0, which is why Scientific Cosmological Agnostics proclaim that we just can't know scientifically what caused the Big Bang. Neither caution nor radical skepticism about the applicability of the laws of quantum physics to conditions at or near the origin of the universe will be popular with many contemporary scientific cosmologists, but perhaps they should reconsider.

Without quantum effects at the beginning of creation, Oscillation Cosmologies are indefensible because they cannot otherwise avoid singularities. With quantum effects, Oscillation Cosmologies are still on shaky ground for reasons already given and for others to follow. We must now examine some of the logical and philosophical implications of these options, even though we have serious doubts about the applicability of quantum laws to an ultracompressed universe.

E. Oscillationism and Thermodynamics

Oscillation Cosmologies presuppose the existence of sufficient mass/energy in all cosmic epochs to reverse their expansion and close them down, no matter how different their contents, laws, constants, and initial conditions might be. Our own cosmic epoch seems to be very much out of place within such an endless series of oscillating epochs. Our universe lacks seventy percent or more of the mass required for reversal. Oscillationists may have a strong faith that the missing mass is really there; but they cannot prove it, and it has now been decisively disproved by the discovery in 1998 that the pace of cosmic expansion is accelerating rather than slowing. This could not be true of a universe that will eventually cease expanding and gradually collapse into a Big Crunch. Oscillation Cosmologists cannot reconcile their robust Omega = 1+ requirement with the paltry Omega = .1 to .3 of our actual universe as we know it.

They might resort to the desperate strategy of contending that either the First Law of Thermodynamics or the Law of Gravity were grossly violated at the beginning of our eccentric cosmic epoch, even though they held in all antecedent universes. They might argue that enough mass or gravity was present in an infinite number of antecedent universes to close them, but our exceptional universe is open because huge quantities of mass/energy or gravitational attraction just disappeared somehow at the beginning of our atypical cosmic epoch. Perhaps so. Perhaps not. If this happened, Oscillationists must explain how and why; but no explanation is readily forthcoming. In an infinite set of successive universes, why is ours an exception to the rules?

If a singularity preceded our epoch, the notion of antecedent universes is meaningless; but without an initial singularity to reshuffle everything drastically, we have no good reason to believe that any physical fundamentals change dramatically from epoch to epoch. Natural laws, including the First and Second Laws of Thermodynamics, would not break down between epochs that are not separated by singularities. Oscillationists have no adequate theory or mechanism to explain an enormous one-time extinction of mass or weakening of gravity. With no initial singularity, the First Law of Thermodynamics, the conservation of mass/energy, weighs heavily against significant extinctions of mass and gravitation between epochs. An implausible theory cannot be saved by appeal to an even more implausible theory.

The Second Law of Thermodynamics, the law of increasing entropy or disorder, also makes serious trouble for Oscillation Cosmologies, even without singularities. Cosmologists like Paul Davies[87] and Alan Lightman *suggest* that it does.[88] Others like Alan MacRobert,[89] Steven Weinberg,[90] Ya. B. Zel'Dovich and I. D. Novikov,[91] James Peebles,[92] and Alan Guth[93] *reject* Oscillationism *outright* because they are convinced that the Second Law of increasing entropy or disorder excludes it. Endless oscillations and divisions are ruled out completely by increasing entropy. S. A. Bludman argued in 1984 that

> because of the huge entropy genèrated in our Universe, far from oscillating, a closed universe can go through one cycle of expansion and contraction. Whether closed or open, reversing or monotonically expanding, the severely irreversible phase transitions transpiring give the Universe a definite beginning, middle and end the ultimate crunch can never be reversed. Nor could it have bounced in the past if it began hot or developed a great deal of entropy in a first contraction.[94]

Most scientific cosmologists believe that the law of increasing entropy holds not only during the initial expansion phase but also during the contraction phase of a universe undergoing gravitational collapse. They have no doubt that the law of increasing entropy would apply continuously through successive cosmic epochs, no matter how these are connected. They may or may not be right, but *if* they are, Oscillation Cosmology is dealt a deadly blow.

Since disorder constantly increases, according to the Second Law, our present universe would be infinitely chaotic if it were preceded by an infinite number of cosmic epochs, as Oscillationism maintains. Because our universe is not infinitely chaotic, an infinite number of antecedent universes did not antedate it, and Oscillationism is dead.

Unless Oscillationists can find a way around the Second Law of Thermodynamics, increasing entropy is a formidable obstacle, not only for them, but also for other Antecedent Universe Cosmologists. But the situation may not be completely hopeless.

John A. Wheeler argued in 1973 that when a collapsing universe shrinks to a singularity, a drastic reprocessing of mass, charge, physical constants, natural laws, and all details of the system occurs; and the ensuing universe starts out on a completely fresh dynamic cycle.[95] Singularities might insure that a universe that ends in maximal entropy would be followed by another universe that begins in minimal entropy, but many unresolved problems are hidden in the obscure notion of "reprocessing."

Quantum Cosmologists wanting to dispense with singularities also speculate that when a cosmic epoch shrinks to quantum size, quantum effects wash out all existing entropy, all information about what went before; and thus each new epoch starts afresh. As Andrei Linde recalls, M. A. Markov developed "a model of an eternal, oscillating universe, which at each cycle of its evolution forgets what occurred before."[96] In Markov's own words, "The state of maximum contraction would play a peculiar role of 'purgatory,' purifying the universe from 'excessive' mass and entropy acquired in a previous expansion and contraction."[97] Robert Dicke and P. J. E. Peebles, commenting on oscillating universes, say that "Experience would suggest the total entropy can only increase, though it certainly is conceivable that in the new physics of the bounce, entropy is eliminated, perhaps lost in black holes left over after the big bang."[98] Notable authorities may be quoted on both sides of the question of whether entropy continues from epoch to epoch, but which view is most plausible?

Many cosmologists are convinced that only an initial singularity, if anything, could guarantee a scramble sufficiently thorough to insure the demise of the Second Law of Thermodynamics from epoch to epoch. Understandably, a cosmic epoch would have to forget its composition and structure if it collapses into a singularity, for at that point there is no space, time, physical causation, physical structures, or natural laws. In singularities, and only in them, the laws of nature are gone, and so are all physical causes, constants, fields, masses, particles, and components. Nothing remains to remember or carry forward the disorder inherited from a preceding epoch. Unfortunately, along with entropy, an initial singularity also obliterates all lawful spatial, temporal, and causal continuity between cosmic epochs In a singularity, both the First Law and Second Laws of Thermodynamics are gone, and nothing remains to insure that enough physical energy or orderliness will persist from epoch to epoch to keep the series going.

Most cosmologists think that complete reprocessing would not occur if a universe collapses only to quantum (or larger) size, and the rest of us may just have to take their word for it. At finite dimensions of compaction, something spatial, temporal, causal, and lawful remains of a maximally compressed antecedent universe; and, despite Markov's doubts, just enough probably remains to carry forward the entropy, the distorting Weyl curvature, of a preceding epoch into a later one. If quantum laws and processes are preserved through the passage from one epoch to another, so are the laws of thermodynamics. If cosmologists can't affirm continuing quantum laws without affirming the laws of thermodynamics, the Second Law of Thermodynamics is a major hurdle for Antecedent Universe Cosmologies that appeal to quantum effects to avoid an initial singularity, for they can't avoid an increase of entropy from epoch to epoch.

The strategy of resorting to total reprocessing in singularities is extremely risky for Oscillation Cosmology. The physics of the Big Bounce without singularities also requires that all antecedent worlds be closed quantum universes that cannot collapse into singularities because of quantum effects like indefiniteness and discreteness; but in an infinite number of diversifying tries, total reprocessing will insure that at least one rescrambled antecedent universe would be open, flat, and/or non-quantum. Infinite diversification through reprocessing would bring about at least one open or flat universe that breaks the series, and/or one non-quantum universe that ends in a spaceless, timeless, lawless, and causeless singularity–with all its problems.

In an infinite number of antecedent diversifying shuffles that actualize all possibilities, quantum laws and quantum effects themselves would be scrambled out of existence at some point, and so would sufficient mass to close an infinite number of antecedent cosmic epochs. From such antecedent universes, no Big Bounce could occur. The endless chain would be broken; and we would not be here; but here we are! Infinite reprocessing might unscramble entropy, but it would also terminate infinite reprocessing. The non-existence of our world is a very high price to pay for avoiding the Second Law of Thermodynamics. Furthermore, since *every* epoch in an infinite series has an infinite number of predecessors, every universe must be nonexistent! The very idea of oscillating universes is unintelligible!

i. Entropy Can't Apply to the Universe as a Whole

Perhaps the Second Law of Thermodynamics does not apply legitimately to the universe as a whole. The philosopher Stephen Toulmin contends that it does not.[99] The preceding argument against Oscillation Cosmologies based on increasing entropy from epoch to epoch applies the Second Law to the universe as a whole. If Toulmin is right, this application is illegitimate and all objections to Oscillation Cosmologies based on thermodynamics are spurious.

Toulmin's argument is complex, but let's try to go directly to the heart of it. He is concerned in part with the practical implications of the idea that the universe is running down; but his primary focus is on the shift of the Second Law away from applied mechanics, where it properly belongs, to cosmological speculation about the universe as a whole, where he thinks it is out of place. Toulmin believes that this shift is illegitimate because the question "Is the universe as a whole a thermally isolated system?" is senseless. The reason, Toulmin explains, is that

> the question how far a given physical system is isolated from its surroundings has a clear enough meaning when asked about any bounded part of

> the universe–being equivalent to the question, to what extent heat exchanges are possible across the boundary–when asked about the universe-as-a-whole, its meaning is completely obscure.[100]

We cannot talk intelligibly about heat exchanges across the boundary of the universe as a whole, Toulmin maintains, because nothing exists outside the boundary of the universe as a whole. The universe has no surroundings.[101] The Second Law of Thermodynamics tells us that disorder increases in thermally isolated systems; it applies intelligibly only when systems have boundaries and surroundings; it applies appropriately only *within* but not *to* the universe as a whole. According to Toulmin, we do not apply the law of gravity to the universe as a whole because we realize that nothing exists outside the universe to be attracted by it. Similarly, the Second Law of Thermodynamics should not be applied to the universe as a whole because it has no boundaries or surroundings.[102]

ii. Entropy Can Apply to the Universe as a Whole

Fortunately, scientists seldom pay much attention when philosophers tell them what they can and cannot do or think. Toulmin's argument is unconvincing. If nothing exists outside the boundaries of the universe as a whole, this means that it is the ultimate isolated system *par excellence*; as Victor Stenger says, "Only the universe is a completely closed system."[103] If the universe as a whole has no surroundings, then it is clearly impossible for anything to affect the constancy of its supply of energy (First Law) or the internal dissipation of its energy and order as it expands (Second Law).

Toulmin's reasons for thinking that the laws of thermodynamics do not apply to the universe as a whole really show that the universe is a perfect example, perhaps the only really perfect example, of a system that can neither gain nor lose energy (once created). Thus, objections to Oscillation Cosmologies based on thermodynamics must be taken seriously after all. If disorder really increases constantly in closed systems, and the universe as a whole is the ultimate closed system, then Oscillation Cosmologies are untenable because they project the entropic universe as a whole infinitely into the past. Our world is not one of infinite chaos; but it would be if the Second Law of Thermodynamics has been operating throughout an infinite past; so Oscillation Cosmologies must be false.

Antecedent Universe Cosmologies clearly affirm that something exists outside a particular cosmic epoch like the one we inhabit, namely, a temporally infinite but imperceptible (to us) larger or older universe. Thus, even on Toulmin's terms, Oscillation Cosmology should have no difficulty applying laws of thermodynamics to our cosmic epoch, which, by hypothesis, is bounded and surrounded by an infinite series of prior cosmic epochs. The real problem is that the surroundings of our epoch transmitted little or no entropy into our Big Bang; so they could not be temporally infinite.

Oscillation Cosmologies are doubly cursed. *Without singularities* between cosmic epochs, the Second Law of Thermodynamics would not break down, and our world would be infinitely chaotic, which it is not. Our orderly cosmic epoch would be impossible if preceded by an infinite number of increasingly disordered universes that dumped their disorder into our Big Bang. *With singularities*, if the First and Second Laws are periodically reprocessed out of existence, and so is everything else that would make oscillating universes go on forever. Even worse, antecedent universes could have no spatiotemporal or causal connections with later universes; no continuous and infinitely comprehensive temporal series could exist within which one universe could be earlier or later than another.

So, what caused the Big Bang? Oscillation Cosmologies answer that it was caused by an influx of energy from a preceding universe or cosmic epoch. With or without singularities, this answer is incredible and indefensible.

To summarize, Oscillation Cosmologies affirm that our universe, our cosmic epoch, was preceded by an infinite number of antecedent expanding/collapsing universes; but this claim lies far beyond the limits of scientific knowledge. Science cannot establish that even one cosmic epoch preceded our own, much less an infinite number of them. Antecedent universes are supernatural beings that fall outside of our system of spacetime or nature, as Robert Jastrow correctly indicates. They are beings that, by hypothesis, exist before time and outside space, even if it is only *our* time and space. We know empirically of no universe other than our own; old fashioned Naturalists are right about that! Scientifically, with respect to supernatural spaces and times, we can only be agnostics or Positivists.

Oscillation Cosmologies have additional flaws. They presuppose that our own cosmic epoch contains enough mass/energy to close it and reverse its expansion process, and that this was true of all antecedent universes, an infinite number of them. However, our own cosmic epoch does not seem to be closed. Its expansion rate is increasing, and seventy to ninety percent of the required mass is missing. Maybe we will find it someday; maybe not. We do not know that we will; and the chances are extremely high that we will not, especially now that we know that the Hubble expansion rate of our universe is increasing.

Oscillation Cosmologies also presuppose that all antecedent universes were bouncing quantum universes, but no known laws of physics would cause or allow crunched-up universes to bounce. Oscillationists appeal to periodic reprocessing to eliminate entropy; but this insures that in an infinite number of antecedent diversifying tries, a prior open or flat universe, a non-quantum universe, or a terminal singularity would have resulted from an endless reshuffling of laws and initial conditions; and we and our world would not exist. Yet, here we are! Since this consideration applies to every epoch in an infinite series, no worlds at all would exist!

Oscillation Cosmologies differ over whether antecedent collapsing epochs begin and end in singularities, but they are damned if they do and damned if they don't. *With* singularities, no lawful spatiotemporal or causal relations could link cosmic epochs because space, time, physical causation, and all the laws of nature break down in singularities. The Naturalistic claim that all spatiotemporal events are caused by other spatiotemporal events would be untrue. The very notion of an antecedent universe becomes incoherent and unintelligible. An antecedent universe is postulated as a physical cause of our universe, but physical causation begins *after* T = 0 and cannot be traced back any further. We cannot infer that the cause of our universe was something physical. Singularities having no magnitude at all are empirically indistinguishable from nothingness.

Without singularities, Planck-size or larger spatiotemporal and causal relations could connect successive epochs; but the laws of nature would not break down between epochs; and without a total meltdown, the law of increasing entropy carries over from epoch to epoch. If our epoch was preceded by an infinite number of epochs, and if chaos increased in each and from one to the next, then our epoch would be infinitely chaotic. It is not; so Oscillation Cosmologies must be wrong. Furthermore, if the laws and initial conditions of each universe are totally reprocessed between epochs, an open or flat universe, a non-quantum universe, a terminal singularity, or the extinction of all energy whatsoever would occur in an infinite number of diversifying antecedent tries that actualize all possibilities; and our universe would not exist; but it does.

Quantum effects rescue Oscillation Cosmologies only at the price of unjustifiable favoritism. Quantum indefiniteness and discreteness would exclude all singularities, both initial and those in collapsing neutron stars and black holes. Initial singularities are excluded only in imaginary time, not in real time, if Hawking is right. Quantum effects support neither Oscillation Cosmologies nor even more eccentric Quantum Cosmologies yet to be examined, for we do not know that other worlds exist(ed) as quantum universes, or that atomic or sub-atomic size universes obey the laws of quantum physics when all the energy within them is drastically compacted in size, pressure, density, temperature, and curvature. With or without an initial singularity, our universe is flat or open, thus radically unlike the infinite number of universes that supposedly preceded it; but what happened to the missing gravity or mass?

Oscillation Cosmologies are initially attractive, but their problems are insurmountable. Antecedent Universe Cosmologies are not serious obstacles to theistic belief. This does not establish the truth of theism, but it clears away much of the rubbish that stands in the way.

Five

BIG FIZZ AND BIG DIVIDE QUANTUM COSMOLOGIES

Quantum Cosmologies add quantum theory to relativity theory and apply them to the universe as a whole. If, as previously indicated, the laws and concepts of quantum physics apply only to tiny, definite, and discrete quanta of energy, the whole enterprise of Quantum Cosmology is doomed from the start. When compacted to the size of a quantum object like an electron, all the energy of an entire universe is just too great for quantum physics to apply; but for the sake of a lively debate, let us assume that it does apply. Even so, Quantum Cosmology in its many forms still has far too many problems to be very plausible.

From the vantage points of common sense and classical Newtonian physics, quantum theory involves many oddities. Two strange quantum effects, quantum indefiniteness and quantum discreteness, may be used as evidence against initial singularities, as previously explained. A third, quantum fluctuations, is used in Big Fizz Cosmologies to account for the origin of bubble universes and for the bubbles within bubbles that give rise to galactic and supergalactic structures. A fourth, plenitude or the actualization of all possibilities, spawns infinitely many universes. A fifth, Quantum Observership, results in an Idealistic Metaphysics according to which the universe exists only because we perceive it.

Quantum Cosmologies modify Standard Big Bang Cosmology by introducing quantum oddities like indefiniteness, discreteness, fluctuations, plenitude, and observership and by applying these to the beginning of the universe. Plenitude from quantum physics spills over into astrophysics, resulting in at least two distinct world-ensemble cosmologies: (1) the Big Fizz view of John A. Wheeler, Stephen Hawking, Alan Guth, and a number of Russian and other cosmologists who proclaim that primordial Superspacetime generates multiple inflationary universes by quantum fluctuations to create all possible worlds; and (2) the Big Divide view, originating with Hugh Everett, III in 1957,[1] which affirms that every universe branches every instant into multiple parallel universes to actualize all possibilities. These world-ensemble cosmologies proliferate worlds to infinity through bizarre Big Fizz and Big Divide processes.[2]

1. Big Fizz Quantum Cosmology

Russian astrophysicists like A. A. Starobinsky,[3] Andrei Linde,[4] and M. A. Markov[5] affirm and defend oscillating universes without singularities, as do other astrophysicists. They also envision other and more fundamental ways of generating universes without appealing to preceding oscillations. They locate their imagined worlds within an infinite Superspacetime that endlessly ejects, then reabsorbs, and ejects again, an infinite number of co-present universes that may or may not oscillate. I shall call this multi-ejection Quantum Cosmology the "Big Fizz" theory of origins. Stephen Hawking and many others are strongly attracted to the Big Fizz position. What does it have to say about the origin of our universe?

A. Mother Spacetime

Big Fizz Cosmology affirms that an infinite number of universes are created by, and co-exist within, infinite Superspacetime, which includes but infinitely transcends the finite spacetime of our own universe. Oscillation and Quantum Cosmologies are not mutually exclusive. Once co-existent universes originate in infinite Superspace through spontaneous fluctuations, they may then oscillate forever after within Supertime, say Big Fizz proponents. Even parallel universes with a finite past can have an infinite future. Oscillation variants are plagued by all the difficulties discussed in Chapter Four, but let us begin with the novel element that Quantum Cosmologies introduce–Superspacetime. Both Oscillation and Big Fizz theories are Antecedent Universe Cosmologies. Universes expand into nothingness in pure Oscillation Cosmologies, but they expand into pre-existing Superspacetime in Big Fizz Cosmologies. *Superspacetime*, not a singularity or an antecedent cosmic epoch, is the ultimate Antecedent Universe in these theories; it produces all Big Bangs, including our own. Maternal Superspacetime is the space beyond our space and the time before, during, and after our time.

According to I. L. Rozental, for whom "metagalaxies" are independent, co-existing universes, including ours, "A multidimensional background space exists, filled with a physical vacuum subject to perturbations. These perturbations give rise to the evolution of objects like the Metagalaxy."[6] Rozental says that "The Universe–eternal and infinite–lives a stormy life reminiscent (metaphorically, of course) of a pot of boiling liquid. Like vapor bubbles, metagalaxies arise, expand, and finally die, giving birth to new metagalaxies."[7] In discussing many worlds cosmologies, John Gribbin calls this infinite multidimensional background spacetime "'super' spacetime."[8] Katsuhiko Sato and his co-authors call it "the original 'mother' universe."[9] Willem B. Drees called it "Mother Spacetime."[10] This maternalistic metaphor is illuminating. Mother or Super spacetime is the primordial progenitor of all particular spatiotemporal universes, an infinite number of them, so Quantum Cosmologists claim.

Pure Oscillation Cosmologies postulate a single infinitely long strand of *consecutive* universes, each of which gives rises to its successor after it collapses. In Big Fizz world-ensemble cosmologies, Mother Spacetime spontaneously generates an infinite number of *co-existing* universes, some of which may also oscillate once initiated. Mother Spacetime is infinitely creative in both space and time, but of what is she composed? Rozental likens her to a pot of boiling water. In more technical terms, how is she depicted? Infinitely transcendent Superspacetime is composed of: the physical vacuum or minimal mass/density of "empty space" as an actualized energy field having a bubbly or fizzy structure that results from spontaneous quantum fluctuations.

John Hick thinks that Big Fizz Cosmology is religiously ambiguous with respect to Theism and Naturalism because "It remains no less conceivable that the super-universe, with ours as one component, is itself the ultimate uncreated reality."[11] However, if the super-universe consists entirely of contingent universes, contingent Superspacetime, a contingent physical vacuum, contingent quantum fluctuations, and contingent laws of quantum mechanics, it is definitely not a necessary being, or what Hick calls "the ultimate uncreated reality." As explained in depth in Chapter Twelve, as far as experience takes us, which is all that we have to go on, any whole composed entirely of contingent parts is a contingent whole, not an ultimate necessary being. This means that as contingent wholes, our universe and any transcendent Superuniverse do not exist self-sufficiently and indestructibly. Certainly, no self-contradiction is involved in denying their existence. Our own universe was definitely created by something and is not everlasting. Contingent wholes exist dependently; their non-existence is possible, so their existence is not self-explanatory. As a contingent whole, Superspacetime itself requires an explanation, one that ultimately comes to rest in a transcendent Necessary Being; but the case for this is yet to be made.

B. The Physical Vacuum and Pure Energy

Mother Spacetime includes or is a physical vacuum.[12] Do not read too much common sense into this notion. As quantum physicists understand the concept, a physical vacuum is not a state of complete nothingness or emptiness. If it were, it would be indistinguishable from a singularity, understood as empirical emptiness; and the creation of the universe from it would be identical with creation *ex nihilo*.

Big Fizz proponents clearly repudiate both creation *ex nihilo* and the identification of Superspacetime with absolute nothingness. A physical vacuum has ephemeral but still real ingredients. Alan Guth writes that "To the particle theorist, the word 'vacuum' is defined as the state of lowest possible energy density."[13] The physical vacuum is "empty space"; but empty space is not empty! It contains no enduring particles like electrons and protons, but it does contain the actualized mass/energy/density of space as space[14] which has not yet been converted to enduring particles; and it contains virtual particles[15] that become actual particles for only a tiny fraction of a second before being annihilated by their antiparticles. It also may contain the "Higgs field" that interacts with electrons to give them mass.[16] Charles Misner, Kip Thorne, and John A. Wheeler thus describe the vacuum state:

> No point is more central than this, that empty space is not empty. It is the seat of the most violent physics. The electromagnetic field fluctuates. Virtual pairs of positive and negative electrons, in effect, are continually being created and annihilated, and likewise pairs of mu mesons, pairs of baryons, and pairs of other particles. All these fluctuations coexist with the quantum fluctuations in the geometry and topology of space.[17]

The physical vacuum of "empty" space is not just vacuous nothingness. "Empty" space itself is not empty; it has a fine-grained bubbly structure. It seethes with particles waiting to be born. It even has its own physical mass/density.[18] Its measurable effects can push two metal plates together after all physical particles have been siphoned away.

The Uncertainty Principle permits very short-lived suspensions of the Principle of the Conservation of Energy, so in "empty space" actual particles and antiparticles are constantly being born. Usually, they immediately lapse back into pure potentiality, or mutually annihilate one another. Occasionally, however, a particle gets away, Big Fizz theorists claim, though this has never been confirmed. Now and then, they say, one of these newly escaped particles or bubbles inflates spontaneously into an entire universe. This, too, has never been confirmed. These metaphysical claims are unabashed conjectures.

C. Forever Blowing Bubbles

Mother Spacetime also consists of bubbles or froth.[19] She spawns a really big fizz, an infinite number of bubbles, interminable primordial foam. At its deepest level, spacetime itself is foamy or granular rather than smooth; but theorists are not clear about whether the bubbles are composed merely of space, or consist in the transient virtual/actual matter/antimatter particles that constantly arise.

Big Fizz Cosmologists recognize that most bubbles do not inflate into entire universes; but, randomly, innumerable bubbles do inflate at the speed of light into self-contained worlds, some of which are mini and some maxi universes. Andrei Linde calls this "chaotic inflation." According to this theory, each mini-universe may subdivide into innumerable distinct mini-universes. As Linde puts it, "Instead of one single big bang producing a single-bubble Universe, we are speaking now about inflationary bubbles producing new bubbles, producing new bubbles, *ad infinitum*."[20] In one way or another, Mother Spacetime gives birth to an infinite number of "child universes," as they are called by Alan Guth[21] and Andrei Linde.[22]

Progeny universes pinch off and become detached from one another, though in some instances wormhole tubes might link them, some speculate.[23] Child universes are so far apart in Mother Spacetime that they cannot act causally on one another, even at the speed of light–except possibly through wormholes.[24] In infinite Superspacetime, an infinite number of co-existent universes could all be infinitely far apart. After birth, most child universes exist in complete causal independence. Relativity theory defines simultaneous co-existence as mutual causal independence. Co-existing postnatal offspring universes live and move and have their being within the womb of Mother Spacetime, the Infinite Universe in the largest possible sense, without influencing one another. Within all-encompassing Mother Spacetime, they have spatiotemporal but not causal relations with one another. Only the Infinite Mother herself, if She is omniscient, can know their relative times and positions within infinite Super-spacetime.

Just be aware that all of this is wild, unconfirmed, and unconfirmable conjecture! Our universe is one such bubble-child, says the theory. Offspring universes begin with their own Bang. According to Andrei Linde, "If one wishes to reserve this name [big bang] for the first 'big bang' (if there was one), one may think about such names as a 'small bang' or 'pretty big bang.'"[25] Viewed from within, each child universe seems to be everything, for it cannot contact its brothers and sisters and is not at all sure that they even exist. After its spontaneous generation by Mother Spacetime, a child universe may expand, contract, expand, contract, forever. Linde describes such an oscillating universe as "an infinite chain reaction of creation and self-reproduction which has no end and which may have no beginning."[26] Linde emphasizes closed oscillating universes, but if Mother Spacetime actualizes all possible universes, she must also include all possible open, flat, and non-oscillating universes.

D. Spontaneous Fluctuations

Supposedly, spontaneous fluctuation is Mother Spacetime's mechanism for world-making. The Uncertainty Principle in quantum theory says that the behavior of entities at the atomic and sub-atomic quantum level is random, indeterminate, and unpredictable. Quantum-size entities (sub-atomic particles or wavicles) cannot simultaneously exemplify velocity and locus and are not bound by rigid causal bonds. Even "empty space" has a tenacious case of the jitters. Unpredictable spontaneous deviations are the rule, not the exception. No hidden variables exist to restore classical causal determinism and reduce physical indeterminacy to mere human ignorance or to uncertainties in the experimental situation. Quantum fields and particles *in themselves* are not fully determinate in position and velocity; they are not fully determined causally by conditions that we cannot find. Quantum entities fluctuate spontaneously and unpredictably. No one can make accurate predictions about when individual molecules of radioactive elements will spontaneously disintegrate, or when particular electrons will jump orbits without passing through the intervening space, though *en masse* vast numbers of individually unpredictable events give rise to statistical regularities.

Big Fizz Cosmology avows that tiny bubbles form in the supercosmic physical vacuum through spontaneous fluctuations. Most bubbles in the primordial fizz are stillborn. Others inflate into child universes, but not all at the same time, as Mother Spacetime might apprehend sameness and time. Universe-producing bubble inflation is creative, spontaneous, random, unpredictable, a-causal, and accidental.

Child universes are almost, but not quite, created *by nothing* as well as *from nothing* in moderate Big Fizz Quantum Cosmologies. More extreme Big Accident Quantum Cosmologies, later considered, will take this final step; but moderate Big Fizz Cosmologists like Linde, Rozental, and Guth do not go this far. They hold that Mother Spacetime, the ephemeral energy of the physical vacuum of empty space, spontaneous energy fluctuations, and the laws of quantum physics, are necessary causal conditions for all Big Bangs.

Causes are either necessary or sufficient conditions, or both together. In the absence of necessary conditions, effects cannot come to be; in the presence of sufficient conditions, effects must come to be. In Big Fizz theory, Mother Spacetime, quantum laws, quantum fluctuations, and the pure energy of the physical vacuum are necessary causal conditions for the spontaneous generation of all universes, including our own. In their absence, neither our universe nor any other would exist. Since its birth was in part spontaneous, our universe did not have a sufficient cause (unless it was the abstract Principle of Plenitude); no antecedent physical conditions were perfectly adequate for its emergence. Given Mother Spacetime, our universe is the ultimate manifestation of spontaneous creativity; but getting an enormous and magnificent universe like ours from nearly nothing may not be quite so easy.

A. Karel Velan indicates that no actual particles or universes ever emerge spontaneously from a physical vacuum alone. Particle accelerators disclose that virtual particles become actual only in the presence of an actualized electromagnetic energy field. If he is right, contingent actualities require additional external actualities for their being, even in a quantum multiuniverse. Velan indicates that an actualized primordial electronic radiation field must exist within the physical vacuum of Superspacetime itself if it is to spawn any universes, including our own.[27] This cosmic energy field, he thinks, is "the fundamental tool of divine power."[28]

After next explaining and exploring Big Divide Cosmology, we will see that it and the Big Fizz suffer from common defects.

2. Big Divide Many Worlds Cosmology

Under the direction of John A. Wheeler, Hugh Everett, III wrote a doctoral dissertation in 1957 titled *The Theory of the Universal Wave Function* in which he developed the Big Divide many worlds interpretation of Quantum Cosmology. This position was later expanded and defended in several of Everett's published articles, and by Bryce DeWitt and others.[29] More recent defenses and critiques are also available.[30]

According to quantum physics, indefiniteness, discreteness, and spontaneity permeate entities at the atomic and sub-atomic levels. In many interpretations, quantum-size entities exist only in an indefinite, indeterminate wave-function state; and they achieve definiteness only if something happens to collapse their wave-function. Prior to this collapse, all possibilities are on an equal footing. If and when its wave-function collapses, a quantum object then acquires definiteness and determination. In the theory of Quantum Observership, presented in a later chapter, wave functions collapse only when physical process are observed and measured; but in the Everett\DeWitt Big Divide theory, wave-functions never collapse.

If wave-functions never collapse, why do we experience a definite and determinate world? Everett and DeWitt gave a peculiar answer. A wave-function is merely an infinite set of potentials or possibilities; and everything, large and small, including the universe, has one. Also, potentiality is identical with actuality in this perspective; so everything's and everyone's infinite possibilities are somehow actualized. How can all the potentialities of every conceivable particle, person, life-form, thing, and cosmos be actualized? Actualization of infinite possibilities is a time-asymmetrical process; any universe divides into an infinite number of additional universes at every moment of time and every point of space. Once created, branch universes sprout branches, which also sprout branches–to infinity. Big Divide theorists do not call attention to them, but divisions must contain dead end branches that go nowhere, since these too are logically possible.

We conscious observers experience a definite world every time we have a sensation or make a measurement or an observation. Conscious observers as such also divide endlessly at every moment into multiple copies, and these too actualize all possibilities. Everett focused primarily on infinite divisions of minds in a "many minds theory," but DeWitt emphasized infinite divisions of both minds and worlds. Every time we conscious subjects divide, the entire universe divides with us. For a fleeting moment, we observe a fleeting determinate universe; but this does not mean that the infinite potentialities of our wave-function have collapsed into *one* concrete actuality within one determinate universe. Instead, it means that we have suddenly and imperceptibly sub-divided or branched into an infinite number of observing subjects in an infinite number of new universes. Definiteness emerges from infinite potentiality because infinite partitions into endless branching universes actualize all possibilities for everything, from quarks to Quakers.

Because "universe" may have more than one meaning, says Frank J. Tipler, Everett's Big Divide interpretation of quantum physics is often misunderstood. "Universe" may mean "all spatiotemporal reality;" but Everett's cosmology does not assert that all spatiotemporal reality constantly branches into infinitely many worlds, Tipler says. In Everett's theory, "universe" refers only to the limited domain of observer and observed; and the theory affirms that "Only the observed/observer system splits; only that restricted portion of the universe acted on by the Measurement operation M splits";[31] but the unobserved portion of the universe does not split.

Tipler's efforts to make an infinitely counter-intuitive theory less so do not succeed, however. Suppose that an observer tries to measure or observe the universe as a whole, the totality of spacetime, instead of just a restricted part of it. Then DeWitt's more extreme interpretation of the Big Divide becomes operational. Tipler himself concedes that "if it is the radius of the Universe that is being measured, the *first* measurement of the universal radius will split the Universe into universes, which collectively have all possible radii."[32] To avoid initial conditions, Tipler affirms that "The Universe consists of *all* logically possible universes";[33] but that is just what the more extreme view espoused by Bryce DeWitt, and occasionally by Paul Davies, affirms all along.

Quantum theorists disagree sharply about whether the indeterminateness found at microscopic atomic and sub-atomic levels carries over to macroscopic perceptible objects like rocks, houses, cats, and people. DeWitt's variety of Big Divide Cosmology offers a peculiar solution to this problem. Schrödinger's cat illustrates both the problem and the weird solution that Big Divide many worlds cosmology adopts to account for relationships between microscopic and macroscopic worlds.

A. Schrödinger's Cat

Erwin Schrödinger proposed a notorious experiment in 1935 to demonstrate how imperceptible sub-atomic quantum events could have perceptible macroscopic effects. First, put a live cat and a bit of radioactive material into a small box with a vial of cyanide that will be broken by a hammer if a Geiger counter detects the radioactive decay of just one atom. Assuming classical physics and a bit of common sense, if an atom decays before the cat in the chamber is examined a short time later, the experimenter will find a dead cat; but if an atom does not decay, the cat will be alive. So far, so good.

However, the indeterminateness of quantum physics creates a profound paradox. Quantum theory, as some interpret it, says that an unobserved atom like the one detected by the Geiger counter is nothing in itself except a superposition or collection of all its potentialities. If we know the sum of its possibilities, we can calculate the probability that an observer will find it in a particular determinate state; but, unobserved, it is largely, perhaps completely, nothing more than an indeterminate set of possibilities.

So, what does quantum indeterminacy predict for Schrödinger's cat? The unobserved decaying atoms to be detected by the Geiger counter are nothing more than the sum of all their possibilities. A negative possibility corresponds to every positive possibility. Possibly, any particular atom decays; possibly not. So, is the cat dead or alive? From the perspective of quantum physics, the macroscopic cat itself is also nothing more than a wave-function superposition of dead and alive states, both before and after the chamber is opened. Is the cat both dead and alive? Has quantum physics disproved the logical principle of non-contradiction, according to which a proposition cannot be both true and false? Has it shown, as some muddled mystics suggest, that Schrödinger's cat is both dead and not dead? Has quantum theory proved the truth of illogical irrationalism?

B. Schrödinger's Cat in Many Worlds

DeWitt's version of Big Divide Cosmology renounces illogicality for its own brand of logically consistent occultism. The Big Divide answer is that the cat is both alive and dead–in at least two different universes. When a trunk universe confronts alternatives, it branches into parallel universes to actualize all of them; so Schrödinger's cat is alive in one universe and dead in another. The dead cat is in an infinite number of states of decay in an additional infinity of universes, and the live cat is instantiated in an infinite number of life-states in yet another infinity of universes. This avoids contradiction, but the price of saving logic is very high! An infinite violation of the principles of parsimony and of the conservation of energy is required to produce an infinity of entirely new universes at every worldly impasse. Needless to say, DeWitt's Big Divide Cosmology is not popular with most physicists. Richard A. Healey understates the situation: "Few working physicists take it seriously."[34]

The paradox of Schrödinger's cat may have less drastic but still perplexing solutions. Less radical interpretations of quantum mechanics say that definiteness is achieved for microscopic events because their wave function collapses into determinateness when perceived or measured by an observer using a macroscopic measuring instrument. More radical DeWitt Big Divide Cosmology affirms that definiteness is achieved because world-division into infinitely many universes actualizes every possibility. The Quantum Observership option says that conscious measurement creates all definiteness; but before turning to it, we must consider problems that are common to both Big Fizz and Big Divide cosmological perspectives.

3. Critique of World-Ensemble Cosmologies

Difficulties for Big Fizz and Big Divide world-ensemble cosmologies focus around A. their lack of empirical foundations and B. their plenitudinal equating of actuality with possibility. These cosmologies are much more other-worldly than traditional theology. Their affirmation of other worlds and the Principle of Plenitude has no empirical basis whatsoever.

A. Lack of Empirical Foundations and Meaning

In many respects, world-ensemble Quantum Cosmology is speculative theoretical science gone absolutely mad. Experience without theory is blind, but theory without experience is vacuous. We human beings have no direct experience of other antecedent or contemporary universes of any kind, including transcendent Mother Spacetime, the infinite froth or fizz in her womb, the co-existent universes she supposedly spawns, or the endless branches that each allegedly sprouts. No matter how ingenious they are, human scientists cannot experience infinitely extended time and space. This much is patently obvious.

Big Fizz Cosmology's physical vacuum in pre-existing infinite Superspacetime is a product of pure theory and brazen speculation, but it is not delivered or confirmed by experience. Blau and Guth concede that "A false vacuum has never been observed."[35] Then they confidently proceed to give a theoretical account of it, as do other Big Fizz Cosmologists. Rozental concedes that it is not possible to check the validity of the Big Fizz creation scenario: "As far as a direct experimental verification is concerned, no approach is in sight";[36] but he anticipates that the development of theory alone will eventually solve the problem of the birth of our metagalaxy.

This is precisely the difficulty. Big Fizz Quantum Cosmology is spun solely out of thin air, ponderous theory, intricate calculations, and audacious speculation. It touches base with experience much too infrequently, and on the most crucial issues, not at all. Theologians conjure up non-empirical cosmologies, and pure mathematicians do abstract calculations, but where is the empirical science in this Big Fizz version of Quantum Cosmology?

When considering contingently existing things, entities that might or might not exist, experience alone can distinguish between actuality and abstruse possibility. Only highly dubious theoretical conjectures support many worlds Quantum Cosmologies. Michael White and John Gribbin say that Stephen Hawking, one influential sponsor of Quantum Cosmology, has never won the Nobel Prize because candidates for the award are considered

> only if a discovery can be supported by verifiable experimental or observational evidence. Hawking's work is, of course, unproved. Although the mathematics of his theories is considered beautiful and elegant, science is still unable even to prove the existence of black holes, let alone verify Hawking Radiation or any of his other theoretical proposals.[37]

Since this was written, the Hubble Space Telescope has located black holes, but no parallel universes. This shows the critical importance, not the unimportance, of balancing theory with experience, for both truth and meaningfulness.

"Meaningful" has many meanings; and empirical meaningfulness is just one of many kinds. Meaninglessness logically correlates with meaningfulness.

No matter what we think of Logical Positivism, some beliefs really are empirically meaningless to us because no human experiences whatsoever could ever verify or falsify them. But claims may be empirically unverifiable or unfalsifiable in at least two ways.

First, the possible referent of a proposition can be inaccessible in principle because we *cannot imagine* experiences in our world or elsewhere that would count for or against it. For Positivism, only sensory experiences count–not religious, moral, or aesthetic experiences. We can imagine sense experiences that would count for or against life in the Andromeda Galaxy, or even in other universes, though we have no practical way to get there; but we cannot imagine how to verify or falsify the proposition that nothing at all exists, because someone must exist to think the thought and make the relevant observation. We also can't imagine observing a thing that both is and is not itself; logic forbids it. No sensory images, the very stuff of positivistic science, correspond with many concepts widely employed in contemporary cosmology; we cannot imagine nothingness fluctuating, or what a singularity would look like. By positivistic standards, much of contemporary cosmology is utterly meaningless. The ultimate limits of imagination are not fixed, however.

Second, claims cannot be verified or falsified if the objects to which they refer are *physically inaccessible to us because they allegedly belong to some other world or universe*. We might be able to imagine what they would look like if they were accessible, if we were there; but since other worlds do not belong to our spacetime system, they are physically unavailable to us; no lawful causal or spatiotemporal relations link them to us. In this second sense, all propositions about other worlds, whether religious or quantum-cosmological, are unverifiable, unfalsifiable, and thus empirically meaningless to us. This second type of empirical meaninglessness is based mainly upon causal inaccessibility. No matter what we do or what experiments we perform while alive in this world, totally other worlds are inaccessible to us. This is true even if we could time travel back to the beginning or our own universe. We can imagine how statements about Heaven, Hell, other worlds, antecedent universes, co-existing world-ensembles, and Mother Spacetime could be verified or falsified if we could only take an all-inclusive God's-eye view that encompasses them all; but we mortals just cannot get there from here (alive) to take a look. We cannot go to transcendent worlds, and entities and messages from them cannot get to here from there. Statements about other worlds are not accompanied by instructions on what to observe in this world to make them either plausible or implausible.

In this second sense of verifiability, Big Fizz and Big Divide world-ensemble cosmologies are perfect examples of vacuous or meaningless metaphysics gone mad. Empirically, we cannot know whether they are true or false. They literally make no sense to us. Alternate spacetime systems and their contents are physically inaccessible to us no matter what we do because they do not belong to our system of space, time, and causation; and they predict nothing about our world that cannot be explained in much simpler and more obvious ways. We can perform no operations, make no observations, that would give us empirical access to what goes on in other worlds.

A sturdy streak of Logical Positivism runs through most interpretations of quantum mechanics, but it is often applied with a highly selective bias. In quantum mechanics, statements about things that cannot be observed or measured no matter what we do–like hidden variables, photons spinning unobserved, and electrons having definite position and velocity–are consistently dismissed as meaningless. Yet, by their own admission, Big Fizz and Big Divide metaphysicians posit the reality of innumerable entities belonging to other worlds that are totally inaccessible to us no matter what.

Reflecting on Big Divide parallel universes, Hugh Everett, III wrote that the "total lack of effect of one branch on another also implies that no observer will ever be aware of any "splitting" process,"[38] and Bryce DeWitt claims that multiple worlds are "mutually unobservable but equally real."[39] By positivistic quantum logic, however, unobservable multiple worlds should be just as unreal or nonsensical as all other states that are totally inaccessible physically, like simultaneous position and velocity for sub-atomic particles; and statements about them really should be regarded not as lies but as gibberish.

Does purely theoretical system-building count as good natural science simply because brilliant scientists are doing it? Not so, suggests Eric Lerner, who correctly affirms that "The ultimate test of scientific theories is observation," and that legitimate scientists use an empirical method. By contrast, "The other method, advocated by mainstream cosmologists and particle theorists, is the deductive method, mathematically deducing how the universe *must* be."[40] Theory and experience are indeed mutually supporting in good science, but many versions of Quantum Cosmology go too far beyond anything experiential. Illustrative of Lerner's point, Heinz R. Pagels says that scientists once tried to deduce the laws of nature from experiment and observation, but "Today this method has been abandoned and physicists do not directly deduce the laws from experiment. Instead they try to intuit the basic laws from mathematical reasoning."[41] Sadly, theory and intuition alone cannot construct a true and meaningful account of contingently existing empirical actuality. Quantum Cosmologists easily confuse eccentric theory with empirical reality, variable private intuitions with public truth, and abstruse possibilities with determinate actualities.

Not only is infinite world-ensemble metaphysics unverifiable and unfalsifiable, but it also flagrantly violates the rational, scientific (albeit aesthetic) criterion of simplicity. As Abner Shimony says, "The continuous evolution of the total quantum state is obtained by Everett at the price of an extreme violation of Ockham's principle, the entities being entire universes."[42] Richard A. Healey argues that the theory of many spacetime systems in Big Divide Cosmology "offers no interpretative advantages" over a theory which affirms that "all but one of the many worlds which emerge from a quantum measurement are merely *possible* worlds." According to Healey, "The intuition behind the argument is just Ockham's razor: the many-spaces version postulates a proliferation of extra entities (spaces, "copy" quantum systems) with no corresponding gain in explanatory power or conceptual clarity."[43] Victor J. Stenger doubts that the "luminaries" who affirm the existence of infinitely many unexperienced coexisting parallel universes actually believe it.[44] He brands the many worlds parallel universe hypothesis as a "bizarre, nontestable notion"[45] and pronounces it to be "uneconomical speculation."[46] The claim that all possibilities are actual, when only one set will do for experience, is empirically untested, untestable, unnecessary, unintelligible, incoherent, and an inexcusable violation of parsimony.

B. Possibility = Actuality, and World-Ensembles

Many Quantum Cosmologists thoroughly confuse mental constructs with reality by identifying possibility with actuality. They catapult conceptually from mere possibilities to the actual existence of infinitely many worlds. Both Big Divide and Big Fizz world-ensemble cosmologists declare that all possibilities are actualized somewhere. Philosophical theologians, influenced by Plato's Principle of Plenitude, once regularly vaulted from possibility to actuality. Today, world-ensemble cosmologists make the same jump with wild abandon.

As Arthur Lovejoy indicated, Plato believed that God would be imperfect if He actually created anything less than everything that He possibly could create.[47] Equating possibility with actuality was a prominent feature of Greek and Medieval theology, and somehow it found its way into today's Quantum Cosmology. As Dennis W. Sciama expressed it in 1993, "All logically possible universes exist in an ensemble of disjoint universes,"[48] and "Everything which is not forbidden is compulsory."[49] Stephen Hawking agrees: "In quantum theory, anything that is not actually forbidden can and will happen."[50] Many serious philosophical questions must be raised about such claims. How do these speculative cosmologists know all possibilities are actualized? What is their observational evidence? Even if microscopic quantum-level events actualize all possibilities, can Plenitude then be generalized to entire macroscopic universes? Where does the energy come from to actualize infinite universes? Is this really natural science? Or is it just sloppy reasoning, wishful thinking?

Quantum physics unearthed a tenuous connection between possibility and actuality at the sub-atomic quantum level with the discovery that individual photons traveling through two nearby pinholes or slits seem to pass through *both* of them. From this finding, some physicists hastily concluded that physical particles do not move continuously along single paths. Instead, they take *every* route possible, that is, an infinite number of routes, to reach their objective.[51] Of course, if a third slit is added to the two slit experiment, the photon will *not* go through all three of them, so there are very strict limits after all, even at the level of quantum events;[52] and Quantum Plenitude cannot be reconciled with the well established quantum truths that electrons can *not* take *every* orbit possible, that particles can *not* take *every* spin possible, and thus that *everything possible can not be actual*!

Furthermore, equating actuality with possibility greatly exaggerates and distorts the "sum over histories" account of quantum processes, according to which most possibilities simply cancel one another out, and only a finite number of paths are actually open to moving particles. But why don't *all* possibilities cancel out one another since for every "possibly so" there is always a corresponding "possibly not"? Possibilities cannot be perfectly identical with probabilities or with actualities. If they were, *every* coherent contingent possibility would be canceled out by its own negation–which is also logically possible. For every possible p, not p is also possible; for every live cat, a dead cat is also possible. For every live you, a dead you is possible. Clearly, not everything possible is actual in our world, so Quantum Plenitude has to postulate an actual infinity of worlds to make a place for the actualization of all possibilities. For many good reasons, we should repudiate Quantum Plenitude!

Big Fizz and Big Divide Quantum Cosmologists really do take seriously the rule that everything actually occurs that is not logically forbidden. They extend this sweeping, hasty, and erroneous generalization from the sub-microscopic level to the macroscopic level of everyday experience, ordinary sense objects, and entire universes. Rozental rejects the leap from the microscopic to the macroscopic,[53] but he still affirms the existence of infinitely many worlds. Leaping hastily from a sub-microscopic to a macroscopic identity of possibility with actuality, Quantum Cosmologists reason that since sub-atomic quantum level entities realize all possibilities (which is false to begin with, as just noted), then *everything* realizes all possibilities. All things that are logically possible are also actual, including universes. From a dubious interpretation of the mysterious behavior of photons, unbridled speculation conjures up infinitely many worlds! Theologians are left in the dust by such preposterous hasty generalizations and spectacular leaps of faith!

According to Andrei Linde,

> The evolution of the inflationary universe has no end and may have no beginning. As a result, the universe becomes divided into many different domains (mini-universes) of exponentially large size, inside of which all possible (metastable) vacuum states are realized. One may say therefore that not only could God create the universe differently, but in His wisdom He created a universe which has been unceasingly producing different universes of all possible types.[54]

How many universes does it take to cover "all possible types"? Obviously, an infinite number.[55] Since possibility equals actuality, an infinite number of physically and qualitatively diverse universes really exist, according to Big Fizz and Big Divide infinite world-ensemble metaphysics.

C. You and I in Many Worlds

You may be thinking: "If possibility is identical with actuality, why ain't I rich?" You are, but only in another "parallel" universe, if that's any consolation. In many worlds Quantum Cosmology, new universes are spawned not only in the primordial fizz, but at every turn of events within every cosmos. The Big Divide version implies that when your past self confronted the possibility of being either rich or poor, the whole universe, yourself included, branched. The rich you entered at least one universe, and the poor you entered at least one other. If you are poor, there is a rich you, and if you are rich, there is a poor you, somewhere in another cosmos. Ain't that grand! Mother Spacetime knows!

Serious problems about personal and cosmic identity abound here. If all possible universes exist that are not logically prohibited, then a real universe exists in which England won the Revolutionary War, the South won the Civil War, and Germany won both World Wars. A universe exists somewhere in which fundamentalistic creationism is true in every detail, and another universe exists in which it is false in every detail. In some universe, Hitler was a saintly born-again religious believer, and Jesus was the Devil incarnate! Surely anyone who believes this sort of thing has been sipping too much bubbly!

Don't just take my word for it. In explaining the Big Divide outlook, Bryce S. DeWitt concedes that

> The idea of 10^{100+} slightly imperfect copies of oneself all constantly splitting into further copies, which ultimately become unrecognizable, is not easy to reconcile with common sense. Here is schizophrenia with a vengeance.[56]

Yet, DeWitt does not repudiate this multiple-world multiple-personality madness! He affirms it! Quantum Cosmology is quantum mechanics gone absolutely nuts! Eric Lerner ridicules and disavows what I call Big Fizz and Big Divide Quantum Cosmologies with these words:

> Some cosmologists, such as Hawking, answer with even weirder ideas: perhaps, they speculate, tiny pulsations in the space around us, even within us, are at every instant giving birth to submicroscopic universes, tiny bubbles of space-time, that then pinch off from our universe to form another universe. From every point, even the tip of one's nose, quadrillions of universes are forming every second. Ours is only one among them, formed presumably from the tip of someone's nose in another, more ancient universe.[57]

Quantum Cosmologists may not comprehend fully the absurd implications of their revitalized Principle of Plenitude that equates possibility with actuality. World-ensemble theorists repeatedly emphasize that even though an infinite number of universes exist, most are unsuitable for life of any form, especially intelligent life. The other side of this coin also needs to be emphasized. Though improbable, many logically consistent and thus possible worlds would be suitable for forms of life that far surpass the excellencies of any life-form on earth. Indeed, some such worlds would approximate, indeed achieve, perfection, no matter how conceived. World-ensemble metaphysics predicts that *everyone's* concept of Heaven is actualized somewhere! So, by the way, is everyone's Hell! And every possible condition in between! And we (or copies of us) are in all of them! We are also in none of them, since that too is logically possible. Does all of this make good sense to you?

The small problem of getting from here to there, from earth to Heaven (or Hell) is really no difficulty at all if everything possible is also actual. Possibly, souls could be transported through wormholes to proper Big Fizz Beyonds. Or, as John Hick suggested, when we die in this world, God could simply recreate us instantly in another world–like being beamed up by Scotty! Or, as Frank Tipler maintains, the immensely complex computers of the future could just emulate our virtual reality, along with that of all other logically possible entities and events. If actualities and possibilities are identical, all of the foregoing possibilities would be actual in some universe: Because possible, therefore actual! Even God(s) and the Devil(s) must exist in all worlds if all possibilities are actual! Yet, if they are contingent beings, they may not exist in many or even in any worlds, for that too is possible! Patrons of cosmic plenitude really have not thought it through very carefully!

Just how ridiculous can plenitude get? In his 1994 book, *The Physics of Immortality*, Frank Tipler contends that billions of years from now, the complex computers of the future will raise us (or our ciberspacetime virtual emulations) from the dead, once computer technology becomes sufficiently complex to create the virtual reality of all logically possible universes, individuals, and their relations.[58] Yet, because he accepts quantum many worlds metaphysics,[59] Tipler need not wait even a second for what he calls immortality. In Big Divide Cosmology, Heavenly *and* Hellacious universes, with natural laws and empirical conditions sufficiently different to make them truly heavenly and hellacious, are created for each of us every instant! Is this not a fitting *reductio ad absurdum* of all varieties of infinite world-ensemble metaphysics?

In sum, Quantum Cosmologies modify Standard Big Bang theory by postulating cosmic or supercosmic-level quantum conditions like indefiniteness, discreteness, fluctuations, the physical vacuum, and Superspacetime. Our universe was preceded by, and is included within, Infinite Superspacetime; and through quantum fluctuations in the primordial physical vacuum, the womb of infinitely many worlds, Motherspacetime endlessly proliferates bubbles and/or branches that inflate or divide into full fledged co-existent child universes. These child universes themselves bubble or branch indefinitely to produce infinitely many additional universes that actualize all possible worlds with and without living things.

Do Big Fizz and Big Divide Quantum Cosmologists really know what they are talking about? This is very doubtful. No one really knows that collapsed universes will exemplify quantum laws or conditions, or that tiny bubbles in transcendent Superspacetime inflate into real universes. What we do know indicates that these things are mere figments of human imagination gone wild.

Other problems are also quite serious. Quantum Cosmologists confuse conceptual constructs and theories with reality. The Principle of Plenitude applied metaphysically affirms that all possibilities are actualized in an infinite number of universes; but this is scientifically groundless and logically incoherent. No empirical evidence whatsoever discloses other universes constantly being created out of a primordial supercosmic fizz, or at every turn of events within or without our own cosmos. You and I could never see it happen. Frank Tipler correctly concedes that "We cannot see the other worlds of the Many Worlds Interpretation."[60] This explicitly acknowledges that other worlds exist only in theory, and that there is no scientific, that is, empirical, evidence for them. So why mess around with them?

We experience no antecedent or co-existing universes, no prior oscillating cosmic epochs, no contemporary worlds in transcendent Mother Spacetime, and no parallel universes branching from our own world at every instant. Timothy Ferris correctly acknowledges that "We have but a single universe to examine."[61] Like the Santa Claus fable, Quantum Plenitude explains nothing for which better, simpler, more empirical, more scientific explanations are not readily available. God might be able to collect empirical evidence for or against infinite worlds metaphysics, but we cannot. Empirically, we can only be Positivists or agnostics about the other-worldly ramifications of Big Fizz and Big Divide Quantum Cosmologies. Even so, we have not come to the end of it. Additional serious difficulties for many worlds metaphysics will be discussed in later chapters, especially in connection with Anthropic Cosmology.

Six

QUANTUM OBSERVERSHIP COSMOLOGY

Conventional non-Everett quantum theory assumes that definiteness in and of the world results from wave-function collapse, and many Quantum Cosmologists try to apply such notions to the universe as a whole. Some Quantum Cosmologists affirm a very intimate relationship between physics and physicists, between observers and things observed. This intimacy is much clearer at the level of microscopic quantum events than at the cosmological level of accounting for the origin of the universe.

Quantum Observership says that if and when a physicist measures for the position of an electron, it then takes a position; if and when a physicist measures for its momentum, it then assumes momentum; since no observer can measure for both simultaneously, it cannot have both together. Sometimes sub-atomic entities behave like particles, sometimes like waves, depending on how observers perceive them. In themselves, apart from being observed, physical entities are nothing; the very concept is meaningless. They have no actuality in themselves; they are only bundles of possibilities.

In Europe, Niels Bohr's "Copenhagen interpretation" of quantum mechanics closely linked observer and observed. In the United States, John A. Wheeler and Eugene Wigner endorsed Strong Anthropic theories of Quantum Observership that emphasize the role of observers in resolving quantum uncertainty and indefiniteness.[1] Many experiments in quantum physics suggest that unobserved sub-atomic particles in themselves do not exist in a single definite state and that the process of being observed somehow affects what they are.[2] The two slit experiment indicates that individual photons seem to pass through two (but not three) separate slits at once. Other experiments indicate that the act of observation significantly affects the physical state being observed.

As the story goes, when two spinning particles are generated together, if one is observed to be spinning in one direction, the other instantly acquires the opposite spin, no matter how far away from the first it might be, or how long it has traveled to get there, or even if the decision about how to measure it is made after its departure. If widely separated particles that originated together are observed simultaneously (a difficult if not impossible feat), they do not have time to communicate; but they still have opposite spins. Quantum mechanics rejects as empirically meaningless the realistic position favored by Einstein that particles that originate together have opposite spins from the very outset and continue to have them while unobserved; whenever they are observed the viewer merely sees what has been there objectively all along. Instead, the correct spin supposedly comes into being because it is being observed; observation itself fixes the direction of spin of the second particle.[3] In Quantum Observership, observation fixes all definiteness.

In John A. Wheeler's theory of Quantum Observership, physical particles in themselves are nothing more than indefinite probabilistic wave function superpositions of all possible states all at once. According to Quantum Observership, quantum systems take on definiteness or "collapse" only when they are observed. Only measurement by an observer using a macroscopic measuring instrument causes collapse. Wheeler illustrates the essential role of measuring observers in quantum mechanics with this anecdote.

> I like the story of the three baseball umpires relaxing over beer one afternoon and comparing notes. One umpire says, "I calls 'em as I sees 'em." The next umpire says: "I calls 'em as they really are." The third one says, "They ain't nothin until I calls 'em."[4]

According to Wheeler, in quantum mechanics "The observer is elevated from 'observer' to 'participator.'"[5] He calls the necessity for observers in the universe the "Participatory Anthropic Principle." According to it, observers are essential for the very existence of the universe and all things within it.

1. Observers Create the Universe

Metaphysical Idealism asserts that only minds and their experiences and activities exist; it denies the objective existence of matter, claiming that material things exist only in being perceived. They do not exist in themselves but only in and as the experiences of observers.

Most physics is realistic and assumes that observers have nothing to do with the objective existence of the physical world. By contrast, Quantum Observership is idealistic and presumes that observers have everything to do with the existence of the physical world. According to this theory, unobserved atomic and sub-atomic entities in themselves bear no resemblance to tiny billiard balls. In themselves, they simply cannot be pictured. They are individuated and determinate only when observed. An unobserved world is nothing more than a set of sum-over potentials for all possible worlds and simply does not exist at all as a definite actuality. Eugene Wigner, who stresses the role of consciousness, claims that a definite world is brought into being only when its potentials are observed and measured by conscious beings.[6] Quantum mechanics thus appears to vindicate John Stuart Mill's dictum that the physical world in itself is nothing more than a set of "permanent possibilities for perception."

In quantum theory as Wheeler and Wigner interpret it, the being of physical entities consists in their being perceived. Wheeler explicitly links quantum mechanics with Bishop Berkeley's "To be is to be perceived"[7] and says that in quantum mechanics "The universe would be nothing without observership as surely as a motor would be dead without electricity."[8] All empirical evidence for the Big Bang is created by observership, and so is the Big Bang itself.[9]

Observers create the world by experiencing it and reflecting upon it. Wheeler says that "The observer is participator in genesis,"[10] and "Observership brings the universe into being."[11] Thus, the world *must* be compatible with the existence of observers, for without observers there is no world. By extrapolating backwards from what we perceive to be happening now, we (and not God) create the past, the entire natural history of the universe from the very beginning. So, what caused the Big Bang? We did, says Quantum Observership!

2. Critique of Quantum Observership

Quantum Observership, as just described, has serious problems that make it extremely implausible as an ultimate explanatory hypothesis.

A. Incompatibility With Cosmic and Biological Evolution

The main problem is that Quantum Observership requires our conscious existence long before we actually exist. Conscious observers like us create the Big Bang, but the Big Bang creates all conscious observers! Fred Hallberg rightly indicates that the theory is logically incompatible with plausible scientific accounts of cosmic and biological evolution, including human origins. Hallberg explains:

> Wheeler emphasizes that life and consciousness entails both biological evolution, and prebiological physical and chemical evolution ["Genesis and Observership," pp. 3, 5]. Yet all these forms of evolution involve very specific interactions among highly individuated molecules. So his story seems to require specific, individuated events and entities before consciousness is present to individuate them.[12]

A more plausible and realistic account says that innumerable physical structures and processes, including our galaxy, solar system, and planet existed for billions of years before we or any other conscious observers came into being. Just because observation affects the definiteness of quantum-level events, it does not follow that they have no definiteness at all when not being observed. Throughout fifteen billion years of cosmic evolution, the unobserved physical world in itself was sufficiently definite to do all the things that Big Bang Cosmology says that it did. The unobserved world evolved either from an initial singularity of nothingness or from minimal Planck dimensions through exponential inflation (perhaps) to an astronomically vast cosmos, unpopulated for eons by any conscious observers. Definite physical forces like gravity, the strong nuclear force, the electromagnetic force, and the weak force, emerged from the original Grand Unification. As the universe expanded and cooled, a primordial soup of radiant energy and sub-atomic particles gave way to discrete atoms, at first mainly of hydrogen and helium. Under the influence of gravity clouds of these gaseous atoms consolidated to form definite galaxies, stars, and planets. Unobserved atoms of heavier elements, manufactured by nucleosynthesis in supernovas, were sufficiently determinate to produce planets. On at least one planet, earth, definite molecules of carbon dioxide, water, ammonia, methane, and so, on eventually combined to produce DNA in living cells. Unobserved living cells increased in complexity and definiteness and united with other cells to form complex multicellular organisms. After billions of years of evolution, some complex organisms became conscious observers. A few even became quantum physicists and cosmologists!

Yet, according to Quantum Observership, only with the advent of conscious observers and measurers does nature in itself acquire any definiteness at all, even with respect to the instant of its origin in the Big Bang. Without a definite world, no conscious observers exist; but without conscious observers, no definite world exists! The idealistic interpretation of quantum physics founders on this contradiction. John Wheeler recognized the problem, but he did not solve it.[13] He only perpetuates the contradiction when he writes, "Beginning with the big bang, the universe expands and cools. After eons of dynamic development it gives rise to observership. Acts of observer-participancy in turn give tangible "reality" to the universe not only now but back to the beginning."[14]

Richard Rorty says that realism, with its concerns for objective truth and objective existence, makes no practical difference,[15] but it does. Unless a highly definite universe existed objectively on its own back to the beginning, we would not be here; our being here cannot give the universe and our biochemical environment a reality that our own existence presupposes. Quantum physics is still in its infancy, but one way or another it must overcome the paradoxical anti-realism of Quantum Observership.

Quantum physicists disagree on whether the physical world is completely indeterminate in itself, or only partly so. The complete indefiniteness of the Quantum Observership interpretation is extreme. In a less radical Critical Realism, the physical indefiniteness of nature is only partial; the sensory modalities of conscious observers add some kinds and degrees of definiteness to things perceived, but not all of it.

Critical Realism can resolve the paradoxes and puzzles of Quantum Observership and still preserve the genuine advances of quantum physics, severed from the pretentious Idealistic Metaphysics with which it is too often associated. If physical things in themselves are individuated and definite to a high degree, but not completely so, conscious observers would not be required to structure the physical world. The sensory modalities and mechanisms of conscious observers may add something, but not everything, to what is given perceptually; observers do not create what exists in itself. A partly realistic position must try to tell us which is which–which properties are objective, and which are mind-dependent.

Realism says that a largely determinate physical world exists in itself apart from conscious observation; we perceive certain sensory properties because physical things actually have them and cause us to perceive them. True perceptions accurately reflect objective properties, but illusions and false ones do not. True beliefs accurately describe objective realities, but false beliefs characterize them incorrectly. Realism comes in many varieties. For Naive Realism, objectively existing things have all of the properties that we perceive them to have; but a long tradition of Critical Realism dating back to the Greek Atomists, and resurrected by the originators of modern natural science in the seventeenth and eighteenth centuries, takes a less extreme view. Critical Realism has many contemporary defenders.[16] It traditionally asserted that "secondary" properties of perceived objects like their color, taste, odor, sound, and perceived temperature are "subjective" or mind dependent; they are added to our percepts by our own creative perceptual modalities and do not exist objectively in nature itself. However, the "primary" physical, spatial, or spatiotemporal properties of things like their size, shape, weight, position, resistance, and motion exist both in our perceptions and objectively in things perceived. Primary properties are the metric or mathematically measurable and quantifiable characteristics of physical things, both in themselves, and as perceived. Our attempts to observe or "measure" quantum-level occurrences always changes their properties, but this does not imply that they have no properties at all when we are not observing them.

In idealistic Quantum Observership, primary qualities are just as mind-dependent as secondary qualities. Idealistic interpretations of quantum mechanics may even border on Positivism, arguing that it is meaningless to postulate the objective existence of any unobserved physical entities and processes in themselves, even their primary properties, because, obviously, we can never observe or even imagine observing such things. As Werner Heisenberg indicated in his *Philosophical Problems of Quantum Physics*,

> In modern physics, atoms possess geometrical qualities in no higher degree than color, taste, etc. The atom of modern physics can only be symbolized by a partial differential equation in an abstract multidimensional space. Only the experiment of an observer forces the atom to indicate a position, a color and a quantity of heat. *All* the qualities of the atom of modern physics are derived, it has no *immediate and direct* physical properties at all, i.e. every type of visual conception we might wish to design is *eo ipso* faulty. Quantum theory made the atom into something inaccessible to our senses or our imagination, unlike objects within our daily experience. An atom or, more correctly, an electron no longer displays 'in itself' ('*an sich*') even the simplest geometrical and mechanical properties but it shows them only to the extent to which they can be made accessible to observation by external interference.[17]

In 1954, Wolfgang Pauli attacked the "hidden variables" of Einstein's physical realism by saying that "One should no more rack one's brain about the problem of whether something one cannot know anything about exists all the same than about the ancient question of how many angels are able to sit on the point of a needle."[18] Quantum Positivism regards all speculation about what things are when we cannot observe or know them as meaningless.

Actually, we should rack our brains about whether the physical stuff of the world existed objectively in relatively determinate physical states for billions of years before conscious observers evolved. This must be so, or we physical, biological, and embodied observers would not exist! Quantum Observership's metaphysical dogma that atomic and sub-atomic physical entities are nothing more than infinite sets of indeterminate possibilities until they are consciously observed must be renounced, along with Idealistic Metaphysics, if we are to have a plausible scientific account of cosmic and biological evolution.

The obvious solution is that quantum entities in themselves are not just infinite sets of potentialities for definiteness. Quantum Critical Realism holds that they contain within themselves a high degree of definiteness inherited from the past, combined with a finite set of potentials or possibilities for further self-determination. Real wavicles are both relatively determinate particles as well as waves, not just totally indeterminate "wave functions" alone.

B. Idealism vs. Realism

Idealistic interpretations of quantum physics must answer this fundamental question: What causes observers to have sensory experiences? For the realistic dualist or materialist, matter does the job. For Bishop Berkeley, God, not matter, produces all our sensations. For Kant, totally unknowable things or realities in themselves (which are not supposed to cause anything because only appearances can be causes) cause our sensory experiences. For the skeptic David Hume, our sensations are of "unknown origin."

Quantum Observership cannot explain how infinitely complex sets of pure possibilities, totally lacking the power of actual beings, can act upon the sensory organs, nerve cells, and brains of living biological observers like us and cause us to perceive one definite and common world. In fact, it cannot account for the objective existence of sense organs, much less cells, and brains! How can the absolutely indeterminate bring about the partly or completely determinate? How can pure possibilities be efficient causes? How can living biological observers have definite sensory organs, nerve cells, and brains if the unobserved physical world has no definiteness in itself? Quantum Observership has no good answers. In Idealistic Metaphysics, nerve cells and brains exist only as relatively infrequent objects of perception, but not in themselves. In light of what contemporary physics has revealed, how should we conceive of the physical or material world?

i. Critical Realism and Quantum/Relativity Matter

Quantum Observership aspires to be anti-metaphysical; but it is both epistemologically anti-realistic and metaphysically idealistic. Thoughts and macroscopic sensory experiences are everything. No actual determinate world in itself exists below the level of macroscopic observables, not even one of primary properties; and there are no objective answers or scientific truths because there are no independently existing objects. In Quantum Observership, no objectively existing world of space/time/mass/energy exists; no atoms, protons, electrons, or other particles or wavicles exist objectively in and of themselves. Since they don't exist at all in themselves, they do not have even partly definite size, shape, position, weight, resistance, velocity, and momentum (which is mass multiplied by velocity). Thus, our questions about independent realities have no correct answers. Correspondence theories of truth and perception are rejected. Concepts, propositions, and sensations have no objective referents; they merely cohere with other concepts, beliefs, and experiences over time; or they fail to do so. Being consists in being perceived, nothing more. Truth consists in coherence with other beliefs, not with correspondence to reality.

Critical Realists, by contrast, think that a real world exists beyond our perceptions, and that it closely resembles the spatially extended and mathematically measurable qualities of our sensations. Scientific and cosmological beliefs refer to this real world. True empirical beliefs describe correctly an objectively existing world of space/time/mass/energy; false beliefs misdescribe it. In modern chemistry, electrons and protons in themselves are often said to lack secondary qualities of color, taste, odor, and sound; but they still have primary properties of size, shape, weight, resistance, velocity, and momentum. Yet, in critically realistic quantum physics, the objective reality of these properties cannot be exactly what they were conceived to be in pre-quantum classical physics.

Before quantum theory, classical particle physics was Newtonian in spirit. The primary properties of subatomic particles were thought to be exactly like those of billiard balls, only smaller. All the way down to the smallest particles, Newtonian matter in itself possessed primary qualities in a fully definite and determinate manner. By contrast, in Quantum Critical Realism, some of the primary properties of quantum matter exist only indefinitely, indeterminately, by degrees, and relative to prehenders or measurers (who need not be conscious beings).

The Whiteheadian concept of "prehension" is fruitful and illuminating in understanding quantum matter. Prehending is the temporal process of grasping something or taking it in experientially. Prehenders are temporally experiencing or prehending subjects or momentary occasions of experience that take data from past events into themselves and actively process this received information in a variety of ways. Perishing past events causally transmit or imbue what they can of themselves into their successors, which actively receive, integrate, and thus prehend that data into themselves. They make their own fleeting creative contribution to ongoing reality and perish themselves, to be pretended in turn by their own temporal successors.

Consider ten important contrasts between classical and quantum matter. They differ significantly with respect to (1) definiteness/indefiniteness of spatial extension, (2) definiteness/indefiniteness of position and motion, (3) definiteness/indefiniteness of location, (4) continuity/discontinuity of existence, (5) causal determinateness/indeterminateness, (6) internal vacuousness/sensitivity, (7) local/non-local causation, (8) absolute/relative spatial properties, (9) absolute/relative temporal properties, and (10)organic holistic unrelatedness/interrelatedness.

(1) *Definiteness/indefiniteness of Spatial Extension.* Classical electrons, protons, and other particles had *fully* definite spatial extension, that is, size, shape, and position, at all times; but quantum particles have merely a *fairly* definite size, shape, and position–but only when they lack definite momentum and velocity, and only when some measurement or prehension (which need not be conscious) somehow takes account of just those properties. When we know a particle's pace of passage, we cannot know its place. When it has a definite place, it cannot have a definite pace. Data about pace and place cannot be prehended simultaneously. Realistically construed, they cannot exist simultaneously. How we or other non-conscious prehenders try to "measure" for them partly determines what they can be.

(2) *Definiteness/indefiniteness of Position and Motion.* Classical physical particles had a fully definite and discoverable velocity and momentum at all times, but quantum particles have merely a fairly definite velocity and momentum at best–but only when they lack full definiteness in size, shape, and position, and only when something prehends, takes account of, or experiences just those aspects of their motion. When we prehend a particle's place, we cannot know its pace; particles cannot be, or be prehended as, simultaneously definite in both ways. The idea is odd, but physical reality just is odd. Things are what they are only interrelationally; but these interrelations exist objectively, not merely within consciousness, says Quantum Critical Realism

(3) *Definiteness/indefiniteness of Location.* Classical physical particles had what Alfred North Whitehead called "simple location," a fully definite, determinate, and independent spatial position or locus at all times, but quantum physical particles behave like "wavicles," being spread or smeared out over small but indefinite regions of space at any given time and at different times. Particle aspects of quantum-level entities that constitute societies of successive events are always accompanied and guided by wave aspects. In two-slit experiments, the waves go through both slits, even though the particles go through only one. String theorists speculate that the smallest and most basic constituents of matter are very short and small looped (in most versions) strings of energy. Every string, *qua* string, in every kind of particle is exactly like every other; and

differences between kinds of particles result from differences in the vibrational frequencies and amplitudes of the tiny strings in their cores.[19] String theory, currently unverified, says that all the observed properties of the physical world can be explained if under different conditions these too-small-to-be-observed strings undulate, ball up, tie knots, disappear and reappear, resonate harmoniously (or disharmoniously) with other strings, and are otherwise sensitive to their environments. These issues need not be settled here; at present, string theory is little more than an intricate mathematical game with little or no empirical grounding. For the moment, we note simply that rapidly and constantly vibrating strings, or smeared out particles and wavicles, could never have the definite independent locus of substantively enduring pointlike classical particles. They may essentially lack other kinds of definiteness as well.

(4) *Continuity/discontinuity of Existence*. Classical physical particles were thought to exist without interruption and to move continuously through the infinitely divisible continua of Newtonian space and time. Electrons, for example, moved smoothly through the infinite number of points into which their supposedly perfect circular orbits are divisible. By contrast, quantum particles exist and move only discontinuously from place to place without passing through the intervening spaces, and the orbits to which they are restricted are broad shells, not perfect circles. When a photon is added to an orbiting electron, the "excited" electron prehends this addition, then jumps up to the next allowable orbit-like region without traversing the quantum-forbidden space that separates allowable orbits. Then, almost immediately, but we can't predict exactly when, it may throw off the photon and jump back down without passing continuously through the intervening space between orbit-like-shells. Something similar happens when an electron or any other particle moves forward within its own orbital shell or quantum-permitted region. Its particle features move forward as so many periodic and discontinuous pulsations of energy, unpredictably here, unpredictably there, somewhere within the tiny region of its orbit-like shell, but never in an exactly circular or linear path. Whitehead, who neglected its wave-like features, compared an electron's orbit to the travels of an automobile that appears only at milestones but nowhere in between![20] Sten Oldenwald says that physical particles play "a hop-scotch game to avoid gaps where space-time doesn't exist."[21] Physical particles endure through time only as so many successively pulsating repetitions and ongoing prehensions of their particular kind of particle/wave/string patterns–electron forms, proton forms, photon forms, and the like.

(5) *Causal Determinateness/indeterminateness*. Causal determinism was believed to be complete for classical physical particles. In rigid conformity to exact and efficacious physical laws, all motions and changes made by classical physical particles followed exactly from necessary and sufficient antecedent formal and physical conditions and were in principle susceptible to absolutely precise mathematical measurement. If the positions and velocities of all particles

were known, everything could be predicted. By contrast, quantum particles, lacking absolute positions and velocities, are not fully and predictably determined by and knowable from quantum laws and antecedent physical conditions. Every mathematically measurable quantity of physical entities is susceptible to unpredictable fluctuations or perturbations. Neither we nor an ace predictor like God can tell or predict when individual particles will change orbits, exactly where they will appear next within broad-band orbital shells, which slits they will go through, the directions in which they will fly when scattered, or exactly which ones will decay and produce atomic radiation. Quantum uncertainty, indeterminateness, and spontaneity pervade all physical reality. By their erratic behavior, vast numbers of individuated quantum-size entity/events create quantum laws, which merely summarize formally and probabilistically both their collective regularities and their individual eccentricities. Formal laws as such exert no efficient causality of their own. The laws of nature are epistemologically useful but ontologically powerless.

(6) *Internal Vacuousness/sensitivity*. Classical physical particles were internally inert, solid, vacuous, and fully actual; but quantum wave/particles are at best only partly determinate internally, and they are environmentally sensitive and reactive. In part, they are indeterminate sets of possibilities in process of becoming fully actual. Each of a wavicle's fleeting pulsations of energy in some unconscious way takes on or prehends a determinate form (electron, photon, and so on) inherited from immediately past pulsations, and each projects this form into the future as it perishes in time. Wave/particles are not internally inert, solid, impenetrable, and vacuous; they take their own immediate constitutions and past and future environments into account in deciding what to do or what to become next. All physical wave/particles experience and are internally sensitive to their external environments, at least unconsciously. They know what kind of wave/particles they have been in the past. As they confront the future, they internally take into account not only their own past but that of their wider environment. What wavicles are and what they do is largely constituted by their relations. Protons, neutrons, electrons, and other particles know when they are and when they are not located within atoms; they know one another's natures and whereabouts; and they behave themselves accordingly. Photons, electrons, and other wave/ particles seem to know whether or not they are surrounded by and partly composed of waves; they respond actively when photon quanta are attached to themselves; they are aware of how their twin particles are spinning. When they are confronted by the environmental prospect of going through either one slit or two, they know the difference in advance and behave accordingly.

Based on earlier work by Louis de Broglie and David Bohm, John S. Bell offers a highly plausible realistic but relational solution to the particle-wave dualism inherent in the two slit experiment:

> A particle passing through just one of two holes in a screen could be influenced by waves propagating through both holes. And so influenced that the particle does not go where the waves cancel out, but is attracted to where they cooperate. This idea seems to me so natural and simple, to resolve the wave-particle dilemma in such a clear and ordinary way, that it is a great mystery to me that it was so generally ignored.[22]

The idea that all elemental particles are accompanied and perhaps even guided by waves is now widely accepted in quantum physics, as expressed in the very concept of "wavicle." Realistic objectivists like David Bohm and B. J. Hiley completely repudiate the wave-function-collapse account of quantum-level definiteness. They contend, probably correctly, that most of the dither about definiteness resulting only from collapsing wave functions just ignores the fact that wavicles are particles as well as waves. Wave collapse conundrums assume incorrectly that in themselves wavicles are nothing more than indeterminate waves and forget that they are also particles bearing their own definiteness.[23] Bohm and Hiley contend that in a two slit experiment, the particle definitely goes through one slit, we know not which, while the accompanying wave goes through both slits.[24]

(7) *Local/non-local Causation.* Classical Newtonian material particles were also externally inert, naturally at rest, utterly incapable of self-initiated motion or of being moved by distant entities. They could be moved only mechanically by other external physical entities with which they were in direct physical contact. Both Newton and Einstein were horrified by the thought of what Einstein called "spooky action at a distance."

But isn't gravity action at a distance? Not for Newton! Gravity, for Newton, was not a physical force; it was a divine force. It was a manifestation of God's direct activity in moving non-contiguous objects lawfully in relation to one another. Through gravity, God moves otherwise immovable material objects that are separated from one another by distance.[25] Later atheistic Newtonians conveniently ignored this feature of classical Newtonian matter/gravity. Einsteinian gravity also involves no action at a distance; it consists of curvature of space emanating directly and with continuity from mutually attracting entities.

In contemporary quantum physics, wavicles are neither externally inert nor internally vacuous. They never exist in inert independence. Action at a distance is allowed, indeed required. Wavicles are definitely not internally vacuous. Information is integral to their internal existence, for they can somehow anticipate the presence of two slits before they get there. They are aware of open alternatives and in some primitive way can choose among them. In quantum non-locality (explained more later), information about what happens to one wavicle is instantaneously transmitted to and registers with another, perhaps at great distances.

(8) *Absolute/relative Spatial Properties*. Classical particles existed in absolute Newtonian space and time, which were supposed to be uniformly the same throughout the whole universe. By contrast, when combined with relativity theory, quantum particles have only a relativistic locus in Einsteinian spacetime. They are constituted by their spatiotemporality. Their primary spatial properties of size, shape, resistance to change, position, and momentum vary immensely, depending on how other events prehend them and on the speed with which they and their accompanying spacetime frames are being accelerated through the universe. A spaceship accelerated to the speed of light would flatten out; its mass would increase to infinity; and it would offer infinite resistance to further increases in speed, which is why acceleration to or beyond the speed of light is so implausible. Approximations to the speed of light would be approximations to infinity. No matter what its speed, a Newtonian spaceship, by contrast, would always have exactly the same size, shape, mass, and inertia; but overwhelming evidence now indicates that ours is not a Newtonian world.

Aristotle defined change or motion as "transition from potentiality to actuality," but this idea never gave rise to a single mathematical formula that empowers human beings, scientists in particular, for prediction and control. Modern mathematical physics began when Galileo replaced Aristotle's qualitative definition with a quantitative definition. Motion or velocity equals the space or distance traveled divided by the time required for the trek: v = s/t. Thus, if we travel a hundred miles in two hours, our speed or velocity is fifty miles per hour; and if we travel sixty miles in three hours, our velocity is twenty miles per hour. This formula, in which velocity varies with differences in time or distance, works perfectly well with everything except the speed of light. The speed of light in an invariant absolute in an otherwise relativistic universe.

Toward the end of the nineteenth century, physicists James Clerk Maxwell and Albert A. Michelson considered the speed of light and made some remarkable discoveries that paved the way for Einstein's insight that light is a universal constant and does *not* vary in speed with variations in distance and time. Because the speed of light is constant, space and time cannot be Newtonian absolutes, so Einstein proclaimed in his "Special Theory of Relativity" in 1905.

Time and temporal processes would proceed at the same pace everywhere and at every speed if Newton had been right. In fact, however, one earthborn twin traveling for an interval on a spaceship at close-to-light speed would be much younger when she returned than the twin who remained on earth. The physical clocks in her spaceship and her own biological clocks would slow down in proportion to accelerated speed. The speed of light can be constant at all speeds only if primary properties of size, shape, mass, resistance, mass, and time itself vary with speed. Classical matter possessed primary properties absolutely; but collectively if not individually, quantum matter possesses them only relative to acceleration. Time slows down and mass increases as velocity increases, and its pace is faster as mass and velocity decrease.

Since the pace of time itself is affected by the speed with which a space-time frame flows through the universe, are our cosmological calculations concerning the age of the universe affected? Would other observers at other speeds get more than, or less than, fifteen billion years for the age of the universe? Does the universe have this age only from our own relative vantage point? Probably so, in answer to all of these questions; but the world of nature manifests much more uniformity than popularizes of relativity often lead us to believe. We have learned in recent years that a uniform microwave background is universally accessible; and its constancy could function as a uniform frame of temporal reference for all space travelers.[26] Light is not the only form of energy with the constancy of the speed of light; all electromagnetic radiation travels at that speed. Quantum non-locality, which we haven't learned to put to practical use, instantly connects distant parts of the universe. The basic structures and laws of nature are the same no matter how fast anyone is going.

Any observers moving at any speed anywhere in the universe should be able to view the overall structure of the universe and determine that it came into being at some point in the finite past. Without being able to correlate their watches and their calculations perfectly, all intelligent beings within the universe should be able to discover the Big Bang and conclude that our universe was created a finite while ago.

The universe-wide instantaneous action at a distance of quantum non-locality restores additional physical meaning to cosmic simultaneity, despite our inability to use non-locality to send instant messages to observers elsewhere in the universe. Still, all intelligent observers scattered throughout the universe should be able to discover *that* the Big Bang happened, though if they could communicate they might not be able to agree about precisely *when* it happened.

(9) *Absolute/relative Temporal Properties*. Classical matter possessed its spatial properties in complete independence of time. Newtonian space and time were completely independent of one another and of the material masses within them. If the temporal duration of an electron were cut infinitesimally thin, it would still fully possess its definiteness of spatial size, shape, motion, mass, and so forth. But the complete independence of time, space, and mass do not apply to quantum/relativity matter. Minimal finite temporal durations, what Whitehead called "specious presents," are required for the existence of any spatial properties at all, and minimal spatiality is required for the existence of any temporality at all. Spaces and times can be sliced infinitely thin by imagination or calculation, but no infinitesimals can actually exist. Nothing can really exist for less time than Planck time (10^{-43} second) or in less than Planck space (10^{-33} centimeters). To exist spatially as matter, as extended stuff, entities must also have some temporal or durational properties. In contemporary physics, time and space are inescapably interdependent, not totally independent as they were in classical physics.

10. *Organic Holistic Unrelatedness/interrelatedness*. Classical physical particles are what they are quite independently of all else. Since they exist contingently, they require causes; but they do not require anything else to be what they are. They can be isolated and studied quite independently of their relations with other things. By contrast, in quantum world-views, both microscopic and macroscopic realities are what they are by virtue of their internal relatedness with other things with which they form interdependent wholes. Nothing requires only itself to be what it is. All properties are relational properties; holistic relations to other things are internal to the constitutions of all realities. All previously identified traits of matter involve organic relational wholes in which they parts are what they are by virtue of the measuring or pretending wholes within which they exist. Smeared out quantum-level entities have only relative but not absolute definiteness of spatial extension. When measured or pretended for definiteness of position and motion they have them separately but not together. Their precise location is relationally definite by degrees. They exist discontinuously in relational spurts. They are partly caused by the externalities that they internalize and are partly self-caused. They are internally sensitive and responsive to their surroundings. They are partly what they are by virtue of local and partly by non-local causation. They are constituted by their spatial and temporal relations, and these condition all their other properties.

Thus, even matter is not what it used to be anymore! Commenting on Gilbert Ryle's characterization of mind/matter dualism as "the ghost in the machine," Paul Davies and John Gribbin tell us that "Today, on the brink of the twenty-first century, we can see that Ryle was right to dismiss the notion of the ghost in the machine–not because there is no ghost, but because there is no machine."[27] Today we see clearly that matter in the classical sense just does not exist at all. A critically realistic theory of perceptual correspondence to objective reality must be adjusted to take account of quantum/relativity physics. It must emphasize both the definiteness within holistic interrelatedness of, and the partial but not total spontaneity, indefiniteness, indeterminateness, and unpredictability of, the depths of nature that physics has unearthed.

Once proper adjustments have been made, matter still exists objectively and retains a great deal of definiteness in and of itself alongside its unpredictable self-creativity. Despite relativity physics and quantum measurement problems, as Victor J. Stenger notes,

> Many properties of matter are fixed and, for practical purposes permanent. They can be determined without their respective measurements interfering with one another. These include rest mass, electric charge, magnetic momentum, and spin. Material bodies possess many unambiguous features that are not the slightest bit ephemeral.[28]

In a similar vein, Holmes Rolston, III indicates,

> We must not be overwhelmed with this relativity of everything, for many things are not observer-dependent. The equations, the basic laws are invariant. Some of the physical constants of nature–the speed of light, the charge on an electron, or the number of atomic shells, the atomic table, chemical reactions, and so on–will presumably be the same for all observers, as will ordinarily be the order of succession of causally related events. Thus, there is considerable objectivity in relativity theory.[29]

A partly or critically realistic interpretation of the physical world is possible if the primary properties of unobserved sub-atomic entities and processes are to some degree interrelatedly determinate apart from consciousness and transmit that determinateness to their successors, including our conscious sensory experiences. In response to those quantum experiments indicating that observation of one photon affects the spin of another photon, Roger Penrose, who declares himself to be a realist,[30] says,

> The best suggestion that I can make at this stage would be for a picture involving some sort of partially formed, partly bifurcating spacetime, where the nature of the spacetime has not been adequately resolved until the second photon observation has taken place.[31]

If the contents of spacetime are partly formed apart from conscious observation, and if they correlate to that extent with conscious perception or prehension, partial or Critical Realism is correct. Quantum Critical Realism, adapted to quantum/relativity physics, still allows enough definiteness in objectively existing nature to account for cosmic and biological evolution. This is probably the best available argument for the truth of Critical Realism.

Most physicists have not abandoned realism for an idealistic metaphysics, despite the advances and challenges of quantum mechanics. Most cosmological theories are to some degree realistic. The Standard Model of Big Bang Cosmology assumes that an objectively existing universe of space/time/mass/energy came into being between ten and twenty billion years ago. The Inflation modification makes the same assumption but tries (successfully?) to make it more plausible. Infinite universe theories like Steady State and Plasma Cosmology affirm that an objectively existing universe of space/time/mass/energy is literally infinite–in itself–not just in our perceptions, thoughts, and theories. Antecedent Universe Cosmologies postulate objectively existing universes that preceded and caused our own. World-ensemble theories postulate an objectively existing Superspacetime that gives birth to many objectively co-existing universes, some perhaps erupting or branching from others. Most quantum and relativity physicists and scientific cosmologists agree that space/time/mass/energy are inseparable; but they do not believe that they exist only in our experiences, thoughts, and theories. Quantum Observership contends that this quasi-realistic approach is all wrong; but for many reasons like those given in the preceding discussion, most physicists think that Quantum Observership is all wrong.

A few dissenting scientists and philosophers of science have subjectivist leanings, at least in their weaker moments. In 1983, a younger Paul Davies expressed antirealism when he wrote that "The commonsense view of the world, in terms of objects that really exist 'out there' independently of our observations, totally collapses in the face of the quantum factor."[32] In some of his moods, the early Albert Einstein surmised that space and time are ultimately unreal and exist only as human illusions. He came to this conclusion, not on scientific grounds, but because he subscribed to a Spinozistic metaphysics. Kantian idealism would give the same results. Later, Einstein vehemently opposed the idealistic, anti-realistic implications of Quantum Observership.[33]

Quantum Observership gives quantum mechanics a robustly anti-realistic flavor, but most masters of quantum mechanics are realists. To the suggestion that we should just follow the rules of quantum physics pragmatically without asking questions about reality or trying to form a picture of reality, Roger Penrose responds: "This seems to me to be wholly unreasonable. Physics, after all, constitutes our best way of groping for the true nature of the real world in which we find ourselves."[34] Bernard d'Espagnat defines Realism as "the doctrine that regularities in observed phenomena are caused by some physical reality whose existence is independent of human observers."[35] After thoroughly examining quantum experiments that intimate anti-realism, d'Espagnat concludes that abandoning realism

> trivializes the entire scientific enterprise. Science is reduced to a set of recipes for predicting future observations from a knowledge of past ones. Any notion of science as "the study of nature" is a phantom. One can imagine a physics grounded on positivistic principles that would predict all possible correlations of events and still leave the world totally incomprehensible. Given the extreme consequences of abolishing realism, one is inclined to cling to this premise.[36]

Reconciling Critical Realism with quantum mechanics generates an odd picture of physical reality, as we saw in contrasting Newtonian with Quantum concepts of matter; but Realists are willing to pay the price. Physical reality just is odd, says quantum/relativity theory.

In 1964, John S. Bell demonstrated a fundamental incompatibility between the natural order of things disclosed by quantum mechanics and Einstein's insistence on "locality,"–the belief that there can be no faster-than-light causality. Einstein called faster-than-light causality "spooky action at a distance." Bell insists, contra Einstein, that "Events at one place propagate to other places faster than light. This happens in a way that we cannot use for signaling. Nevertheless it is a gross violation of relativistic causality."[37] Many quantum theorists believe that non-local causality is the key to reconciling quantum theory with scientific Realism. Bernard d'Espagnat insists that messages really *can* travel faster than the speed of light.[38] Roger Penrose affirms both that there are faster-than-light correlations between entangled events and that photons really *can* be (partly) in two places at once.[39] David Bohm and Basil Hiley[40] defend realistic and deterministic interpretations of quantum mechanics that allow for non-local faster-than-light connections between events. Just before sending the final version of this book to my publisher, I discovered that John A. Jungerman explains non-locality clearly and affirms it unequivocally on the basis of now abundant experimental evidence.[41]

Not all realists accept non-locality. In his very thorough but somewhat technical defense of quantum realism, Henry Krips rejects non-locality in favor of hidden variables.[42] In his very readable discussions of the topic, a skeptical Victor J. Stenger contends, "Now, after a series of precise experiments, the issue has been decided: hidden variables that are both local and real are ruled out" and "nonlocality exists only in theory" and cannot be confirmed experimentally.[43] The question of non-local causation does not have to be decided here, but the case for it is much stronger than Stenger suggests. There is probably something to it, and it must be considered carefully by anyone who wishes to understand the interconnectedness of things.

Realistic quantum theory emphasizes the objective reality of fields as well as of waves and particles. Fields are objectively existing, invisible, colorless, regional sets of physical habits or dispositions with ill-defined borders; they require no medium of actualized waves or particles for their causal efficacy and reality; but they bear both information and energy; and these influence the wavicles within them. They inform, give form to, their components. Specialized regions, perhaps all regions, of space itself–some more than others–bear physical and formal properties that structure particle/wave events within themselves.

Finally, a rigid causal determinism *appears* to reign supreme at the level of macroscopic entities encountered in everyday experience to which Newtonian physics applies; but at the level of quantum systems, both indeterminateness and indeterminism are the rules. Quantum events in themselves and in our perception of them are neither fully determinate nor fully indeterminate with respect to their primary spatial and temporal properties. They are neither fully determined by nor left completely undetermined by their causal antecedents. Quantum Critical Realists are convinced that spontaneity is objectively real, not just an expression of human ignorance. Einstein was wrong when he decreed that "God does not play dice with the universe!" The universe plays dice both with itself, and with God.

ii. Observership and Causation

A plausible scientific account of human origins must affirm that relatively definite physical processes existed objectively for billions of years before conscious observers evolved; yet, according to Quantum Observership, conscious observers contribute *all* definiteness to what is observed, and nothing unobserved exists at all. Quantum Observership must avoid this logical predicament, which it cannot do; it must also explain the magical causal process by which conscious observation affects or creates physical entities and processes. Is all seemingly physical causation just another example of spooky action at a distance without a physical medium? Is energy really exchanged between observer and observed? Do conscious observers as such project photons or radiant energy onto other photons and particles to collapse their infinite potencies into definite actualities of position or momentum? If so, could the projected photons or radiation be detected? Would this energy have enough definiteness of its own when not being observed to do its work? Or, as idealist Jonathan Edwards held, does God cause everything directly, while events and beings within the world, including human observers, cause nothing? Or do human observers without God simply create photons, electrons, other particles, and the universe itself *ex nihilo* in the very acts of looking for and finding them? How does this happen? Do conscious observers create the Big Bang in the very act of perceiving or conceiving the evidence for it, in something like the way in which God, in some versions of fundamentalistic "Creation Science" directly creates the fossilized bones of dinosaurs that never existed? If so, is Quantum Observership any more plausible than magical Creationism?

iii. Ambiguities Involving "Observer" and "Measurement"

In Quantum Cosmology, the meanings of "observer" and "measurement" are unclear. Quantum physics appears to divide the universe decisively into (1) observers (2) measuring instruments, and (3) observed or measured quantum events; but these are not sharp distinctions. What is an observer? Must all observers be conscious entities? Are observers themselves composed of quantum events? Are their instruments composed of quantum events?

In Quantum Observership, observers must be conscious–like human beings. Perhaps a conscious animal, a dog, or even Schrödinger's cat would do, but this is not always clear. What if a person instead of a cat had been in Schrödinger's box? Why couldn't the observer be an omni-observant God who, presumably, would always confer as much definiteness upon the world as it needs? Most physicists prefer not to appeal to God's existence or to consciousness to solve physical problems. God might provide a perfect solution to the puzzles of Quantum Observership if most quantum physicists were not biased against Theism. Yet, conscious divine, human, or animal observers may not be necessary at all for conferring definiteness upon quantum events. We cannot simply equate "observer" with "consciousness."

In realistic interpretations of quantum theory, observers need not be conscious beings. Anything responsive to its environment will do. According to Werner Heisenberg, "It does not matter whether the observer is an apparatus or a human being."[44] John S. Bell affirms that "The only 'observer' which is essential in orthodox practical quantum theory is the inanimate apparatus which amplifies microscopic events to macroscopic consequences."[45] Observers may be unconscious measuring machines like Geiger counters or photographic plates.

Definiteness-conferring observations need be nothing more than unconscious physical processes taking account of or interacting causally and prehensively with others. Scientific instruments functioning as unconscious observers are themselves composed of quantum-level physical processes. No sharp line separates microscopic and macroscopic processes and observers, even though our senses are generally responsive only to quantum-level events *en masse*. Victor J. Stenger indicates that detectors "need not be limited to the sensory apparatus of human beings or their scientific instruments. The term 'detector' can also encompass the particles in the environment surrounding the system."[46] Thus, one physical process pretending or interacting responsively with another is all that is required to confer physical definiteness. If all physical processes interact sensitively with other physical processes, the universe in itself, quite apart from conscious observers, would have all of the exactitude that Critical Quantum Realism attributes to it. As Abner Shimony puts it:

> When a physical variable which initially is merely potential acquires a definite value, it can be said to be *actualized*. So far, the only processes we have mentioned in which potentialities are actualized are measurements, but in a non-anthropocentric view of physical theory the measurement process is only a special case of the interaction of systems, of special interest to scientists because knowledge is thereby obtained, but not fundamental from the standpoint of physical theory itself.[47]

Heisenberg, Bell, Stenger, Shimony, and other Quantum Critical Realists think that there is nothing special about either conscious observers or measurements. Popularizers of quantum mechanics like Paul Davies[48] contend that definiteness is achieved at the level of microscopic wave-function events only as they collapse when measured by conscious observers using macroscopic measuring and recording instruments. Schrödinger's cat really is in a dual alive-dead wave-function state until a human observer reads the Geiger counter. John Bell, by contrast, maintains that notions like "observer" and "measurement" are so obscure that they cannot be fundamental for physics, and that there is no sharp line of demarcation between the microscopic and the macroscopic.[49] Elemental reflection will show that Bell is right. John Wheeler himself eventually came around:

> Let us not invoke either "consciousness" or "observer" as prerequisite for what in quantum mechanics we call the elementary act of observation. What counts as "observation" for the purposes of quantum mechanics is the irreversible act of amplification. It may not be clear how much amplification is required.[50]

The quantum-level domain seems to confer degrees of definiteness upon itself merely through causal interactions between quantum events, occasionally at speeds as fast as light, but usually not. Physical causation is always temporalistic and prehensive, but it is not rigidly deterministic.

A significant degree of both definiteness and indefiniteness could exist in quantum events if, as Process Philosophy affirms, partly indeterminate events in their present moment of immediacy prehend, take account of, and in that sense measure, immediately past events that become fully definite only as they perish and are prehended by their successors. Demarcations between event-durations need not be as sharp as process thinkers previously assumed, as explained later. The present occasion in any causal cone of spatiotemporal events possesses a high degree of internal freedom, self-creativity, and indeterminateness, mixed with degrees of definiteness or data inherited from the past. Events acquire their fullest definiteness as they complete themselves, perish into the past, and are perceived or prehended (usually unconsciously) by their partly-self-creating successors. In physical terms, wavicles are always highly definite in themselves; but they acquire a more complete but still very similar definiteness as they perish in time and are succeeded and "observed" by their immediate temporal successors.

Quantum theorists recognize that a measurer must be something *spatially* distinct from the object measured, but *temporal* distinctness must also be emphasized. All causation from space to space is temporally ordered. Also, according to Process Philosophy, all mentality is spatially ordered and extended; and no totally disembodied mentality exists anywhere. All events at every level of existence have both mental and physical poles, but physical poles are both spatially extended as well as causally responsive and efficacious. Even God is embodied in the world, or some world, and is not a purely incorporeal disembodied spirit.

If all immediately successive events are observers or prehenders of their predecessors and receive forms and data from them, the unconscious world of nature in itself has sufficient definiteness to exist in itself and to evolve complex conscious observers like us; and each relatively independent and partly self-creative present moment at every level of complexity has sufficient indefiniteness to allow for ubiquitous freedom and creativity.

In Process Philosophy, the partial indeterminateness of present events extends to all levels of reality, no matter what their degree of organization and complexity. Atomic and sub-atomic uncertainty is not the same thing as free will and creativity in immensely complex human subjects, but both are manifestations of a universal principle of creativity; and human creativity supervenes upon physical indefiniteness. Newtonian mechanistic determinism excluded human freedom, but quantum level uncertainty makes room for it without being identical with it. Natural processes become more and more fully determinate as decisions are made and information is synthesized in the present, as events perish into the past, and as their successors perceive or prehend them. Temporalistic entities receive as much as they can from their predecessors and transmit what they can of themselves into their successors. These transmitted forms and data give continuity, memory, and relative self-identity through time to chains or temporally ordered societies of consecutive spatiotemporal events, ranging from quark-pulsations to streams of human consciousness.

Our own conscious present moments of receptivity, partial self-creativity, relative independence, subjective immediacy, and self-enjoyment are partly indefinite, but not infinitely indefinite. Degrees of definiteness and indefiniteness are equally real at all levels of natural complexity. Photon wavicles definitely cannot collapse into electrons; electron pulsations cannot collapse into proton or neutron pulsations; and streams of human consciousness cannot degenerate directly into streams of canine or bovine consciousness. Each partly determinate particle/wave includes a very limited set of unactualized possibilities. Where will it jump to next? Where exactly will it go when scattered? Through which slit will it pass? When will it "decay" into a free particle?

Events become determinate partly by inheritance, partly by decisions of the moment among open possibilities, and partly by perishing in time to be prehended by their successors. Quantum indefiniteness versus measurement is nothing more mysterious than causation between immediately past and present spatiotemporal occasions. That in itself is mysterious enough!

To summarize, Quantum Observership argues from unusual interpretations of perplexing experiments in quantum physics to an Idealistic Metaphysics that denies the objective existence of space/time/mass/energy. No physical objects exist without conscious physicists or observers of some sort, says Idealism. In itself, the physical world is merely an infinite set of indeterminate possibilities for perception; and the determinateness of the physical world consists in and is caused or created by its being perceived or observed.

Quantum Observership is incompatible with cosmic and biological evolution. It generates the unresolved paradox that observers create the Big Bang and the evolutionary process, but the evolutionary process and the Big Bang create all observers. It cannot explain the magical causal procedure by which observers confer determinateness on physical entities and processes, including those in the distant past. As in Kantian Idealism, Quantum Observership reduces space, time, energy, and causation to forms of experience that have nothing to do with things in themselves or objective realities. Most modern scientists reject Quantum Observership and its insistence that space/time/mass/energy exist only in human consciousness; they strongly prefer, with good reason, the partly realistic theory that the physical world exists objectively, independently, but relationally. Yet, the relations need not be with conscious observers.

Finally, the notions of "observer," "instrument," and "measurement" cloak serious ambiguities. The referents of all of these concepts are partly composed of quantum-level events. If observers are allowed to be unconscious physical objects, including measuring instruments, and if no sharp divisions exist between the macroscopic and the microscopic or between minds and matter, and if measuring is merely causal interaction, the mutual interconnectedness of effects and causes, then in the absence of conscious observers, the physical world can prehend, observe, measure, and confer sufficient determinateness upon itself to account for the objective reality of the Big Bang and for the evolution of nature prior to the emergence of consciousness.

A Kantian/Copernican idealistic revolution in epistemology is just as anthropocentric as the geocentric theory that placed humankind at the geographical center of the universe. A realistic nonanthropomorphic natural science affirms that we have our being and our becoming within and as a part of a vast, natural, objectively existing, vibrant universe of interdependent, partly determinate, partly indeterminate pulsations of space/time/mass/energy. For billions of years, this independently existing system of nature possessed a high degree of definiteness in itself; unconscious prehensive relational responsiveness is ubiquitous. In our part of the Milky Way, no conscious observers existed before animals, including our own most primitive evolutionary ancestors, emerged from the primordial terrestrial slime. We exist within and as a part of the totality of objective spacetime. Nature does not exist merely within us as mind-dependent conscious perceptions to which nothing objective and determinate corresponds. The objective existence of highly (but not totally) determinate space/time/mass/energy makes a very practical difference. Without it, we would not be here at all. With it, Quantum Observership is wrong.

Seven

BIG ACCIDENT QUANTUM COSMOLOGY

Big Accident Cosmologists grant the objective existence of space/time/mass/energy; but, they contend, it all adds up to nothing. All opposing forces within our closed universe balance out perfectly; and the net result is zero. If a perfectly balanced universe is closed, the net energy in that universe is nil. As Alan Guth puts it, "The Universe is the ultimate free lunch."[1] According to the "free lunch" interpretation of quantum physics, our universe is a delicate but accidental balance of negative and positive forces that sums up to nothing. This perfect balance is called "perfect symmetry."

1. The Universe as a Big Accident

If symmetry is perfect on a cosmic scale, the total amount of energy in the universe is actually zero. Does this mean that nothing caused the universe? If our universe is an absolute zero, absolutely nothing seems required to cause it! Is our universe such an ultimate absolute accident? Is it a nothing that is caused by nothing for no reason or purpose at all? Extreme Big Accident Cosmology answers affirmatively. This cosmology is advocated by Quantum Cosmologists like Edward P. Tryon,[2] Peter Atkins,[3] A. Vilenkin,[4] Victor J. Stenger,[5] Quentin Smith,[6] and a few others[7] for whom the origin of our universe was indeed a stupendous accident, having no cause whatsoever.

Cosmologists who take this final step are not Antecedent Universe Cosmologists because they acknowledge no antecedents at all, no oscillating predecessors, not even Mother Spacetime and the unstable energy situation of the physical vacuum, not even really empty space itself. They view our world as an absolute accident, requiring neither necessary nor sufficient conditions for its existence. They believe that our universe was created not only *from* nothing but *by* nothing. As Edward P. Tryon put it, "Our universe is simply one of those things which happen from time to time."[8] Tryon was only a forerunner of Big Accident Cosmology who suggested that our universe may be produced by a quantum fluctuation of "the vacuum of some larger space in which our Universe is imbedded,"[9] so he actually presupposed the pre-existence of something after all. Renunciation of *all* antecedent conditions by real Accidentalists like Peter Atkins and Victor J. Stenger is much more complete. As Atkins expressed it,

> In the beginning was nothing. Absolute void, not merely empty space. There was no space; nor was there time, for this was before time. The Universe was without form and void.
> By chance was a fluctuation, and a set of points, emerging from nothing and taking their existence from the pattern they formed, defined a time.

> The chance formation of a pattern resulted in the emergence of time from coalesced opposites, its emergence from nothing. From absolutely nothing, absolutely without intervention, there came into being rudimentary existence.[10]

And as Stenger put it,

> I picture the origin of the universe as follows: in the beginning there was a void more empty than a perfect vacuum, empty not only of particles and fields but of space and time as well. It had perfect symmetry and zero energy. It was as much nothing as nothing can be. A fluctuation in that void then occurred, generating our universe and perhaps countless others very different from ours.[11]

Quentin Smith, another Big Accident Cosmologist, rejects Tryon's "vacuum fluctuation in empty space" account of the origin of the universe,[12] along with the principle of universal causation. He affirms that "The most reasonable belief is that we came from nothing, by nothing, and for nothing."[13] That is Big Accident Cosmology in a nutshell!

Extreme quantum Big Accident Cosmology proposes that the creation of our universe is the ultimate chance happening, a totally spontaneous quantum fluctuation in, from, of, and by pure nothingness. How plausible is the view that nothing caused the Big Bang?

2. Critique of Big Accident Quantum Cosmology

The fundamental affirmations and presuppositions of Big Accident quantum cosmology are empirically vacuous and should be dismissed as utterly meaningless by empiricist standards. "Absolutely nothing caused the Big Bang" presupposes that "Absolutely nothing once existed," but no conceivable experience could ever directly verify this affirmation. Any confirming or disconfirming experience would exist and would thus falsify the claim. *No examples* of absolute non-existence, or of causation by non-existence, could ever be given directly in any conceivable experience. Also, we cannot reason inductively about such things since we have no instances with which to start.

Aside from being utterly unintelligible experientially and inductively, Big Accident quantum cosmology is troubled by three extreme affirmations that make it utterly implausible as an explanatory hypothesis. Big Accident theorists are committed to the preposterous claims that: A. The universe exists in such perfect symmetry that its net energy equals zero. B. Natural quantum laws exist and function in a state of absolute nothingness. C. Causality must be totally abandoned at the point of ultimate origins. But why are these claims so preposterous?

A. Perfect Symmetry and Zero Energy

Our universe was created *out of* nothing *by* nothing, *for* nothing, and *is* nothing, according to extreme Big Accident quantum cosmologists. All the forces operating in the universe exist in such perfect balance or symmetry that the net energy of the universe is absolutely zero. Supposedly, the kinetic energy of the initial explosion is perfectly balanced by the counter-tug of gravity; positive and negative electrical charges of material particles all ultimately cancel one another out, as do matter and antimatter. Everything in the universe is so arranged that absolutely everything adds up to absolutely nothing.

Victor J. Stenger explicitly links zero energy with *creation by nothing*, arguing that "Since the universe has zero total energy, no energy was required in its production."[14] This extreme "caused by nothing" hypothesis would be false, however, if the universe contains the slightest bit of positive energy, if symmetry is in any way imperfect. Does the empirical evidence support perfect symmetry? The average density of all positive and negative electrical charges may be zero, but assuredly we have not totaled them all. Even so, at least three important facts about our world count heavily against the theory that all manifestations of energy are so equally balanced that they sum up to zero: i. We live in an open universe; ii. Matter decisively triumphs over antimatter; and iii. The kinetic energy of the initial bang is more powerful than gravity.

i. We Live in an Open Universe

First, only a closed universe can have zero energy, but overwhelming evidence, given in Chapter Three, indicates that our universe is open. Andrei Linde refers to "The well-known fact that the total energy of a closed Universe is zero, being a sum of the positive energy of matter and the negative energy of the scale factor *a*."[15] But Linde's "well-known fact" is not really a fact about our universe. Free-lunchers do a lot of wishful thinking!

At this stage in the development of science, the most reasonable view is that our universe is open. Entities involved in the Hubble expansion have enough escape velocity to keep expanding forever; the Hubble rate of cosmic expansion is increasing, not decreasing; and gravitational and kinetic energy are decisively out of balance. After all identifiable dark matter is counted, seventy to ninety percent of the mass/energy required to close the universe is missing. Recall earlier discussions of the claim that sufficient mass exists to counterbalance the expansion of the universe. The missing mass just isn't there.

Many prominent scientific cosmologists believe that the universe is open and will expand forever, precisely because not enough matter or mass exists to close it, balance it, or even slow it down.[16] Some cold dark matter exists; but prospects are dim that sufficient missing mass will ever be located to close down the expansion of the universe.[17] This is especially obvious now that we know that the rate of Hubble expansion is increasing, not decreasing. Oddly, in his most recent book, Victor J. Stenger insists that we live in an open universe, one lacking enough mass to close it;[18] but he inconsistently affirms that enough invisible dark matter exists to balance exactly the kinetic and rest energies of the mass of the universe.[19] Stenger concedes, significantly, that what he means by "zero energy" actually contains a small positive amount of energy![20] So, where did that tiny bit come from? Stenger offers no answer.

If enough mass/energy exists to register at .99999 instead of a measly .1 to .3 of Omega at 1, the universe would still be open. Exactly at Omega, expansion would proceed very slowly forever; but the universe would never contract; a spent world would endure forever. Omega plus an additional scrap of mass/energy is required for reversal. In the extremely unlikely event that our measurements might some day approximate these near-closure figures, margins of error would always prevent us from knowing for sure that the universe is closed.

Without sufficient mass/energy, gravity loses; kinetic energy wins; and the average net results are greater than zero. Victor J. Stenger claimed in his 1988 book that the universe is balanced between open and closed;[21] yet, an impressive seventy to ninety percent of the evidence cannot be found. Stenger concedes this in his 1995 book where he announces that the universe is open,[22] but he does not acknowledge that only a closed universe could be a Big Accident. Believing in a closed universe, or even a delicately balanced one, is too much like believing in Santa Claus and Ptolemaic epicycles. Abdus Salam concedes that "At the present time, measurements do not appear to sustain" the claim that "the mass of the universe adds up to zero;" and without this, "We shall discard the whole notion of the universe arising as a quantum fluctuation."[23] Yes indeed!

ii. Matter Prevails over Antimatter

Second, our material universe exists precisely because matter and antimatter do *not* exist in perfect symmetry. We live in a material world because matter triumphed over antimatter near the very beginning, or because our universe was created from the outset with a massive imbalance of matter over antimatter. Stenger defends a hidden symmetry of matter and antimatter, despite the complete absence of evidence for it and substantial evidence against it. His only evidence is the non-empirical Principle of Plenitude. All possible states are actual, he insists. Symmetry is broken and matter prevails in our domain, he concedes, but in other domains (other universes, presumably composed of antimatter) in infinite Superspacetime, things balance out.[24] But how does he know that? We have no empirical access to any Superspacetime containing antimatter universes, much less an infinitude of it! And Superspacetime, if it exists, is not sheer nothingness.

We do *not* know that any, much less that all, possibilities are actualized in other domains or universes somewhere in Superspacetime. This *a priori* metaphysics cannot be verified or falsified. Empirically, the claim is utterly meaningless, permanently without experiential content for us. An implausible hypothesis–creation of, by, out of, and for absolutely nothing–cannot be defended successfully by appeal to theories that are utterly senseless.

In this world, the only one that we know to exist, mass/energy is definitely not zero. As A. Zee wrote in 1986,

> Ever since Dirac's ideas about antimatter were experimentally confirmed, people have speculated that the Universe has an equal amount of matter and antimatter, segregated into domains. More precisely, it was asserted that all conserved quantum numbers of the Universe should be zero. In particular, since electric charge Q is known to be zero to a high degree of accuracy, it seems "aesthetically appealing" that baryon number B and lepton number L should also be zero. Unfortunately, the weight of the observational evidence is against this supposition.[25]

After thoroughly reviewing the empirical evidence, Gary Steigman concluded as early as 1979 that "The Universe is not symmetric and contains little, if any, antimatter."[26] In 1991, John D. Barrow concurred:

> Although particle accelerators produce matter and antimatter in equal abundances quite routinely and there is a democratic relationship between the two, we see no antiplanets, no antistars, no antigalaxies, and there is no evidence of any antimatter in the cosmic rays that come from outside our solar system. Nor do we see any evidence of the wholesale annihilation of matter and antimatter, which would erupt anywhere in the Universe, where the two came into contact. Thus, for some mysterious reason, there exists a form of cosmic favoritism.[27]

In 1993, after describing his research team's extensive probing during the 1970s for antimatter particles with sensitive instruments carried by high altitude balloons, George Smoot wrote that "During all the years, we found not a single convincing sign of cosmic antimatter.[28]

Perhaps, however, a perfect symmetry of matter and antimatter existed "in the beginning," even though little or no antimatter exists today. As Heinz R. Pagels suggests, "The present matter-antimatter asymmetry of the universe does not reflect the original state of the primal fireball, which could have perfect symmetry."[29] Even so, free-lunchers must explain how the original perfect symmetry of matter was broken and overcome, how enough matter to comprise our universe survived an initial perfect balance of matter and antimatter, why so little antimatter survived, and how this all adds up to nothing.

A widely accepted explanation is that a surplus of matter over antimatter resulted because the earliest physical particles decayed at an irregular pace, as permitted by the weak force.[30] This answer is very problematic. Why did the weak force permit asymmetrical decay into matter instead of antimatter? Why didn't the weak force exemplify perfect symmetry? Emerging from perfect symmetry, why were these primitive particles asymmetrically inclined? Why weren't they all annihilated by their own perfectly symmetrical antiparticles?

This widely accepted explanation pushes the asymmetry back one step but does not account for it. No one has ever observed the primitive particles that supposedly decayed into a surplus of matter; they are purely hypothetical explanatory constructs with no empirical status whatsoever. Most seriously, if symmetry was so perfect originally, all of the original primitive particles should have been annihilated by their own primitive anti-particles. The conventional explanation of why asymmetry exists in a perfectly symmetrical universe is unverified, incoherent, and unintelligible.

In 1998, A. G. Cohen, A. De RúJula, and S. L. Glashow indicated that primordial matter/antimatter symmetry would elevate the gamma ray background and distort the cosmic microwave background far above observable quantities. They argue that although small pockets of antimatter might exist here and there, empirical evidence excludes a patchwork universe composed of widely separated regions of matter and antimatter. After reviewing the evidence, they conclude that "A matter-antimatter symmetric universe is empirically excluded."[31]

Even if equal quantities of matter and antimatter existed originally, this would still not prove zero net energy for the universe as a whole. When material and antimaterial particles collide and explode, they do not leave behind zero energy or absolute nothingness, as the "free lunch" Big Accident theory predicts and requires. Instead, they leave a residue of gamma radiation, which is a definite and positive form of mass/energy.[32] In fact, residues of primordial gamma radiation may now have been found;[33] but the early mutual extinction of matter and antimatter did not result in zero mass/energy. Clearly, the primordial annihilation was asymmetrical, a fact that perfect symmetry cannot explain.

iii. The Bang Overpowers Gravity

Before concluding that the universe is really not a free lunch after all, let us consider another argument for this conclusion. If matter/antimatter symmetry does not add up to zero, perhaps matter-gravity symmetry does. In explaining Stephen Hawking's commitment to the zero-energy free lunch theory, Michael White and John Gribbin declare that "If all the matter in the Universe could be collected together at a single point, its negative gravitational energy ($-mc^2$) would exactly cancel out all the positive mass energy ($+mc^2$) of all the matter."[34] This argument for a zero energy universe completely ignores the stupendous kinetic energy of the Big Bang itself, against which gravity is fighting a losing battle. It was written before we discovered that the rate of

Hubble expansion is increasing. It assumes incorrectly, as Hawking does (at times), that the universe contains enough mass/energy to close it.

Surprisingly, Hawking himself says in his 1988 *A Brief History of Time* that "The present evidence suggests that the universe will probably expand forever";[35] but his no boundary model of the universe as a globe that begins at the North Pole, expands to the Equator, then shrinks to the South Pole, implies that the universe will not expand forever because an open universe that expands forever would have no South Pole. Which is the real Hawking?

In *Stephen Hawking's A Brief History of Time: A Reader's Companion*, published in 1992, Hawking says, "I predict that the universe in time will come to an end at the big crunch."[36] In his 1993 book, *Black Holes and Baby Universes and Other Essays*, Hawking defends *both* the openness and the closedness of the universe and concludes that he is hedging his bets "by predicting both ways."[37] He concedes that observation discloses only ten percent of the mass required for closing the universe.[38] His argument for a closed universe is purely conjectural, theoretical, and *a priori*.

Peter Coles and George Ellis wrote in 1994 that "The primary reasons for the widespread belief in a critical density of matter are theoretical." They emphasize that

> this is indeed an experimental question, where theory–no matter how dear it may be to us–will eventually have to bow to the experimental evidence. It may be that the theoretical prejudice in favor of the high-density models will one day be confirmed; if so that will be a great triumph for theory. However, at present the weight of evidence if anything favors a low-density universe.[39]

After examining both theoretical and empirical evidence for a closed universe, Coles and Ellis affirm that "The amount of dark matter for which there is compelling direct evidence is a long way short of closing the Universe,"[40] and that "No strongly convincing case can be made for a critical-density Universe, and on the balance of the evidence, an open Universe should be preferred." They conclude that "Those cosmologists who take it for granted that we live in a high-density Universe and there seem to be many may turn out to be profoundly mistaken."[41]

Consider also this argument against critical density symmetry. If ours is a critical density closed universe, it must be no older than eight billion years, some astrophysicists indicate. Yet, many galactic structures in the visible universe are older than an eight billion-year-old universe itself! Thus, if critical-density cosmologists are right, they must be wrong! Only an open universe allows enough time to account for the age of all its structures.[42]

Trusting experience, the most reasonable thing to believe, though not absolutely certain, is that we live in an open universe. For the net energy of the universe to equal zero, *all* forms of energy, not just a few, must cancel out, including the cosmic contests between matter and antimatter and between gravity and the residual kinetic energy of the Big Bang. They do not balance out. The universe is really something after all! Perhaps it also has a real cause, more than zero existence, and a real purpose!

B. Quantum Natural Laws Operating in Nothingness

Without realizing or admitting it, Big Accident Cosmologists actually presuppose the existence of *something* when they imagine the creation of the universe out of and by absolutely nothing. They presuppose the existence of definite laws of nature, specifically those of quantum mechanics, operating on absolutely nothing in a state of absolute nothingness. Our Big Bang and the resulting universe were created, they claim, by quantum fluctuations within nothingness in accord with the laws of quantum physics. According to Stephen Hawking, "It is possible in the quantum theory for the ordinary laws of science to hold everywhere, including at the beginning of time."[43] Quentin Smith concedes that the Godless quantum cosmology to which he subscribes,

> represents the universe as beginning about 15 billion years ago in accordance with a physical law. The universe is described as beginning from nothing in accordance with some law. Here 'nothing' does not mean the quantum-mechanical vacuum (which it often means in quantum cosmologies, such as Tryon's), but *literally* nothing, i.e., the absence of all concrete objects (mass, energy, spacetime).[44]

Nothing existed. Yet quantum laws were there. So were quantum fluctuations allowed by these laws. Nothing to fluctuate existed, yet nothingness lawfully fluctuated! Quantum laws existed and produce spontaneous fluctuations within absolute nothingness, says Big Accident cosmology; and this supposedly accounts for the origin of our universe![45] But none of this makes good sense, for *something* (laws and fluctuations) cannot exist in or as *absolutely nothing*; and no physical laws exist when no physical entities exist. The position is logically incoherent, and it confuses formal with efficient causes. It confuses laws with efficacious energy.

i. The Incoherence of Something in Nothing

Big Accident Quantum Cosmology cannot be formulated coherently. The difficulties are partly with the fluctuations, partly with the laws. Supposedly, fluctuations in nothingness make bubbles of nothingness that inflate into entire universes of nothingness, one of which is ours. But how can absolutely nothing fluctuate? What could be the difference in meaning or reference between

"Nothing fluctuated" and "There were no fluctuations"? How could nothing fluctuate at all? How could it fluctuate lawfully? How could "Nothing fluctuated" explain the origin of the universe, especially when the fluctuations themselves were nothing?

Recall that less extreme inflationary theorists derive our universe from the pre-existing "empty space" of Superspacetime, and that this "empty space" has its own physical density and mass/energy. Particles and antiparticles are spontaneously generated by unpredictable fluctuations in a primitive actualized energy field, and an occasional particle that escapes annihilation supposedly inflates into a full-fledged universe. Although primeval inflation within transcendent Superspacetime is very far removed from experience, the quantum-fizzy nature of empty spacetime *within* our system of nature is well established; and it is not pure nothingness.

As for quantum laws in absolute non-being, Heinz Pagels acknowledged the incoherence of the idea of "laws in nothingness." He wrote,

> The nothingness "before" the creation of the universe is the most complete void that we can imagine–no space, time or matter existed. It is a world without place, without duration or eternity, without number–it is what the mathematicians call "the empty set." Yet this unthinkable void converts itself into the plenum of existence–a necessary consequence of physical laws. Where are these laws written into that void? What "tells" the void that it is pregnant with a possible universe? It would seem that even the void is subject to law, a logic that exists prior to space and time.[46]

If laws are there in nothingness, it is not pure nothingness!

ii. Laws Are Only Formal Causes

Most seriously, in contemporary natural science, physical laws merely describe the statistically average habituated behaviors of actual physical entities, and they change if and when these habits change. Laws are merely formal causes, not efficient causes imposing external limits on what natural entities can do. No abstract laws can exist when nothing else exists for them to describe. Where nothing exists in spacetime, there are no physical laws. No formal causes can operate when nothing has no form, where no habituated energy-laden actualities exist. The actualized energy field of "empty space" has a habitual case of jitters, and so do all the particles that emerge lawfully from it; but it is not sheer nothingness. Quantum laws are finitely probabilistic and predictive, but no finite predictive probabilities exist to be calculate in a state of absolute nothingness. Either primitive physical actualities with habits existed within Big Accident's alleged nothingness, or no quantum laws were there to permit or describe accidental probabilistic world-creating quantum fluctuations of energy-laden realities. If quantum laws and primitive physical actualities existed in primeval nothingness, it was not absolute nothingness.

Additional important questions remain unanswered. Do the adduced pre-world quantum laws and primitive physical actualities come from somewhere and thus depend on something else for their existence, or are they self-existent, eternal, and necessary beings? If so, by virtue of what do they have this metaphysical status? Why do the laws of quantum physics obtain in nothingness when so many other physical laws–like those of Newtonian mechanics, or of relativity physics alone–are logically possible? Who or what selects the relevant laws? How meaningful is the presumption that totally disembodied laws exist anywhere, much less in absolute nothingness? Big Accident Cosmology has no good answers. It is too much ado about nothingness!

C. Total Abandonment of Causation

Some cosmologists admit it, some do not; but the spontaneous creation of our universe occurred only because necessary causal conditions for its creation were met. In that sense, our universe definitely had a cause. Most Quantum Cosmologists are convinced that these necessary conditions include transcendent Mother Spacetime, the physical vacuum, the primordial fizz, an actualized primordial field of energy, and the laws of quantum mechanics.

Extreme Quantum Accidentalists repudiate all of this; they conjecture that our universe is an absolute accident that just popped into being out of pure nothingness devoid of all causal conditions whatsoever. An absolute accident has no causal conditions at all. Accidentalists deny both necessary and sufficient conditions for the existence of our universe. As Stenger says, "The simplest hypothesis that so far seems to explain the data is that the universe is an accident."[47] What caused the Big Bang? "Nothing!" answers extreme Quantum Accidentalism.

If the Big Bang that created our world was an absolute accident requiring no causal conditions at all, it was the most stupendous accident that ever happened, the greatest miracle of all time, but without a Divine miracle worker. And to think that some people have trouble with God as a miracle worker! Big Accident Cosmologists insist that their creation-by-nothing scenario is all very plausible; but upon closer examination, we see that this is not so.

i. Necessary but Not Sufficient Causal Conditions

Paul Davies claimed in 1983 that quantum physics abandons causality completely and "permits events to occur without causes in the quantum world."[48] An element of truth is in this, but not the whole truth. Spontaneous fluctuations do occur in the quantum world; sufficient causal conditions for spontaneity do not exist. Nevertheless, quantum fluctuations always presuppose and require *neces-*

sary causal conditions. Davies concedes as much: "Recent discoveries in particle physics have suggested mechanisms whereby matter can be created in empty space by the cosmic gravitational field, which only leaves the origin of spacetime itself as a mystery."[49] So both spacetime and a gravitational (or some kind of) energy field are necessary causal conditions for the creation of matter; but how can gravity exist without physical mass? Davies finally concluded that quantum gravity "would allow spacetime to be created and destroyed spontaneously and uncaused in the same way that particles are created and destroyed spontaneously and uncaused."[50] Still, this presupposes quantum gravity, not pure nothingness; but what would quantum gravity attract (or repel) in absolute nothingness? How could it operate in and on nothing? As Einstein insisted, no gravity exists without mass/energy and spacetime. One thing leads to another!

Currently, gravity is best treated within classical relativity physics, which allows for definiteness and continuity in gravitational effects; but no workable theory of quantum gravity is available.[51] In addition to quantum laws, quantum gravity has its own necessary conditions like gravitational particles–as yet undiscovered gravitons and gravitinos that presently exist in theory only. It also requires mass and spacetime, for gravity is identical with spatiotemporal mass and curvature, according to relativity theory.

Also, in a quantum universe, gravitational particles should display their own quirky quantum effects–indefiniteness, discreteness, spontaneity, scattering, and so forth; but in well-established domains of particle physics, these quantum effects always presuppose their own conditions. When an adequate theory is developed, quantum gravity cannot and will not exclude necessary causal conditions. It will not abandon causation absolutely.

ii. Freedom Has Necessary Conditions

To turn to another problem, perhaps experience discloses occasional exceptions to the principle of universal causation–like human freedom; and maybe the origin of the universe falls under one of the exceptions.

Freedom and creativity exemplify self-originated spontaneity. This means that things can and do happen without sufficient causes, but not that anything ever happens without necessary causes. This is clearly at odds with the suppositions of extreme Quantum Accidentalists. Contingent realities–things that might or might not be–never come into being when causal grounds, necessary causal conditions, are completely absent. A very decisive empirical consideration supports this: *All experience illustrates it.* Experience universally confirms that necessary conditions are required for the existence of every contingent entity that comes into being, everything whose non-existence is logically possible, including creativity and acts of free will.

An experienced-based philosophical perspective can make a place for freedom and spontaneity. Both at macroscopic human and microscopic quantum levels, creative freedom is incompatible only with sufficient but not with necessary causal conditions. Quantum theory does not dispense with necessary causal conditions, and neither does human freedom. All experience confirms that contingently existing things always have causes, including quantum effects and free choices. Spontaneous quantum fluctuations cannot and do not occur in the absence of necessary causal conditions like Mother Spacetime, or our spacetime, the physical vacuum, an actualized energy field, and quantum laws. Freely made human choices do not occur in the absence of necessary conditions like desires, interests, and some awareness of open possibilities. Without the presence of *something*, free choices cannot be made. So freedom is no exception to the rule that everything has a cause. Some causal conditions are necessary for the occurrence of everything that happens. Quantum physics and human freedom presuppose only that things can happen in the absence of totally sufficient causal conditions.

iii. Applying Causation to World-Origins

"Everything that comes into being has a cause" was often defended by rationalistic philosophers like Descartes and Jonathan Edwards as an intuitively certain *a priori* or self-evident truth. Quite recently, our knowing this principle *a priori*, without appeal to experience, was vigorously defended by William L. Craig[52] and attacked (successfully in my view) by Quentin Smith.[53] To be sure, I, too, find it intuitively obvious that everything that comes into being has a cause; but I cannot rule out the possibility that this powerful intuition is psychological or empirical, not *a priori*, that it results from (is caused by) its universal confirmation in experience, the last court of appeal in factual matters.

Quentin Smith presents powerful critiques of the empirical argument for the principle of causation and of its application to the creation of the world out of nothing around fifteen billion years ago. The origin of the universe was caused by absolutely nothing, he thinks, because this causal principle is not self-evident and does not apply to the origin of the universe. "Every thing that comes to be has a cause" is well confirmed by experiences of happenings *within* the world, he admits; but it has no relevant application to the origin of the world itself.[54] We have *no* experience of world-causation as such; all relevant experiences pertain only to causation within the world, Smith insists. His view resembles Kant's contention that we have no experience of causation by things in themselves; all relevant experiences pertain only to causation between appearances. Yet, Kant had to explain the existence of appearances causally.

Actually, we have no direct experiences of world-origins from singularities, from "empty space," from quantum tunneling, from God, or from absolute nothingness. So Smith's argument cuts decisively against his own solution to the problem of world-origins. Admittedly, uncertainty creeps into any theistic or non-theistic account of world-creation by God or by any other transcendent entities like antecedent universes or Superspacetime. Smith's a-causal Accidentalism is as fatal to Antecedent Universe and Big Fizz Cosmologies as it is to Theism. Uncertainty pervades all of science and philosophy. We are never absolutely sure of the truth of the premises from which we argue; but some premises are still much more plausible than others. Without reiterating Craig's astute responses to Smith, I will offer two important objections of my own to Smith's atheistic account of world-creation from, by, and for absolutely nothing.

First, even Smith does not really believe it. He definitely presupposes the existence and operation of quantum laws, spontaneous fluctuations, quantum tunneling, and a singularity composed of infinitely compressed physical mass–all in alleged nothingness. Instead of a disembodied God, Smith appeals at rock bottom to the ultimate reality and causal efficacy of disembodied conceptual constructs–quantum laws, fluctuations, tunneling, and a sizeless, timeless, and imperceptible "physical" singularity. Without them, Smith has no explanation of the origin of the universe from and by nothingness. Even without his singularity, Big Accident Quantum Cosmologists always presuppose the antecedent reality of quantum laws, perturbations, tunneling, or something within absolute non-being; but they pronounce their theory only at the price of incoherence. Their nothingness is not real nothingness; it is really something after all!

Second, recall Smith's argument that our knowledge of causation, drawn from *within* the world, cannot be applied to the *origin* of the world because we have never experienced world-origins. But Smith's Accidentalist account of origins is vulnerable to exactly the same objection. In quantum physics, he claims, particles can come into existence spontaneously under conditions of quantum uncertainty, and the law of energy conservation does not apply to these circumstances. Once, such a particle inflated into our universe. But note carefully that Smith's theory of world-origins depends entirely upon an analogy with experienced quantum fields and processes *within* our world. As he puts it, "There is observational evidence, albeit indirect, that this uncaused emergence of energy or particles (notably virtual particles) frequently occurs."[55] This could be true only *within* our system of spacetime, if Smith is really serious about appealing to experience. We have no "observational evidence" of world-origins.

Smith's position is implausible for many reasons. No one has ever seen a quantum particle inflate into an entire universe. All our empirical or observational knowledge of quantum effects is based upon experiences of quantum events *within* our world. If knowledge of universal causation drawn from within the world cannot be applied to the origin of the world, then knowledge of quantum effects drawn from within the world cannot be so applied either. Also, physicists have proved that only a closed universe could arise through quantum tunneling,[56] the process by which virtual particles become actual; but ours is not a closed universe, as earlier demonstrated. Thus, Smith cannot make his case for thinking that our world is a quantum-induced Big Accident.

We have *no* experience of world origins through spontaneous quantum fluctuations in absolute nothingness. All human experiences of quantum phenomena pertain only to minute quantum events *within* our system of spacetime. If we cannot extrapolate from the universality of causation within the world to world origins, neither can we extrapolate from quantum effects within the world to world origins. *All* experienced quantum effects, including those in virtually empty space, presuppose not only quantum laws but also *our* space, not transcendent Superspace, and definitely not pure nothingness. Empirically, the spontaneous origin of actualized particles under conditions of Heisenberg uncertainty occurs only under vacuum conditions *within our system of spacetime*, never in absolute nothingness; and few *if any* such particles endure for more than a fraction of a second to violate the Principle of Conservation.

Smith actually concedes that all known quantum tunneling, by which, in theory, particles momentarily break the barrier separating potentiality from actuality, takes place *within* our system of spacetime.[57] All known tunneling and quantum effects presuppose necessary causal conditions like our spacetime, vacuum (low energy) conditions, an actualized energy field, and perhaps even observers or prehenders (not necessarily conscious). From nothing, nothing comes, even in quantum physics.

Smith's own account of world origins is incompatible with his argument against "Everything that comes to be has a cause." It presupposes and thus grants that we *can* extrapolate from what is known within our spacetime to the origin of the whole of it. But if we can apply what we know about quantum laws and effects to the origin of our universe, we can also apply the universally confirmed principle of causation to world origins. That rebuts Smith's No-cause Big Accident theory of the inception of the universe. Unless everything that comes into being, including the universe, has a cause (presupposes necessary conditions), no Quantum Cosmology ever gets off the ground. Big Accident's misunderstanding of quantum physics can be extended to the whole of nature only if explanatory principles drawn from the parts can be so extended; but that lets causation back in under the wire. *Any* account of the origin of our universe must draw upon analogies with what happens within our universe, for that is all that we know.

Stephen Hawking maintains that "It is possible in the quantum theory for the ordinary laws of science to hold everywhere, including at the beginning of time."[58] If, as Smith and Hawking suggest, the laws of quantum physics do not break down at the origin of the universe as a whole, then neither does the law of universal causation, nor the law of increasing entropy. If any explanatory principles can be extrapolated from parts to the whole, surely the universality of causation is one of them. Contra Smith, a quantum universe presupposes necessary causal conditions. But, deep down, Smith really believes that anyway.

The *ad hominem* retort that "You do it too!" will not satisfy extreme skeptics. Still, anyone, whether an atheist or a theist, who attempts to answer the question, "What caused the Big Bang?" must employ concepts and principles of explanation drawn from within the universe and extend them to the origin of the whole. We could just abstain or quit trying to find a cause for the Big Bang, but the inclination to try is almost irresistible. Anyone who decrees that it simply cannot be done merely blocks the path of inquiry.

The causal principle that "Everything that comes to be has a cause," at least as necessary conditions, must be distinguished from what Victor J. Stenger calls the principle of "causal precedence," that "Cause always precedes effect." Stenger argues that at the elementary level of quantum interactions, "Cause and effect are not always distinguishable" because time is reversible. The irreversibility of time's arrow, he maintains, is an emergent property that exists only at the macroscopic level of everyday experience and common sense.[59] The solution to the two slit experiment problem, he holds, is that a particle goes "through one slit to the detector, then *back in time* to the source through the second slit and finally forward in time once more through either slit to the detector."[60] He concedes that he is "in a minority" on the issue of the reversibility of time![61]

That everything which comes into being has a cause would still be true even if we cannot always tell the difference between cause and effect, or *if* cause, effect, and time are reversed on the quantum level. Stenger, another Big Accident Cosmologist, also affirms that "everything can have come from nothing" in the beginning[62] through quantum fluctuations in the spacetime vacuum. We now know that this actually involves the antecedent reality of necessary causal conditions like Mother Spacetime, the physical vacuum, actualized energy, habituated actualities, quantum laws, and quantum effects. In their absence, spontaneous fluctuations cannot and do not occur. All contingencies have causes.

D. No Contingency Without Causation

By definition it is true that if something exists contingently, it is causally derived from or dependent on something other than itself; but a mere definition cannot settle the substantive question of the causal dependence of the universe on God. Can a wedge be driven between the various elements that conventionally define the notion of contingent existence? Could an existing entity that endures for only a finite span of time have no cause at all even though it is possible for it not to exist? The most extreme Big Accident Cosmologists think so. They contend that the universe popped into being within the finite past as an absolute accident, requiring neither necessary nor sufficient causal conditions. If true, the universe could be contingent (having possible non-existence and finite duration), yet–in another sense–neither contingent (causally dependent) nor necessary (impossible non-existence, self-sufficient, everlasting, uncreated, and indestructible reality).

Big Accident Cosmologists appeal to quantum theory, which really does not support their position because quantum mechanics renounces only sufficient but not necessary causal conditions. Every quantum fluctuation has necessary prerequisites like the laws of quantum physics, spacetime or Superspacetime, the minimal pure energy of a physical vacuum, and the presence of an actualized energy field. They aspire to do so, but Big Accident Cosmologists cannot avoid these necessary conditions for quantum fluctuations, primordial or not; but primordially where and how did these conditions originate? They are not metaphysically necessary beings because as wholes they are composed entirely of contingent, dependent parts. This point is well developed in Chapter Twelve, but let us now note that Big Accident Cosmology falls prey to the first two premises of the Cosmological Argument From Contingency there presented. All wholes composed of contingent beings are themselves contingent; and no necessary entities exist within our system of nature.

To summarize, extreme Big Accident Cosmology takes quantum fluctuations to their ultimate extreme and posits their reality where nothing exists to fluctuate. Our universe was created out of absolutely nothing, was caused by absolutely nothing, is absolutely nothing, and exists for absolutely no purpose. It is the Ultimate Accident. Extreme Accidentalism assumes that our universe is composed of zero energy, and that since the universe is nothing, nothing is required to create it. But a zero energy universe is very doubtful–especially ours. The kinetic energy of the primordial explosion that initiated our universe is not perfectly counterbalanced by gravity, and matter clearly prevails over antimatter in the only universe that we really know to exist. A zero-energy universe exists only in theory, but experience clearly shows that our universe is really something after all!

Extreme Accidentalists always presuppose something as a necessary ground for the universe, even if nothing more than the laws of quantum physics and spontaneous fluctuations. They fail to distinguish adequately between necessary and sufficient causal conditions. If the Big Bang lacked a sufficient cause, it does not follow that it had no necessary cause. All experience, including quantum physics, supports the causal principle that some causal conditions are necessary for everything that happens or comes to be, including the Big Bang. Scientific Cosmological Agnosticism correctly indicates that empirical science cannot identify the transcendent necessary cause of the Big Bang.

Big Accident Cosmologists have wild imaginations, but their theory is logically incoherent and cannot be squared with the facts. Like all other atheistic Quantum Cosmologists, they are driven to preposterous extremes in order to avoid God. A contingent universe cannot exist in the absence of everything; but what is the something upon which it depends for its being? What caused the Big Bang? No atheistic cosmologies examined thus far, quantum or not, give an adequate answer; but a plausible answer is forthcoming!

Eight

ATHEISTIC ANTHROPIC COSMOLOGY

We live in a remarkable universe. Among possible universes, the fact that our universe is compatible with and supports our existence makes it extraordinary. Any universe in which intelligent creatures like ourselves could exist would be a fabulous universe, for lifeless universes could be produced in an infinite number of ways, but only a few highly contrived ways can produce life-sustaining ones. Some cosmologists claim that there is only one way to make a life-sustaining universe.[1]

Except for the Greek and Roman Atomists, most pre-modern thinkers assumed that some kind of special relationship exists between humanity and the universe, that humankind is made for the universe and the universe for humankind. Teleology means purposiveness. Western philosophers and theologians traditionally believed that we live in a purposeful universe, that teleology is an important and conspicuous feature of nature.

Because seventeenth and eighteenth century mechanistic materialists vigorously attacked cosmic teleology, non-teleology became a fundamental presumption of modern science. To exorcize Aristotelian final causes and all other purposes from nature, modern natural science aspires to explain everything in terms of formal causes (natural laws), efficient causes (energy transfers) and material causes (spatially extended entities). Naturalism, as explained in Chapter Two, made anti-teleology a fundamental metaphysical principle. Humanistic Naturalists think that we and similar organisms have purposes, but not the whole of nature, and not some purely fanciful supernatural ground or cause of nature.

Teleology on a small scale inescapably reappears in natural and social sciences like biology and psychology, despite the domination of scientific orthodoxy by naturalistic metaphysics. Teleology also resurfaces on a larger scale in recent cosmology as the Anthropic Principle, so named by the physicist Brandon Carter in 1974. Carter was not the first to notice that the universe is fine-tuned for the emergence of human life, but he first christened this the "Anthropic Principle."[2]

1. The Anthropic Principle and Cosmic Purpose Without God

Cosmologists are again finding purpose in the universe, but we should not jump to the wrong conclusion. Most of them do not wish to revitalize and embrace a new version of the religious Argument from Design for the existence of God. In fact, most Anthropic Cosmologists are thoroughly atheistic and naturalistic and aspire to show how there can be cosmic teleology without God. Most of them make two fundamental claims: We live in a purposive universe that is exquisitely fine-tuned for the production of intelligent life, and God is not required to explain such conspicuous cosmic purposiveness.

In their 1988 book, *The Anthropic Cosmological Principle*, John D. Barrow and Frank J. Tipler thoroughly develop Anthropic Cosmology, explore its history, and defend it against many challenges.[3] Additional prominent contributors to the development of Anthropic Cosmology include Brandon Carter, P. C. W. Davies,[4] Robert H. Dicke,[5] Freeman Dyson,[6] Stephen Hawking,[7] B. J. Carr and Martin Rees,[8] John Leslie,[9] and John A. Wheeler.[10]

Anthropic Cosmologists do not want to overturn the Copernican revolution and reposition mankind in some privileged position in the center of the universe. They believe that we can have a special place in the universe without being at its physical center. As Barrow and Tipler express it, "Although we do not regard our position in the Universe to be central or special in every way, this does not mean that it cannot be special in *any* way."[11] Brandon Carter remarks that "Our location in the Universe is necessarily privileged to the extent of being compatible with our existence as observers."[12]

The purpose of the universe, says Anthropic Cosmology, is to produce complex intelligent forms of life, like human life. The word "anthropic" is a bit misleading, suggesting that we human beings are the only complex, intelligent, living things in the universe, that the universe is designed to produce only us; but these intimations are not really intended. This would distance Anthropic Cosmology too far from what many astrophysicists misleadingly call "the Copernican revolution." Copernicus himself did not doubt that God created the universe for mankind, even if our earth orbits the sun rather than *vice versa*.

Anthropic Cosmologists agree that other complex intelligent life forms may exist on planets in other solar systems. After all, a hundred and fifty billion stars exist in our Milky Way; at least a hundred twenty-five billion other galaxies of equal or greater complexity exist in the observable universe; and intelligent life is very likely to exist elsewhere. The basic chemistry for life is widespread. At the moment, neither the existence of life outside our solar system nor the degree of its prevalence elsewhere have been confirmed; but many planets orbiting other suns have now been located, and many cosmologists are convinced that life is prevalent throughout the universe.[13]

Anthropic Cosmology tends to be excessively anthropocentric only in the sense that it exhibits a definite bias toward *intelligence*. It assumes that other forms of life less intelligent than ourselves have little if any intrinsic worth, and that intelligence as a value epitomizes even if it doesn't exhaust our own worth. To avoid these errors, a broader Biopic Cosmology is needed, one that recognizes the great intrinsic worth of an immense variety of terrestrial and possible extraterrestrial forms of life, one that is not biased against the non-cognitive dimensions of human and non-human nature. Human life stands at the apex of complex, intelligent, affective, volitional life on earth. Yet, we differ from non-human animals only in degree, not in kind. Degrees of intelligence and many other traits that make life worth living for its own sake are shared with other terrestrial animals.

According to the Biopic Cosmology advanced in this book, the purpose of the universe is to produce "an immense variety of forms of experience, love, loyalty, enjoyment, responsibility, initiative, creativity, achievement, and satisfaction, even at the price of conflict."[14] These desirable traits enrich and exist only in the lives of concrete conscious individuals, the proper locus of intrinsic worth. Valuable individual lives need be neither anthropic, that is, humanoid, nor carbon-based; but carbon, hydrogen, oxygen, and nitrogen have properties that greatly favor the origin and evolution of life.[15] These elements exist abundantly throughout the universe. For all we know, the purpose of the universe may be fulfilled only on our earth; but this seems unlikely.

Spectroscopic analysis discloses that the basic chemistry of the universe is the same throughout; and elements and compounds necessary for the formation of carbon-based lives are widely distributed throughout the universe. Carbon-based living things require hydrogen, oxygen, nitrogen, phosphorus, and traces of many other elements; but these too are found in ample quantities sufficient for life throughout our galaxy and in myriads of others. Many other special conditions may be necessary for life, so we are admittedly uncertain about the prevalence of life throughout the cosmos. Non-carbonaceous life-forms are possible, even if carbon specially favors life. Complex non-carbonaceous life-forms are improbable; their existence has not yet been confirmed; but if any exist, the universe is even more suitable for the production of complex and valuable conscious living things than we commonly suspect.

Considering mainly carbon-based life, our universe manifests a huge number of "extraordinarily finely tuned coincidences,"[16] as Barrow and Tipler put it, that seem designed intentionally to create life as we know it. More details of this fine-tuning for life will be given in the next chapter; but we must first examine several meanings of the Anthropic Principle and note that most Anthropic Cosmologists favor only those meanings that exclude Divine foresight, planning, and purpose.

Barrow and Tipler say that the Anthropic Principle has at least three meanings, the first two of which were recognized by Brandon Carter: (1) The Weak Anthropic Principle says nothing more than that we would not be here unless the universe were compatible with our existence. (2) The Strong Anthropic Principle affirms that the universe *must* produce human or intelligent existence. This "must" generates Theistic, Quantum Observership, and Infinite World-Ensemble (Big Fizz and Big Divide) interpretations. (3) The Final Anthropic Principle says that we exist for the sake of a final Omega Point. Each version has its weaknesses, and the whole enterprise of Anthropic Cosmology is highly controversial. Atheistic Anthropic Cosmologists accept either the Weak Principle, the second or third interpretations of the Strong Principle, or the Final Principle, according to which God's non-existence is only temporary.

They generally reject the theistic interpretation of the Strong Principle that is defended in this and following chapters.

After the Weak Principle is given a metaphysical underpinning, it is indistinguishable from the world-ensemble interpretation of the Strong Principle. Because it adds nothing to it and subtracts nothing from it, the fatal flaws of the former are also ruinous to the latter. Only the Weak World-Ensemble Anthropic Principle, the Strong Quantum Observership Principle, and the Final Anthropic Principle are viable options for Atheistic Anthropic Cosmology; but do they hold up under critical analysis?

2. The Weak and Strong Anthropic Principles

The Weak Anthropic Principle affirms nothing more than that we would not be here if the universe were not compatible with and supportive of our existence.[17] Brandon Carter's formulation of the Weak Anthropic Principle says that "What we can expect to observe must be restricted by the conditions necessary for our presence as observers."[18] As Stephen Hawking put it, "We see the universe the way it is because we exist."[19] The Weak Anthropic Principle is not exactly a tautology, though it is occasionally denounced as such. Nevertheless, it is singularly uninformative. It tells us nothing more than that we are here only because the universe is compatible with and supports our existence. It gives no reasons why.

In its purest form, the Weak Anthropic Principle offers no reason for the compatibility between the universe and ourselves. Anthropic Cosmology becomes interesting and informative only when someone tries to explain *why* we live in a universe that is compatible with and supportive of our existence. Part of the answer is very obvious: if the universe were otherwise, we would not be here asking the question, and Anthropic Cosmologists would not be here concocting the answers. This is so patently obvious and unilluminating that Weak Anthropic Cosmologists usually take further steps. They advance from Weak to Strong. They offer a metaphysical underpinning for the Weak Anthropic Principle–an infinite worlds metaphysics; but this converts the Weak Anthropic Principle into an infinite universe interpretation of the Strong Anthropic Principle. These two options, having become one, will shortly be evaluated together.

As Brandon Carter formulated it, the Strong Anthropic Principle says that "The Universe must be such as to admit the creation of observers within it at some stage."[20] The emphasis here is on the word "must," but what Carter meant by this is unclear. It suggests that no universe can come into being that lacks intelligent observers altogether, that some observational selection principle excludes universes inhospitable to our kind of life. What could this selection principle be? Barrow and Tipler consider three possibilities.[21]

First, the universe may have been designed deliberately by Divinity, by some "Supercalculating Intellect," as Fred Hoyle expressed it, who intended to create and sustain finite, conscious, intelligent beings. Atheistic Anthropic Cosmologists like Barrow and Tipler reject this option in favor of the second or third. They are determined to give us teleology without Theism–at least until Omega comes in all its glory.

Second, observers may be necessary to bring the world into being, as claimed by Niels Bohr's Copenhagen interpretation of quantum mechanics and John A. Wheeler's theory of Quantum Observership (discussed and refuted already), and by the Participatory Anthropic Principle.

Third, infinitely many worlds co-existing in Superspacetime would necessitate our existence because they actualize all possibilities. Worlds containing intelligent life are possible worlds, so a few such worlds will be actual if all possibilities are actualized somewhere. We just happen to live in one of these. Some gloves will fit given an infinite number of gloves. No Supercosmic Intelligence is required to explain why we live in a universe in which astonishing cosmic coincidences conspire to produce and support our existence. The Principle of Plenitude insures the existence of infinitely many worlds; supposedly it explains everything, although it really explains nothing. This atheistic infinitely many worlds metaphysics must now be examined carefully. After finding it wanting, the Final Anthropic Principle will be examined and dispose of in the next chapter. Theistic options will fill the concluding chapters

Atheistic versions of the Strong Anthropic Principle usually appeal to the existence of infinitely many worlds to explain why, without God, we live a universe that is exquisitely designed to support conscious, intelligent, sensitive life. Given an infinite number of possible universes, most of which are doubtless incompatible with life, why do we live in one that supports life? According to the metaphysical Principle of Plenitude, all possible universes, an infinite number of them, must actually exist. Possibility is identical with actuality. Given an infinite number of diverse universes, at least a few of them will support life accidentally; and we just happen by chance to be in one that does. No God planned it. The shoe fits; but if an infinite number of different shoes exist, at least one is bound to fit. This is almost self-evident; but it is false!

At least four infinite universe cosmologies would serve the metaphysical purposes of Atheistic Anthropic Cosmology. First, as in Plasma Cosmology, a single universe may be infinite in space and time and contain an infinite number of relatively isolated metagalaxies, most of which are hostile to life, but a few will be randomly life-supporting. We just happen by chance to live in a supportive metagalaxy.

Relatively isolated metagalaxies belong to a single spatiotemporal universe presumably because they continue to have causal contact with other metagalaxies along their edges; but these edges may be so far removed from particular observers like us that we cannot detect them. The relative isolation of metagalaxies cannot be complete because, if complete, this option is indistinguishable from Big Fizz or Big Divide world-ensemble cosmology. John Leslie points out that many worlds theorists disagree about whether many worlds interact causally.[22] If worlds do interact causally in Big Fizz or Big Divide theories, they no longer differ from positions that affirm relatively but not completely isolated metagalaxies in a single universe. This is the old "same universe" quandary all over again.

Second, as in Oscillation Cosmology, an infinite succession of consecutive universes with different laws and initial conditions might exist, being separated temporally rather than spatially from one another. In most cosmic epochs, laws and conditions would be hostile to life; but within an infinite number of diversified tries, a cosmic epoch will occasionally come along that supports life. Periodically, in an infinite number of diversified successive universes, one will be suitable for habitation by conscious, intelligent, living beings like us. We just happen by accident to live in such a one. No observers, astronomers, philosophers, or ordinary people inhabit most of the others.

Third, in Big Fizz world-ensemble cosmology, an infinite number of spatially co-existing universes with different laws and initial conditions are promiscuously spawned by Mother Spacetime. Some of these may then oscillate, so this metaphysics may be combined with the preceding. Infinitely many co-existing worlds are completely separated from and have no causal contact with one another in infinite Superspace. Given an infinite number of structurally diverse contemporary universes, most will be incompatible with life; but a few will support life. By chance, we just happen to exist in one of these. In most of the others, no observers, no scientists, and no inquirers wonder about the purpose of the universe.

Fourth, in Big Divide many worlds cosmology, every universe branches profusely and indiscriminately into new and otherwise causally isolated parallel universes at every turn of events. All possibilities for every reality are actualized, and it takes an infinite number of universes to make it all happen. When parallel universes face a choice between life and no life, they divide; and at least one universe containing life is created. Given an infinite number of branches, some will be life-supporting. By pure chance we live in a life-supporting offshoot. Our own universe is constantly sprouting new universes that actualize every possibility open to every point of space and every instant of time, but most universes are uninhabited.

By appealing to one or more versions of infinite worlds metaphysics, Atheistic Anthropic Cosmology tries to account for the life-supporting purposiveness of our universe without resorting to an intelligent and purposive God. It offers teleology without theology, a universe fine-tuned for life purely by accident. But which is easier to swallow, an unseen transcendent infinite God, or an unseen transcendent infinity of worlds? Should an intelligent person affirm infinitely many worlds that don't know what they are doing, or an infinite God who knows what he is doing? The following considerations should facilitate a more informed decision.

3. Critique of Infinite World-Ensemble Teleology

A. Non-Empirical Status

Atheistic interpretations of the Weak and Strong Anthropic Principles do not explain purpose in the universe until they trot out an infinite worlds metaphysics. Unfortunately, we have no direct experiential or legitimate inferential access to even one other world, much less an infinite number of them. Given an infinite number of universes, Atheistic Anthropic Cosmologists claim, there can be cosmic purpose without God; but there is no good reason to give this, whether it be an infinite number of distant metagalaxies, antecedent universes, disconnected worlds co-existing in Superspacetime, or worlds branching from our existing universe. If quantum theory rejects as empirically meaningless the objective existence of unobservable quantum states, it should also refuse to proliferate unobserved and unobservable quantum and non-quantum worlds *ad infinitum*.

Barrow and Tipler acknowledge the non-empirical status of an infinite number of metagalaxies and oscillating universes, but they seem blind to the non-empirical status of the infinite world-ensemble (Big Fizz) option that they embrace. Eric Lerner's Plasma Cosmology postulates an infinity of loosely connected metagalaxies. Barrow and Tipler find the same postulate in the publications of G. F. R. Ellis.[23] After briefly explaining his position, Barrow and Tipler say that "It is hard to evaluate this idea any further, but one thing is certain: if it is true then it is certainly not original."[24] The theory is indeed hard to evaluate because it is nothing more than sheer fantasy!

Barrow and Tipler are much clearer about the non-empirical status of Oscillation Cosmology, remarking that "It is far from being testable."[25] They fail somehow to see that this is true also of their own infinite world-ensemble metaphysics. If any version of an infinite worlds metaphysics is true, we merely human mortals could never know it; but we have no good reasons for thinking that it is true.

Oscillation Cosmology is not testable, according to Barrow and Tipler. Inconsistently, they later suggest that it actually makes a testable prediction, one that they are unwilling to accept. Given an infinite number of oscillations, if the basic laws and constants of nature change with each bounce,

> Sooner or later the geometry would be exchanged for a noncompact structure bound to expand for all future time. The Universe should currently be 'open' destined to expand forever since this state will always be reached after a finite series of oscillations.[26]

Barrow and Tipler do not realize that this concession is utterly devastating to Oscillation Cosmology. *Every* cosmic epoch in an infinite oscillating series is preceded by an infinite number of antecedent universes persisting collectively through an infinite past. An infinite amount of time is quite long enough for "sooner or later," so *every* universe in such a series should be open. For this reason alone, the very idea of an infinite number of antecedent universes is completely untenable.

Our universe's being open or closed is testable, in principle capable of being decided. All we have to do is find out how much mass/energy the entire universe contains. Unfortunately, this is no easy task; but, given what we now know, as explained earlier, the available evidence strongly indicates that we live in an open universe. Because of their theoretical biases, Barrow and Tipler opt for a closed universe. They recognize that an infinite series of oscillating diversified universes eventually achieves a geometry that destines a terminal member to expand forever; but they still insist that our universe is closed in order to secure their Final Anthropic Principle; only a closed universe can achieve what they think is its ultimate purpose–the Omega Point.

A closed universe is implausible indeed if grounding the Final Anthropic Principle is the best reason that can be given for it, but more about that in the next chapter.

B. The Principle of Plenitude

Weak and Strong World-ensemble Anthropic Principles presuppose the validity of the Principle of Plenitude, or the Principle of Fecundity, as Robert Nozick calls it,[27] which says that all possible worlds are actual worlds. Here is the heart of the Atheistic Anthropic position on the cause of our life-producing Big Bang. The existence of our universe is required by the Principle of Plenitude–an abstract, non-empirical, disembodied, supercosmic normative principle which necessitates that all possible worlds must be actual worlds. But what grounds and drives this supercosmic compulsion to actualize all possibilities? What gives it the power to actualize its ideal? Atheistic Anthropic Cosmologists have no clear answer except that it is not Divine, at least not yet!

The ancestry of the Principle of Plenitude is unquestionably theological. Plenitude is definitely not an empirical principle or a verified discovery of empirical natural science. Arthur Lovejoy, who coined the phrase, classified it as "metaphysical theology."[28] Lovejoy traced the principle back to Plato's conviction that the actual world copies every one of, and every possible combination of, the eternal forms *because* the richer reality is, the better or more perfect it is, and *because* the Divine Demiurge would not be good, perfect, complete, and divine unless he actually is or creates everything that he possibly could be or create.[29]

As Lovejoy interpreted it, the Platonic Principle of Plenitude requires the existence not only of every possible grade or kind of being, but also of every possible individual at every level, since the forms combine to constitute individuals. Plenitudists have wavered on the issue of whether plenitude requires merely that all possible kinds of things be actual, or whether all possible individuals of every kind are also required.

The Greek Atomists hypothesized the existence of infinitely many worlds within the infinite void of space, but Plato and the Medieval theologians who followed him did not derive infinitely many distinct universes from the Principle of Plenitude. They maintained that for both God and the creation to be perfect, God only had to create representatives of all possible compatible kinds, degrees, or levels of being within this one world. So how many continuous grades of being are there? An infinite number, says Lovejoy, though he admits that neither the Greek philosophers who espoused plenitude nor the Medieval theologians who embedded it into Christian theology were always fully aware of its implications; they often combined it with incompatible qualifications.

The levels of being that the Medieval theologians actually identified were very finite; and they, like Plato, subscribed to incompatible premises that prevented infinite plenitude from really adding up to infinity. God in absolute self-sufficiency needs nothing, not even to create, they held; and God's being and goodness are in no way enriched by creation. God creates only what He freely chooses to create, even though Plenitude requires Him to create absolutely everything. Further, a morally good God creates only a morally good world, even though plenitude necessitates the creation of every possible world, no matter how nasty. The good world that God created contains only those kinds of being and those individual beings that are at least roughly harmonious with one another on the whole, but plenitude requires that every possibility be actual, whether harmonious or discordant, whether good or bad.

Traditional Christian theologians through the centuries so restricted the Principle of Plenitude that it did not readily translate into infinite creation. Only God is actually infinite, they held; the created universe, though in some vague sense complete and perfectly ordered from the beginning, is finite. God's plenitude is in himself–an absolute plenitude of being in which there are no unactualized possibilities; but Divine plenitude does not entail creating a truly infinite world or an infinite number of worlds. In himself, quite apart from any and all creation, God is pure being, everything that a perfect being could possibly be, they thought; but God creates only what he wills to create; and he willed to create a finite world which, in the light of modern cosmological knowledge, seems rather paltry.

Before Copernicus and the dawning of modern cosmology, the Christian theologians who espoused plenitude believed that God created a multi-storied finite universe with the earth at is center, surrounded by a finite number of concentric celestial spheres that separate us from God's Heaven. In the closest celestial sphere, a finite number of "wandering stars"–the planets, understood by many to be embodied angels–rotated around the earth. Our sun and moon were classified as planets. In the outermost spheres were the "fixed stars"–

those we now know to have relatively fixed positions in our Milky Way. Many Medieval theologians thought the fixed stars to be either embodied angels or moved by angels who exist on a higher link in the chain of being. All heavenly bodies were mistakenly thought to rotate around the earth in perfect circular orbits. Non-circles are imperfect, and God created nothing imperfect. God sets the outermost celestial sphere into motion, and everything between there and the earth is moved by other motions in the next highest sphere. Despite plenitude, also affirmed, this great chain of being manifests no infinite regress (thus, God exists) and no infinite continuum of grades of being.

From the beginning of Christian cosmology, and through the Middle Ages, the Principle of Plenitude applied most conspicuously to God in himself; it really did not apply to the created world–even when it did! Christian theologians primarily emphasized the infinite plenitude of Divine Being, not the infinite plenitude of creation. Plato thought that the supreme form of The Good required that all the forms be actualized or copied within creation; and his creative and beneficent Demiurge executed this requirement.

For St. Thomas Aquinas, God in himself is pure being, pure actuality in whom there are no unactualized divine possibilities; but this did not translate into infinitely many created universes, partly because no created things made any internal difference to (had any real or internal relations with) God, and partly because a good God would create only a good world, not all possible evil worlds. Aquinas considered the hypothesis that God created infinitely many worlds because his infinite power[30] or goodness[31] seems to require infinite creativity, but he rejected it for reasons not always very clear or defensible. Still, something that superficially resembles infinite plenitude of creation does appear in Aquinas. He held that the perfection of the finite world requires God to create representatives of innumerably many species or grades of created beings; but Aquinas never says *all* or *infinitely many* species. Many things that might exist do not, he insisted, and only those species were created that harmonize on the whole with the existence of other species, especially humankind.[32]

Still, we might wonder, as did Lovejoy, why an infinite God, who is everything that he could be, did not create everything that he could create, and why infinitely many degrees of created beings could not exist within one infinite universe. Why and how could God's creativity be limited when God himself is pure, infinite, limitless, actual being in whom there are no unactualized potentialities at all?

By affirming repeatedly that God wills to create only that which the divine intellect proposes under the form of goodness, Classical Theologians like St. Thomas Aquinas easily avoided the difficulty that some possible worlds would be so horrible that a morally virtuous God would never create them. The perfection of unqualified plenitude of creation is clearly incompatible with the perfection of moral goodness or righteousness. A benevolent God would not create all possible worlds. The problem of theodicy, addressed in more depth in Chapter

Eleven, says that whether a good and intelligent God could have created this world of woe is seriously doubtful; but much worse worlds are possible, thus actual somewhere, under the unqualified aegis of the Principle of Plenitude.

Historically, the Principle of Plenitude did not originate with empirical natural science. It sprang from a very peculiar but highly influential ancient Greek way of conceiving of divine perfection. Amazingly, despite its lack of empirical origin or justification, the principle is alive and well today among scientific-minded cosmologists. To avoid God, many contemporary Anthropic Cosmologists affirm either a single infinite spatiotemporal universe or an infinite number of universes in time and/or space. Just what differentiates them depends upon how universes are distinguished, as articulated earlier.

The infinite worlds metaphysics of today's Weak and Strong Anthropic Principles is nothing but metaphysical theology without God; yet, without God the whole rationale for it is lost. Unless Plato's ideal of divine perfection is first accepted, we have no good reason to affirm any version of the Principle of Plenitude or the infinity of this or any other universes. Also, many theists have very different concepts or ideals of divine perfection; and Chapter Ten will show how divine perfection may be conceived without Plenitude.

Today's world-ensemble cosmologists presuppose that the disembodied Principle of Plenitude is the ultimate, final, and efficient cause of our life-producing Big Bang, just as it is the ultimate cause of every other universe, an infinite number of them. What caused the Big Bang? The Principle of Plenitude!

Yet, we have no good reasons to believe that disembodied, abstract principles are anything more than impotent Aristotelian formal or final causes. The main difference between Aristotelian and Platonic approaches to metaphysics is that Platonists think that universals, including abstract principles, can and do exist and exercise efficient causation without being located or embodied in actual entities. By contrast, Aristotelians are persuaded that universals exist only in actual individuals and that abstract principles can be causally effective only through embodied conscious individuals who understand and act upon them. On this issue, experience always favors the Aristotelian approach. Experience never supports Platonism. That is quite enough to make the Aristotelian view, but not the Platonic, rationally warranted.

In his brilliant book, *Universes*, John Leslie gives a highly persuasive Anthropic Argument from Design for the existence of God; but by "God" Leslie does not mean an actual personal conscious mind who understands and acts knowingly and deliberately upon abstract principles. Instead, he defines "God" as nothing more than an abstract Neoplatonic principle. For him, God is

> *the creatively effective ethical requirement that there be a good universe or universes.* Or again he is the Principle that the ethical need for a universe or universes *is itself responsible for the actual existence of that universe or those universes.*[33]

But who or what has this need? Nothing; no one!

If "there is," as Leslie says, an ethical need for infinitely many universes, his Neoplatonic God is nothing but the abstract Platonic Principle of Plenitude, except that unrestricted plenitude entails both bad and good universes, since it supposedly produces *all* possible universes. In response to the objection that his theology provides no mechanism by which ethical needs produce their effects, Leslie explains that "Neoplatonism is the view that ethical needs *are themselves* creatively effective, unaided by any mechanism."[34]

No Aristotelian or Whiteheadian metaphysician would be convinced! Alfred North Whitehead's "Ontological Principle" says that

> The reasons for things are always to be found in the composite nature of definite actual entities–in the nature of God for reasons of the highest absoluteness, and in the nature of definite temporal actual entities for reasons which refer to a particular environment.[35]

If Aristotle and Whitehead are right, disembodied normative principles cannot be efficient causes of anything, much less of universes. Only definite actual entities like human beings or an embodied personal God can understand and act knowingly on normative principles to make them efficacious. Principles themselves are only formal causes; they cannot act by themselves. Abstract principles have no needs at all, especially no ethical needs for life-supporting universes. This Aristotelian/Whiteheadian position is universally confirmed by experience. This is really all that needs to be said for it–and against Platonism.

M. A. Corey protests that Process Theists cannot affirm that God unilaterally determined the initial conditions of the universe so that they would be suitable for the later evolution of complex forms of life because these initial conditions themselves would have been free to resist the divine will, and some would have done so.[36] This spurious objection also treats abstractions–the laws of physics, the constants of nature, and the aggregate quantity of initially undifferentiated mass/energy in the nascent universe–as if they were concrete individuals capable of making choices; but no process thinker would so regard them. In fact, since the reasons for things must always be traced back to individual entities, and since God is the only individual entity capable of determining the initial conditions that govern and limit all lesser individual entities, we could and should expect God to set the initial limits for all creation. Disembodied, unindividuated abstractions can neither resist the will of God nor actualize *any* possibilities, much less *all* of them. Impersonal aggregates like door knobs and just-created, pure, undifferentiated, grandly-unified energy have no consciousness or freedom to resist God's will, just as impersonal principles have no power to accomplish anything by themselves. Ethical needs for good universes, or for all possible universes, must exist in some actual entity, if they exist at all; but these needs are not compatible with one another.

C. Infinity = All Possibilities

Let us grant for the sake of the argument that the Principle of Plenitude might be the efficient cause of whole universes, either temporally as an infinite linear series of oscillating worlds, or spatially as an infinite number of co-existing worlds, or both. Fully deployed, this principle demands the existence of an infinite world or number of worlds in which every possible grade of reality, every possible individual entity, and all possible qualities and relations are actualized. In physical terms, this principle necessitates the realization of every possible quantity, quality, kind, and combination of mass/energy. Each tiny variation calls for a whole new universe in which everything else in the universe has a new relationship with that variation, no matter how minute. Oscillation Cosmologies express the temporal alternative; and co-existing universes in Big Fizz and Big Divide Cosmologies express the spatial option, which may be combined with the temporal.

Logically, a *single* temporal series of oscillating universes can never absolutely fulfill Plenitude of Creation, for an infinite number of spatially co-existing universes are also possible; but neither can Mother Spacetime's Big Fizz, for possibly only a single temporal strand of oscillating universes exists. Many possible worlds and combinations of them exclude other possible worlds and combinations. Absolute plenitude of creation is utterly unintelligible!

We have no good reasons to believe in any other worlds, much less infinitely many of them. More seriously, even if infinitely many universes exist, infinity as such does not explain why we live in a life-supporting universe, despite what most atheistic cosmologists assume. Why must an infinitely prolonged spatial or temporal series of worlds actualize *all* possibilities, or even all logically compatible possibilities? Plato's God and the God of classical western theology supposedly, but not always consistently, actualized all possibilities knowingly and deliberately; but would infinity alone do so in the absence of Divine Agency, as Atheistic Anthropic Cosmologists contend? No, for infinite numbers alone do not necessitate or entail endless categorial, individual, qualitative, and relational diversity, despite what most atheistic philosophers and astrophysicists think. We must take a careful look at this last great unchallenged dogma of unwarranted metaphysics.

By assuming finite matter enduring for infinite time, David Hume tried to escape the theological implications of the Argument from Design by arguing:

> Instead of supposing matter infinite, as Epicurus did; let us suppose it finite. A finite number of particles is only susceptible to finite transpositions: and it must happen, in an eternal duration, that every possible order or position must be tried an infinite number of times. Innumerable revolutions produce at last some forms, whose parts and organs are so adjusted as to support the forms amidst a continued succession of matter.[37]

What insures that Hume's "innumerable revolutions" will all be different, instead of just being endless repetitions of lifeless sameness? *Finite* matter existing for infinite time cannot by itself insure infinite diversity. A universe composed of a single hydrogen atom enduring from everlasting to everlasting would fulfill Hume's suppositions. It is a logically possible world, but clearly it would never produce any diversity at all. Neither would a universe of two hydrogen atoms, or three, or four, or many more. All these are logically possible Humean universes. At exactly what point would the addition of finite materiality result in the actualization of infinite diversity? None!

Suppose that *infinite* quantities of matter exists for infinite time and/or throughout infinite Superspace. Would this insure infinite diversity? Would it necessitate that every possible world, relationship, quality, and individual be actualized? For some theologians, God intentionally makes it so, as Lovejoy points out; Divine perfection deliberately diversifies universes and thereby insures the existence of every good kind and degree of being, but not all possible imperfections. Perhaps divine perfection, Pure Being for and in whom no unactualized possibilities exist, deliberately diversifies universes and guarantees an infinite variety of actual individuals, qualities, relationships, and degrees of being. Hume's disproof of the existence of a purposive God tacitly appeals to a metaphysical principle that requires God, who can supply diversity when sheer numbers cannot!

For Atheistic Anthropic Cosmology, what insures that every possibility is actualized in infinite time and/or space? No one, including David Hume, really knows that time or space are infinite, but let us grant this groundless assumption for the sake of the argument. Infinite time and/or space alone would not insure that every finite combination of individuals and forms will inevitably be actualized because: i. This presumption confuses two very different things–numerical spatiotemporal infinity, with an infinite variety of classes, individuals, qualities, and relationships. ii. It clearly lacks a Principle or Agent of Diversification. iii. It definitely confuses *infinite* possibilities with *all* possibilities. iv. It mistakenly assumes that infinity would eventually use up all lifeless universes. v. And the very idea of actualizing all possibilities is logically incoherent.

i. Infinity Is Not Infinite Diversity

The assumption that all possibilities must be actualized in infinite time or space may confuse different orders of infinity. Georg Cantor discovered different orders of infinity, some of which are richer than others. The members of a denumerably infinite set can be put in one to one correspondence with the set of whole numbers, but a nondenumerably infinite set is so rich that its members exceed the set of whole numbers. According to Cantor's theory of transfinites, spatial or temporal infinities are denumerable sets. However, the set of all possible individuals, qualities, and relations is nondenumerably rich.

Denumerable sets cannot exhaust or use up nondenumerable sets. Denumerably infinite spacetime cannot actualize a non-denumerable infinity of potential individuals, qualities, and relationships. There are just too many of them; but even if this error is not committed, other more serious difficulties plague infinite worlds metaphysics.

ii. An Agent of Diversification Is Needed

Most seriously, infinity alone does not insure diversity. In infinite worlds metaphysics, lifeless worlds could simply repeat themselves endlessly; and probabilities insure that they would. An infinity of spacetime alone does not guarantee life, the rich environmental order required for its support, or its harmony and goodness on the whole. Endless accidents, or random spontaneous fluctuations alone can insure the actualization of nothing more than an infinite number of garbage universes. Mere infinity cannot guarantee that any universe would ever be life-supporting. For this, an informed choice must be made between lifeless and life-supporting universes, and between good and bad ones; but nothing in the concept of sheer infinity insures any kind of diversity, goodness, or intelligent selection.

Perhaps many worlds metaphysics does not require an infinite number of universes. Some cosmologists suggest that finite but very large number of universes would be sufficient to make a life-supporting universe just happen occasionally. The trouble is that neither finite nor infinite numbers as such insure any diversity at all; if infinity can't do it, neither can finitude. Universes, finite or infinite in number, can be identically lifeless. This is possible and probable. Neither finite nor infinite numbers provide or insure diversification. For that, some additional cause, an agent who selects for diversity, is required.

Life-supporting universes are infinitely improbable. Atheistic Anthropic Cosmologists regularly concede that for every single way to get a life-supporting universe right, there are infinitely many ways to get it wrong.[38] Something more than mere spatiotemporal infinity is required if conditions essential for life are to converge in some actual universe. The Teleological Argument for the existence of God, developed in a following chapter, says that to get it right, this informed choice must be made by a Superintelligent Being who comprehends infinite errors and blind alleys and chooses against them.

Alfred North Whitehead recognized the necessity for a Divine Principle or Agent of Limitation or Concretion who picks or selects desirable universes to be actualized from unfathomable numbers of undesirable possible worlds.[39] Nothing in the mere concept of infinity insures *any* diversity whatsoever, much less the right kind of diversity needed to support highly complex living beings who can live worthwhile lives. An individuated, actualized, and embodied intelligent principle or agent of qualitative selection and diversification of Divine proportions is necessary to make any numerical set of universes exemplify the right kind of qualitative conditions to support worthwhile and complex forms of life.

Neither David Hume nor today's Atheistic Anthropic Cosmologists provide for an intelligent, or even for an unintelligent, agent of qualitative limitation and diversification. Atheistic Anthropic Cosmologists dismiss the Strong Theistic Anthropic Principle much too quickly.

iii. *Infinite* Possibilities Are Not *All* Possibilities

Infinite Worlds Metaphysics definitely confuses *infinite* possibilities with *all* possibilities. "All" includes "infinite," but "infinite" does not entail "all." God or Mother Spacetime could create an infinite number of good universes without creating a corresponding number of evil universes. Plenitude of creation requires both; but plenitude as a perfection is radically different from and incompatible with the perfection of moral goodness, which adds responsible and purposive moral selectivity to the requirement to produce infinite actualities.

As the ancient Stoics and nineteenth-century Friedrich Nietzsche contended, infinite time (or space) could simply repeat an identical set of mass/energy configurations, individuals, qualities, and relations an infinite number of times in an Eternal Recurrence of sameness. They supposed that life would participate in Eternal Recurrence, but what would guarantee this? An everlasting or an eternally recurring universe composed of a single hydrogen atom would result in no life or meaningful diversity; neither would an eternally recurring universe composed of an infinite number of hydrogen atoms (and no others). Both are logically possible universes that could eternally recur. That *all* universes are merely hydrogen universes is logically possible. "All" does not translate into "infinite diversity"!

Despite incessant claims to the contrary, a monkey banging on a typewriter for an infinite amount of time would *not* necessarily write the *Bible* or all the works of Shakespeare. Infinite time alone could not and would not prevent a secretarial monkey from banging on just one key–forever. Its doing so forever is logically possible, and if all possibilities are actual, it must be so! If the monkey is bored (teleology), it might strike a diversity of keys. At first, it would produce only trash. In an infinite amount of time, it would produce only an infinite amount of trash. Garbage in, garbage out–forever.

The secretarial monkey universe actually raises more questions than it answers. Who or what created the monkey? Who created the typewriter? Does only one monkey exist? Did it have parents? Does it have a navel? What does it eat? Does it ever take a break? Why does it type rather than doing something else? How does it live forever? Has its universe been around forever? How so? Some things that sound possible in the abstract are not very plausible in the concrete. The example requires a cosmology, which requires a metaphysics, which requires a theology!

Those who assume that an infinitely diverse temporal or spatial series (which requires an Agent of Selection and Diversification) will exhaustively actualize *all* possibilities simply fail to understand the concept of infinity. Suppose that an infinite number of diverse individuals, qualities, and relationships were or are actualized during an infinite past or in endless space. This would not mean that *all* possibilities are actualized; even infinity does not use up infinity; an infinite number of universes would still remain to be actualized! Infinity subtracted from infinity still leaves infinity, and "infinite" does not mean "different" or "all."

If probability is to be our guide, without God, all past, present, and future worlds would probably be lifeless worlds. The existence of an infinite number of antecedent or contemporary universes does *not* explain why we live in a universe in which incomprehensibly complex and remarkable concurrences conspire to generate and support our existence and that of innumerable other inherently valuable forms of conscious life. An infinity of antecedent or co-existing universes would likely be nothing more than endless variations on inexhaustible themes of lifelessness. Life-supporting worlds are infinitely improbable, and rationality rejects infinite improbability.

iv. Infinity Would Not "Use Up" Lifeless Universes

Those who think that life-supporting universes will inevitably occur in an infinite amount of time (or space) seem to assume that certain sub-sets of infinity like lifeless universes would eventually be used up, and then a life-supporting universe would necessarily come along; but this is not true at all. This approach treats infinity as if it were finite. It assumes that infinite sets can be *used up* in infinite time or space; but infinity, especially nondenumerable infinity, cannot be actualized exhaustively in any amount of space and/or time. Neither qualitative infinity, nor sub-sets of it, can be totally depleted in numerically infinite time or space. Sub-sets of infinity–all possible lifeless universes–contain just as many members as infinity itself, so they can never be exhausted or used up so that something else can come along. Even after an infinite amount of time, an infinite number of lifeless universes would remain to be created. Each one could and probably would be exactly like the last unless some infinitely wise and intelligent Divine Agent of Limitation and Diversification knowingly and deliberately selects just the right conditions for life-affirming qualitative diversity.

Those who think that an infinite number of universes will actualize all possible qualitative diversity may believe that universes will be randomized, and that infinite randomization will produce infinite qualitative diversity and eventually use up all possible lifeless universes. However, randomization itself needs explaining and has its own necessary conditions. What selects and causes these conditions? Why does infinite randomization not result in infinite repetition?

The concept of randomness makes sense only against an established background of law and order in an actually existing universe. As Alan Guth recognizes, "'randomness' is ill defined. The word 'random' does not become meaningful until the probability rules are stated."[40] But no established finitistic probability rules exist for pure sets of all possible universes.

Given the orderly background of the actual finite laws of physics, cards in a deck can be randomly shuffled; but with no actual laws of physics, no actual cards, and no actual shufflers, no random shuffle can occur. Randomization of actualities makes no sense in a realm of undifferentiated possibilities devoid of any and all background order; and randomization alone could not use up infinity to insure that randomly produced lifeless universes are not repeated endlessly.

Given the orderly background of quantum Superspacetime and the laws of quantum physics, quantum spontaneity might generate randomness; but what guarantees the existence of Superspacetime, the ubiquity of quantum laws within it, and the exclusion of all possible non-quantum universes? Why does Superspacetime exemplify quantum instead of non-quantum laws? Alternatives are logically possible! Quantum-foamy Superspacetime cannot use up all possibilities, for a uniformly featureless Newtonian/Kantian absolute Superspacetime is logically possible also.

A teleological explanation for quantum universes is readily available and highly plausible. A creator God wanting to make free, responsible, and co-creative creatures would choose quantum laws, not Newtonian laws, for created universes in order to make room for creaturely freedom, responsibility, and creativity. God would just make spontaneously creative creatures ranging from sub-atomic particles to conscious animals and human beings. Quantum laws actually presuppose their existence, for laws are products of the average behavior of concrete entities, not efficient causes that antedate and constrain their behavior. This is also true of the laws of sociology and psychology.

v. Actualizing *All* Possibilities Is Incoherent

Finally, the very idea of actualizing all possibilities is logically incoherent. Possibilities are both negative and positive. Possibly, all universes are life-supporting; possibly, none are. Possibly, the physics of every world is Newtonian; possibly the physics of every world is quantum/relativity; but these universalized possibilities cannot both be actual. Possibly the epoch that preceded ours was Newtonian, or it was open or flat; but then quantum effects in an antecedent singularity or in antecedent nothingness would not account for our world. (They don't anyway.) If all possibilities were actual, we would not be here, for it is possible that neither we nor our world should exist. Yet, we would be here, for it is possible that we and our world might exist. Possibly, all worlds are good, possibly created by a benevolent God; possibly, all worlds are bad, possibly created by a malevolent demon. A negative possibility corresponds with every positive possibility; and *all* possibilities include both! *All* possibilities just cancel out everything!

The Principle of Plenitude of creation is logically unintelligible. If modified to say that all "consistently combinable" possibilities are actual, who or what does the selecting? Different combinations are possible in different universes and between universes. Contradictions indicate mutually exclusive alternatives, but they do not select one or the other to be actualized. To settle this, a choice must be made. Not an abstract Principle, but an *Agent* of Selection and Diversification is required for this because disembodied abstract principles are impotent.

Postulating multiple universes does not account for our life-supporting universe, as it seemed to do for Schrödinger's cat that was alive in one universe but dead in another, for many possibilities that would apply to *all* universes are mutually exclusive. For example, it is possible that all universes are life-supporting or that none are, that all are quantum or that none are, and so on. If the "consistently combinable" qualification is accepted to begin with, Schrödinger's cat poses no problems. Unlike the Medieval theologians, Quantum Cosmologists seem to be very serious about the actualization of *all* possibilities–logic be damned. But logic says the same thing of Quantum Cosmologists!

D. Infinitely Many Life-Sustaining Universes

Suppose we knew that an infinite number of worlds exists, and *all* of them are good, interesting, and life-sustaining. This too is logically possible. Would this tell us anything about cosmic or supercosmic teleology? Anthropic Cosmologists themselves concede that life-supporting worlds are rare exceptions, not the rule, among possible universes. Low probabilities are possible but highly unlikely. Possibility should not be confused with probability. That each member of an infinite series of oscillating or co-existing universes handsomely supports valuable life-forms is logically possible but not very probable. Yet, if *all* possibilities are actual, this must be true; but it must also be false!

If a large finite or even an infinite number of universes were life-supporting and none were lifeless, what would be the most plausible explanation? Atheistic interpretations of the Anthropic Principle explain the remarkable life-supporting coincidences of our universe by postulating an infinite number of antecedent or contemporary worlds, and by equating numerical infinity with infinite diversity. Given an infinite number of qualitatively diverse shoes, at least one will fit accidentally, they say, even though it is logically possible that none will fit, or that all will be exactly alike–all size eight when you wear a ten, or all left feet. But suppose that an infinite number of shoes *all* fit! No mere chance depletion of infinite diversity could explain it, for there would be no relevant diversity; no lifeless universes would be used up. Why would each world support life when endless lifeless worlds are logically possible? Only a Divine Agent of Limitation and Diversification could explain it.

Abstract principles are not efficient causes; they are efficacious only when embodied in and acted out by concrete decision-making individuals. Only a concrete individual agent of Divine proportions could select the right conditions for worthwhile life even once, much less an infinite number of times. If life-supporting universes are infinitely improbable subsets of a nondenumerable infinity of possible individuals, qualities, and relations, these subsets would be actualized only if a deliberate and informed choice is made between them and infinite subsets of lifeless universes. A following chapter on the Teleological Argument for the existence of God concludes that God made the choice.

Just one life-supporting universe is infinitely improbable, and so are an infinite number of them. To be sure, we do not know that *any* other universes exist, much less an infinite number; but one interesting argument for the existence of an infinite number of life-supporting universes should be considered, even if it is finally rejected.

If the constants, initial conditions, and laws of nature did not break down in the final collapse of an infinite series of oscillating universes, they would presumably be the same in every successive universe. Conservation laws, along with all others, would carry over from epoch to epoch. The First Law of Thermodynamics would be in effect from one epoch to the next, and the amount of mass/energy in each cosmic epoch, no matter how many, would be identical; no mass/energy would be lost from epoch to epoch. Conservation of mass/energy would be metaphysically everlasting.

Collectively, the harmoniously integrated laws, constants, initial conditions, energy densities, and so on, of our own universe produce sufficient qualitative richness and diversity for complex living creatures to exist. If these do not break down at the end of any cosmic epochs, they would be the same in every previous and succeeding cosmic epoch. With identical laws, physical constants, quantities of energy, and initial conditions, each preceding universe would be generally life-supporting like ours (with a fly in the ointment to be explored shortly).

An infinite series of exclusively life-supporting quantum universes would not destine every cosmic epoch to be a detailed rerun of its predecessors. This might be true of deterministic Newtonian universes, but not of indeterministic Heisenberg universes. If quantum laws always obtain, each of us would not live our lives over and over again in an endless series of Eternal Recurrences. If, like ours, all previous cosmic epochs include the laws of quantum physics, then quantum indeterminacy, fluctuations, and free choices made by complex creative agents in each epoch would preclude detailed duplication. Only general life-supporting conditions could be replicated by the laws, constants, initial conditions, and quantity of mass/energy of all earlier and later relevantly similar universes. But if endless universes are as rich in life as our own, how, why, and by whom were life-supporting conditions chosen for all, when infinitely improbable even for one, given all possible universes from which to choose?

The Second Law of Thermodynamics is the fly in the ointment for an interminable string of life-supporting universes. If entropy or disorder increases from one cosmic epoch to another, then chaos increases from epoch to epoch. Because more chaotic, no universe could be qualitatively similar to its predecessor after all; each would have to be much less ordered than its immediate antecedent. An infinite number of earlier increases in chaos would result in pure chaos today, and too much disorder is incompatible with complex valuable life. Since our present universe is not pure chaos, either the Second Law does not carry over from epoch to epoch, or there have not been an infinite number of prior cosmic epochs, or a Divine Being intervenes periodically to bring order out of chaos, or God just created our universe out of nothing. But which is it?

E. Infinite Time and the Inverse Gambler's Fallacy

Ian Hacking brilliantly argued in 1987 that Oscillation Cosmologists like John A. Wheeler commit the Inverse Gambler's Fallacy.[41] The gambler makes a logical mistake that usually costs him dearly. He comes into a game of dice and asks how many previous rolls occurred before he arrived. If many previous rolls produced no double sixes, the gambler assumes that the chances for non-twelves have been used up; and he concludes that his chance of rolling a twelve are very high; he bets heavily on twelve–and loses. The problem is that no matter how many previous rolls, the gambler's chances are exactly the same when he starts to play as on the very first roll. Once the relevant statistics about habituated entities are available, the odds do not change. The *odds* are the same on *every* roll! This is what gamblers, both cosmic and non-cosmic, do not understand.

Hacking's essential point is easier to grasp if we consider flipping a coin. With only two sides, the odds for flipping heads or tails are equally fifty/fifty. In a fluke situation, a gambler learns that tails have come up forty-nine times in a row and assumes that the odds favoring heads on the next flip are forty-nine to one; so he bets on heads–and loses. All along, the odds are just fifty/fifty. In the inverse situation, learning that heads just turned up, the gambler bets that tails appeared on the previous forty nine tosses. He bets on that–and loses.

The inverse gambler's fallacy begins with the information that an improbable double six was just rolled. From this, the gambler infers that it must have been preceded by a large number of rolls with no double sixes. If no dies were cast since he entered the room, he waits to see how the roll comes out. If it is a double six, he infers that many previous rolls must have produced different results. He then bets that many previous rolls resulted in no double sixes–and loses. In actuality, once the odds are established and known, the statistical chances of rolling a double six are just as good (or as bad) on the next as on the millionth roll. The number of rolls makes no difference to beating the odds, and a single current roll makes no difference in estimating the number or kind of previous rolls.[42]

Applied to Anthropic Cosmology, the inverse gambler's fallacy says that "The universe has been around for ever so long, so it is not in the least surprising that it should have got into its present orderly state. Given an old enough universe, we would expect our order to arrive eventually by mere chance."[43] This Humean reasoning is fallacious, says Hacking, because "There would have been no ground for believing we have an old universe, except that it explains the present order. But it does not explain the present order. So there is no ground for believing in an old universe."[44]

Thus, the supposition that our orderly life-supporting universe occurs late in an infinite sequence of non-life-supporting universes does nothing whatsoever to explain its existence. There is no such thing as "occurring late in" or "using up" an infinite temporal series. In an infinite series, *every* universe, no matter where positioned, is preceded by infinite time and an infinite number of antecedent universes. Cosmological gamblers cannot beat those odds! And if an open universe inevitably occurs after a finite or even an infinite number of oscillations, then *every* universe will be open, and the very notion of infinite oscillations is incoherent.

All Atheistic Anthropic Cosmologists commit the inverse gambler's fallacy when they assume that infinitely many preceding worlds would account for our life-supporting universe. Gamblers should learn that the odds against winning are the same on *every* throw of the dice; and cosmologists should learn that the odds against life-supporting universes are eternally the same–infinite–no matter how many antecedents. We lack absolute certainties, but rationality bets against infinite improbability. Rationality bets on a God who knows and cares.

E. Infinite Space and the Inverse Gambler's Fallacy

Does the *spatial* option for the Weak and Strong World-ensemble Anthropic Principle commit the inverse gambler's fallacy? Big Fizz Cosmologies affirm that if infinitely many worlds co-exist in infinite Superspace, as Carter, Linde, Barrow, Tipler, and others believe, then somewhere in that vast simultaneous infinity, some worlds would support life, not by Divine design, but just by beating the odds. Ian Hacking thinks that the infinitely many simultaneous worlds theory does *not* commit the Inverse Gambler's Fallacy, for time is not a factor; this theory just deduces its conclusion directly from the Principle of Plenitude. If all possible universes simultaneously co-exist, and if our ordered world is a possible universe, then our ordered world will exist. Here "Everything in this reasoning is deductive. It has nothing to do with the inverse gambler's fallacy."[45] Our life-supporting world is just deduced from the truth of the Principle of Plenitude. But plenitude is not a truth!

Our well-founded doubts about the Principle of Plenitude are just as cogent when applied to infinitely many worlds co-existing in Superspace as to an infinite temporal succession of cosmic epochs in Supertime. Plenitude is not an empirical, scientific truth. It originated with a peculiar theological concept of Divine Perfection, and there is no other plausible rationale for it. It cannot be employed to avoid Theism. Applied to co-existing worlds, Plenitude confuses orders of infinity and is logically incoherent. The probability is infinite that lifeless worlds would occur endlessly in mindless Superspacetime. Without some intelligent and deliberately selective Divine Agent of Limitation or Diversification, infinitely many co-existing worlds would probably all be lifeless. This is possible as well as highly probable, and if possibility equals actuality, it must be so–even if it cannot be so! World-ensemble theories fail to provide an Agent of Selection and Diversification. Infinite Superspace cannot actualize *all* possibilities, for no matter how many contemporary lifeless universes exist, an infinite number of lifeless universes will remain to be actualized.

G. Faith vs. Reason

In discussing the relation between the Weak Anthropic Principle and world-ensemble metaphysics, Fred W. Hallberg wrote:

> So all those "other" universes would exist unobserved (Gale 1981, 168). The existence of our special life-enhancing universe would be the inevitable result of chance within this larger ensemble of universes. Of course, this entire supposition of an ensemble of universes is a purely speculative idea beyond any conceivable scientific determination. An equally valid alternative supposition, that our universe expresses an *intention* that life and consciousness be realized, also takes us beyond what could be conceivably determined by scientific experimentation.[46]

Hallberg thinks that very different attitudes toward the world would be appropriate, depending on which explanation of life-supporting cosmic coincidences we accept. He concludes that "Neither choice is more factual, or realistic, than the other" and that "The weak anthropic principle limits reason in a way that leaves room for faith (Kant [1787] 1958, 29)."[47]

Hallberg is right about the limits of science, but perhaps he restricted reason (conceived more broadly as philosophy) prematurely. Theism has more than an equally valid claim with world-ensemble Atheism on our confidence. Hallberg does not challenge the assumption that infinite number equals infinite diversity. Consequently, he does not realize that life-affirming diversity within infinitely many worlds requires an Agent of Selection and Differentiation of Divine proportions. Chapters Ten through Twelve of this book will show that belief in God is rationally warranted. Theistic faith (confidence) need not be blind or irrational. All rational beings, including cosmologists, should take the theistic option much more seriously.

To summarize, Atheistic Anthropic Cosmology concedes that our universe is exceptionally fine tuned for intelligent life forms; but this can be explained, its adherents claim, without appeal to God. The universe is compatible with our existence because we would not be here if it were not, says the Weak Anthropic Principle. But *why* is the universe compatible with our existence? The Weak and non-theistic versions of the Strong Anthropic Principle explain our good fortune by appeal to infinite worlds metaphysics. Yet, no infinite worlds metaphysics is verifiable directly or inductively. As an explanatory hypothesis, it is derived historically from an unscientific theological Principle of Plenitude, which affirms that God would be less than perfect unless he actually creates everything that he possibly could create. The assumption that an infinite number of worlds in time and/or space would actualize *all* possible individuals, qualities, and relations is not true, partly because a lower is not equivalent to a higher order of infinity, partly because infinite sets cannot be used up, partly because infinite sets need contain no diversity at all, and partly because the notion of actualizing all possibilities is logically incoherent. Lifeless worlds could and most likely would be repeated infinitely, just as monkeys banging typewriters would most likely produce garbage infinitely. Infinite numbers do not translate automatically into the existence of infinite diversity. Only a super-calculating life-loving God could select conditions for one or more life-supporting universes.

Infinite temporal oscillations do not improve the odds against a life-supporting universe, according to the Inverse Gambler's Fallacy. An infinity of worlds, whether successive or co-existing, is deduced from the indefensible Principle of Plenitude. An infinite plenitude of universes does not make good sense and is supported by no good reasons because abstract, disembodied, normative principles are not efficient causes, and the notion of actualizing all possibilities for all conceivable universes is logically incoherent. Atheistic Anthropic Cosmology does not successfully account for what caused our Big Bang.

Nine

THE FINAL ANTHROPIC PRINCIPLE

According to another version of the Strong Anthropic Principle, the Final Anthropic Principle, our universe *must* exist. Why? Because it helps to create God. Our universe, together with infinitely many others, insures that God will one day come into being. This is its ultimate purpose and reason for being.

1. The Omega Point as the Purpose of the Universe

John Barrow and Frank Tipler develop and defend what they call the "Final Anthropic Principle,"[1] according to which the purpose of our presently Godless universe is to bring about God or the Omega Point. Only gradually do they identify the two, but this is the end result of their reflections, especially Tipler's. From the future, the Omega Point creates the world, but only after the world creates the Omega Point (if that makes any sense). God creates the world only after the world creates God.

According to Tipler, who identifies himself as an atheist,[2] God's non-existence is true now, but it will be false at the end of time when Theism becomes true. Tipler thinks that a godless universe will one day create God. *The Anthropic Cosmological Principle* culminates with the Omega Point. Tipler most fully develops the idea that the forthcoming Omega Point = God in his *The Physics of Immortality*, 1994, where he contends that the Omega Point will raise us (or virtual cyberspace computerized emulations of us) from the dead billions of years from now and give our emulations eternal life.[3]

Every step in the futuristic pseudo-scientific eschatology of Barrow and Tipler is highly conjectural, unverified, and improbable. Over billions of future years, they believe, biological human life will perish; but it will re-embody itself in computerized robots that will gradually spread throughout the cosmos. Some day computerized humanity will exit planet earth and our solar system in space ships. Self-replicating humanoid or android computers will come to inhabit all of the Milky Way, then move on from there to conquer all other galaxies.

Barrow and Tipler remind us that eventually our solar system will cease to support biological life. Our sun has already burned half of its energy and has only five billion years to go. In four billion years, it will expand as a red giant to incinerate all its planets. Eventually, all stars/suns will exhaust their nuclear fuel, and all planets everywhere will become biologically uninhabitable in a cosmic heat death.

After biological human life as we know it is extinct,[4] intelligent life itself will not end, Barrow and Tipler claim. They define life as "information processing,"[5] which, by definitional fiat, makes computers both alive and intelligent. (Yet, we must note, so much of the fullness of human reality and value is missing!) They insist that "Intelligent *machines* can be regarded as people. These machines may be our ultimate heirs."[6] Eventually, we will construct complex robotic computers that can reproduce themselves and survive both the final cold of a closed universe's vast expansion and the final searing heat and pressure of its terminal contraction. In due time, all matter will decay, including protons and magnetic monopoles, so living computers will have to reembody themselves in radiant energy toward the end of the expansion phase. Toward the end of the contraction phase, they will be embodied in matter denser than iron and will endure long periods of hibernation with only fleeting moments of information-processing. Complex intelligent computers will still continue to communicate with one another and to gather information–until they take possession of, and make use of, all the mass/energy in the universe. Fat chance!

Similar life or information-collecting processes that include all possible histories are going on in infinitely many co-existing worlds in an all-inclusive Superspacetime. The Final Anthropic Principle agrees that the Principle of Plenitude as objectified in infinite worlds metaphysics caused our Big Bang. As our universe approaches its final singularity, it will merge with infinitely many other worlds that include all possible histories. Collectively, they will possess an infinite amount of information. The final Supersingularity in which infinite worlds merge in Superspacetime is the Omega Point. The whole meaning and purpose of our universe, of every universe, is to generate the Omega Point. Objectively, when the Omega Point is reached, "This is the end;"[7] but subjectively it will not be the end.

The Omega Point will be subjectively immortal because the pace of events will be slowed down so much that time will *seem* endless; (and presumably the then-Omniscient Omega will be too stupid to realize that it isn't). Barrow and Tipler tell us that "A modern-day theologian might wish to say that the totality of life at the Omega Point is omnipotent, omnipresent, and omniscient!"[8] In *The Anthropic Cosmological Principle*, they do not actually say that the Omega Point is God; but since it exemplifies the traditional defining attributes of God, it is not misleading to say that according to the Final Anthropic Principle the purpose of the universe is to make God. Tipler explicitly calls the Omega Point "God" in later writings.[9] Like Samuel Alexander, whose *Space, Time, and Deity* appeared in 1920, Barrow and Tipler predict that some day the universe will create God. Unlike Alexander, they also contend that God created the universe, since the future (God created by the universe) creates the past (God creating the universe). Anthropic Atheism results in Omega Point Theism in the very far distant future; Tipler's God "exists mainly at the end of time."[10]

2. Critique of the Final Anthropic Principle

The incredibly conjectural predictions proffered by Barrow and Tipler make their Final Anthropic Principle (FAP) highly problematic. Not inappropriately, one critic, Marvin Gardner, renamed their principle the "Completely Ridiculous Anthropic Principle (CRAP)".[11]

A. Unfounded Assumptions

Not one of the following claims made or presupposed by Barrow and Tipler are known to be true, and most of them are either meaningless, blatantly false, immensely improbable, or logically incoherent.

1. Human beings (or our robot computer descendants) will one day travel to all inhabitable planets in the Milky way and eventually to every galaxy in the universe. This presupposes that:

a. Human minds and bodies can either survive the weightlessness and other adversities of very long-distance space-travel; or they can be perfectly encoded in lightweight computers that can survive such rigors.

b. Human beings will invest heavily in space-research and travel in the future.

c. Cheap and abundant sources of energy for space travel will be available in the future.

d. We and our biological descendants will long survive the enormous genocidal propensities of our species, our unpredictable adventures and misadventures with nuclear energy, our incredibly short-sighted environmental destructiveness, our dabbling with bioterrorism, and our propensity to overpopulate the earth.

e. Many planets throughout the Milky Way and the rest of the cosmos are inhabitable and will provide suitable habitats for our computerized robot descendants.

f. If inhabited, the occupants of other planets will be receptive to computerized humanoid aliens.

g. The native bacterial, viral, chemical, physical, and social occupants of other inhabited planets will not be devastatingly hostile to and destructive of computerized humanoid aliens.

The list could go on and on.

2. Life is nothing but "information processing," which implies that automobiles and all other machines are alive.[12]

3. Machines (information processing computers) are people; they are (or will be) just as conscious, intelligent, and valuable intrinsically as biochemical, carbon-based human beings.

4. We and our biological descendants can be persuaded that 2. is true, will come to care about the long-term destiny of android computers, and will recognize that the whole meaning of our existence depends on what happens to merged computerized robots billions of years from now.

5. Our universe, and each member of the oscillating set to which it belongs, contains enough mass/energy to close it, so its expansion phase will halt,

and the terminal Big Crunch will coalesce into the Omega Point. This is also true of infinitely many co-existing worlds and/or oscillating sets of worlds in Superspacetime. The Omega Point will result from the merger of an infinite number of universes that actualize all possibilities.

6. Intelligent computers can survive and function in the extremely hostile physical conditions that will prevail toward the end of the universe's expansion and contraction phases. At the end of time, they and the information they encode can survive the infinite density, timelessness, and spacelessness of the ultimate singularity, the Omega Point.

7. Computers can and will eventually capture and make use of all the mass/energy in the universe and manipulate its evolution.

8. Processes like 1–7 above are going on in infinitely many worlds co-existing with ours in infinite Superspacetime. Infinitely many worlds actualize all possibilities.

9. Near the final state of the Superuniverse, infinitely many computer-captured universes will merge with ours to form the Omega Point. Their contraction phases will somehow coincide with that of our world or its ultimate oscillating successor. Infinitely many universes can and will find ways to contract and merge that preserve and are not destructive of information about every detail of their existence.

10. The duration of the Omega Point will be objectively finite, but it will be subjectively immortal–or too slow and stupid to know that it is mortal!

11. At the end of all spacetime, the Omega Point, God, will become omnipotent, omnipresent, and omniscient.

The possible and the probable are not identical, and the logically incoherent is not even possible. None of the above claims are known to be true, and most if not all of them are highly improbable, blatantly false, unintelligible, or otherwise implausible. No attempt will be made here to criticize these claims in depth, but many of these presumptions have been decisively refuted in earlier discussions. The absurdity of this position speaks for itself.

Let us consider only two of many insuperable obstacles to the realization of *Star Trek*, *Star Wars*, and all fanciful futuristic human space-travel scenarios–the time involved in space-travel, and the incredible quantities of energy required. As Timothy Ferris indicates,

> The stars are just too far away: A spacecraft capable of traveling a million miles per hour–and this would be a stunningly fast ship, one that could fly from Earth to Mars in less than an hour–would take nearly *three thousand years* to reach Alpha Centauri, the nearest star."[13]

Alpha Centauri is 4.3 light years from us; it actually consists of three stars closely encircling one another; it probably boasts no habitable planets, for any that try to form would be ground to bits by this encirclement.

In his 1994 book, Tipler concedes all of this, then opts for travel beyond Alpha Centauri to two stars resembling our sun that are 11.3 and 10.7 light years away as the first feasible extra-solar destination. Habitable asteroids might exist along the way, he hopes. Tiny microcomputers could be transported to these stars at 90 percent of the speed of light in only twelve years or so, he claims.[14] But how and when will we build spaceships that can travel at 90 percent of the speed of light? Surely not in twelve years. Probably never.

Building space ships to carry tiny microcomputers is one thing; building them to carry human beings is another. This is where the available energy problem gets serious. Physicist John A. Jungerman tries to bring Trekkies back to sober realism about this. Flying the starship *Enterprise* to the nearest star at only half the speed of light (much faster than in Ferris's example above) would take eight years one way, but accelerating it to that speed would be absolutely prohibitive. At four million tons of mass, the energy (fuel) required to accelerate the *Enterprise* to half the speed of light for the trip would be

> the energy equivalent of about 10^{16} tons of TNT, or ten thousand trillion tons! The nuclear arsenals of all countries contain about twenty billion tons of TNT equivalent. So to put this into perspective, the fuel required for the acceleration to half the speed of light would be the energy equivalent of about five hundred thousand times all the nuclear arsenals of the world.[15]

Jungerman further indicates that when the *Enterprise* gets to the nearest star, an equal amount of energy will be required to slow the ship down and land it. And this says nothing of fuel required for the return trip! Fictional space warps that allow for overcoming these obvious limitations of space-travel remain "a science fiction dream," Jungerman cautions.[16]

Ferris, Tipler, and Jungerman discuss only space-travel within our galaxy. Intergalactic travel, the real stuff of science fiction, is conspicuously less feasible. The light we see from the "nearby" Andromeda Galaxy has been traveling (at the speed of light) for over two million years to reach us. If you plan to vacation some day in the Andromeda Galaxy, forget it! Even if you would not age much while traveling at almost the speed of light, assuming you could attain that speed, you would still take over two million earth years to get there. We cannot conceive of the obstacles that are likely to arise or the fuel required for such a trip. The only intelligent space travelers likely to migrate to other solar systems or galaxies will be human-made computers; they won't be us or our biological descendants. Even if a chosen few elite specimens of biological humankind eventually travel to and survive in extraterrestrial environments, most ordinary people (the likes of you and me) will be left behind to suffer the fate of this fragile earth, whatever that fate may be. If *we* don't make it here, *we* won't make it anywhere!

Some futurists paint a much rosier picture of our prospects for space travel. Whether it will ever happen or not is really quite irrelevant to the central issue, which is: Upon what does living a meaningful life depend? Are our lives meaningless if the ultimate destiny of our species is tied to our fragile planet earth, or to our solar system? Are our lives meaningless if neither we nor our biological descendants will ever colonize other planets, first within and then beyond our solar system, and then beyond our galaxy? Is every person's life meaningless unless he or she makes some contribution to space travel and the "terraformation" and colonization of other planets? Some astrophysicists think so, as indicated next.

Suppose, as seems likely, that neither we nor our biological descendants will ever get beyond our solar system, or that even if we do only a tiny number of human beings will ever colonize other planets. Does this mean that most of us live meaningless lives? Suppose that the earth ultimately dies a "heat death" and that this is true of all the other planets that a few elite human beings might ever inhabit. Will it all be for nothing? The answer is "yes" if the worth of our lives depends on endlessly perpetrating our species or contributing to some ultimate transspecies objective. The answer is "No" if our lives have *intrinsic* meaning and worth here and now.

B. The Meaning and Value of Human Life

Philosophers identify at least two types of value or goodness–intrinsic, and instrumental or extrinsic. An intrinsic good is an end in itself, valuable for its own sake. An extrinsic good is an efficient means to some other goal or value beyond itself. Extrinsic goods have desirable consequences. Systemic goodness–the value of concepts, ideas, and formalities of every description, was recently added to this traditional duality of goodness by Robert S. Hartman.[17] In considering the value of individual human lives, we must decide whether we have intrinsic worth, extrinsic worth, systemic worth, or some combination of all three.

The word "meaning" may have many different meanings, but it is something conceptual, something systemically good, as is "a meaningful life." Meaningful lives and valuable lives are intimately related. All valuable lives are meaningful, but they are not merely conceptual. Let us specify that human life is meaningful if we can conceptually comprehend and wholeheartedly affirm its intrinsic, extrinsic, and systemic worth and can understand how human values are supported by broad social, physical, psychological, cosmological, and metaphysical structures and environments. Support makes little sense unless threats and dangers exist, so these too must be factored into any conceptual scheme that captures the meaning of human life. Metaphysical support for human life will be covered in following chapters. Valuable individual human lives are the concrete realities to which our concepts of "meaningful lives" make reference.

i. Human Existence as Merely a Means to Something Beyond

Advocates of the Final Anthropic Principle make some commonplace but highly questionable assumptions about the value and meaning of human life; and they incorporate their flawed assumptions into their interpretation of the FAP. Barrow and Tipler suppose that the meaning and value of human life (of all kinds of life, for that matter) lie *entirely* in its future consequences. This implies that human lives here and now are nothing more than instrumental goods, that our own lives have no inherent or intrinsic meaning and worth at all. The following excerpt expresses this key axiological assumption presupposed by the Final Anthropic Principle. According to Barrow and Tipler,

> We *know* space travel is possible. We argued that even interstellar travel is possible. Thus once space travel begins, there are, in principle, no further physical barriers to prevent *Homo sapiens* (or our descendants) from eventually expanding to colonize a substantial portion, if not all, of the visible Cosmos. Once this has occurred, it becomes quite reasonable to speculate that the operations of these intelligent beings could begin to affect the large scale evolution of the Universe. If this is true, it would be in *this* era–in the far future Near the Final State of the Universe–that the true significance of life and intelligence would manifest itself. Present-day life would then have cosmic significance because of what future life may someday accomplish.[18]

If the "true significance" of life and intelligence is manifested only in the very distant future, this means that it has no true significance here and now. If present-day lives have significance only *then* or as means to *then*, they have no significance *now*, except as extrinsic goods. The end to which we are mere means lies billions of years in the future, the Omega Point. But the promise of the Final Anthropic Principle is as hollow as the sign in the bar that says "Free Beer Tomorrow," for tomorrow never comes. Biological human life will be totally extinct for billions if not trillions of years before it has any "true significance."

Tipler makes the purely extrinsic or instrumental worth of human life even clearer in a later essay where he asserts that (1) "Value is something connected with life, and thus if value is to remain in the universe, life must persist indefinitely."[19] (2) A universe "in which life (and hence intelligence) and all its works disappeared forever would in my judgment be ultimately meaningless."[20] (3) The laws of physics ultimately doom the human species to extinction.[21] (4) "Humankind's place in the scheme of things is that of an intermediate link."[22] (5) The future of life belongs to computers, not to DNA (biological) based life.[23] In his 1994 book, Tipler adds that the intelligent computers of the future will eventually raise us–or rather, computerized virtual cyberspace emulations of us–from the dead; and we will be identical with these virtual emulations.[24] Do you see any problems about personal identity here? Does the Final Anthropic Principle make you feel exploited as a mere means to ends beyond yourself, ends that you do not embrace?

ii. Enduring Grand Objectives

Many science-minded cosmologists feel and express deep anxiety about the meaning and value of human life because, like Barrow and Tipler, they presume that our lives are utterly worthless unless they contribute significantly to the achievement of some Enduring Grand Objective located in the far, far distant future. Unlike Barrow and Tipler, many skeptical or pessimistic cosmologists are convinced that no Enduring Grand Objective like the Omega Point will ever exist. So they despair.

Cosmological pessimists fully comprehend that humankind is destined to perish some day in a hostile cosmic environment, and they are obsessed by an awareness that our lives are extraordinarily tiny and brief within the vastness of cosmic spacetime. They conclude that human life has no value or meaning, and neither does the universe. This sort of cosmological pessimism was well expressed in an often-quoted excerpt from Steven Weinberg's book, *The First Three Minutes*. According to Weinberg, "Whichever cosmological model proves correct, there is not much comfort in any of this." Reflecting on the beautiful and supportive earthly environment in which we live, Weinberg comments,

> It is very hard to realize that all this is just a tiny part of an overwhelmingly hostile universe. It is even harder to realize that this present universe has evolved from an unspeakably unfamiliar early condition, and faces a future extinction of endless cold or intolerable heat. The more the universe seems comprehensible, the more it also seems pointless.[25]

All twenty seven contemporary cosmologists interviewed by Alan Lightman and Roberta Brawer for their *Origins: The Lives and Worlds of Modern Cosmologists* were asked to comment on this quotation from Weinberg.[26] Their responses are fascinating! Clearly, many contemporary cosmologists share Weinberg's cosmic pessimism, and he did not change his mind in his more recent *Dreams of a Final Theory*, 1992.[27]

iii. Human Insignificance in the Grand Scheme of Things

What should we make of all of this? The universe and our own existence are likely to seem pointless if we make any or all of the following assumptions:

1. No Enduring Grand Objective will ever exist.

2. Genuine value and meaning correlate with enormous physical size and immense temporal duration or permanence. Little things have little worth!

3. Compared to the universe, individual human lives are indeed small and brief, and all human (and nonhuman) life in this world will eventually be extinguished.

Although 3. is true, 2. is not; and 1. is irrelevant to the question of the meaning and value of human existence.

Biological human life is indeed spatially tiny and temporally short when compared with the vastness of the spatiotemporal cosmos. Like Weinberg, Stephen Hawking also contrasted the immensity of the universe with "insignificant creatures like ourselves." [28] In the 1990 BBC television program *Master of the Universe*, Hawking said that

> We are such insignificant creatures on a minor planet of a very average star in the outer suburbs of one of a hundred thousand million galaxies. So it is difficult to believe in a God that should care about us or even notice our existence.[29]

Many individuals who reflect on our position in the universe seem to equate worth with size. Even the theologian John Hick proclaims that "a naturalistic conclusion" is strongly supported by the "sheer size" of the universe and "humanity's correlatively minute place within its spatial and temporal immensity."[30] Victor J. Stenger acquiesces: "So small is humankind" and "So vast is the universe," he laments "The insignificance of humanity is almost impossible for most humans to accept."[31] "Surely the universe does not care about human existence,"[32] Stenger bemoans. Frank Tipler, who locates ultimate meaning in the Omega Point, argues that we humans live "at an exceedingly early time" in the history of the universe, that "Most of life is in the future," and that "It is our relative insignificance in time, not space, which is the real challenge posed by modern cosmology for traditional religion."[33] Temporalistic versions of the insignificance argument focus on how little of the cosmic time line is occupied by living things.[34]

Cosmologists with this pessimistic mind-set often assume that attributing great worth to individual human beings is merely a human judgment, therefore untrustworthy; but note carefully that the judgment that we do *not* have great worth is also merely a human judgment! The real problem is that the connection these pessimists assume between size, duration, and significance is all very wrong, both cosmologically (spatiotemporally) and axiologically (valuationally). Cosmologically, Anthropic Cosmologists demonstrate, human life could never evolve in a universe very much smaller or younger than our own. Temporally, producing complex life requires billions of years for generations of supernovae to come and go and billions more for complex forms of life to evolve locally. Spatially, a stable and hospitable solar/planetary system requires vast separations between heavenly bodies if life-budding, life-building planets are to avoid being torn out of their orbits by the gravitational effects of passing stars or other planets.[35] Spatial versions of the insignificance argument center on how tiny our bodies are in relation to the vastness of space, temporal versions on how briefly we endure within the whole of time.

Axiologically, our species and our individual lives are not insignificant simply because of our relatively limited physical size and duration. Value and meaning in and for our lives do not correlate with or depend upon great magnitude and/or permanence. Our species has existed for only a fraction of the age of the universe. As adults we typically live only seventy years or so, weigh between one and two hundred pounds, are only about six feet tall, and are less than two feet wide. There are exceptions! By comparison with the totality of time and space, we are tiny and trivial indeed; but that is the wrong comparison, a childish comparison, uninformed by the study of value theory. Philosophers tend to be scientifically naive; but scientists doing cosmology tend to be philosophically naive!

iv. The Intrinsic Worth of Human Existence

Philosophers deeply ponder the question of whether anything has intrinsic worth, is valuable in and of itself; and after careful consideration they usually find immense intrinsic worth in human existence. Few philosophically astute value theorists would agree with pessimistic astrophysicists that our species-typical spatiotemporal limits are incompatible with immense intrinsic worth, for they reject the premise that value depends on immense size and duration. We don't have little worth just because we occupy very little spacetime!

If and when we are not blinded by cosmological magnitude, we can readily appreciate the great intrinsic worth of even small human infants, to say nothing of larger adults. When loving an infant, we realize that within broad limits smallness is no obstacle to great significance. Teilhard de Chardin was right (in part) in affirming that significance depends on the complexity of consciousness.[36] It also depends on the uniqueness of that consciousness, and on properties experienced and actualized by and within that consciousness. Philosophers who reflect deeply on the question usually conclude that nothing has much intrinsic worth, if any, apart from consciousness; but more than mere, pure, or complex consciousness as such is required for life to have meaning and value.

Physically we are tiny, and our lives are short compared to the cosmos as a whole. Yet, we complex, conscious, and unique human beings are immensely if not infinitely valuable or significant in and for ourselves. Our great inherent worth does not depend upon our size, our duration, or upon what we produce in the near or distant future. This is true also of all but the simplest animals, but that is another story, too long to tell at present. We are immensely complex conscious individuated creatures, despite our spatiotemporal limitations. We can even comprehend the Big Bang and our place within the universe! Human consciousness and the physiological conditions that support it are incomprehensibly complex. Nothing else in the world known to us matches the intricacy of the human brain. Intrinsic value depends heavily on the enormous complexity and activity of embodied consciousness; and all but the most unfortunate human individuals have it or eventually achieve it. Great intrinsic worth depends on complexity of individuated consciousness; it also requires just the right kinds of intricacy, activity, and contents. But intrinsic value is quite independent of vast physical size and endurance. Some size and endurance are essential for spatiotemporally embodied individuated consciousness, but not vastness.

Intrinsic value is complex. Complex individuated consciousness as well as just the right activities and intensional objects of consciousness are required for great intrinsic worth. Intrinsic human goodness or worth is a synthesis of unrepeatable, unique, individuated consciousness with many additional repeatable concretized universals that are dynamic, conceptual, emotional, affective, and volitional. Many philosophers and other thoughtful persons realize that brain-grounded individuated consciousness can be enriched positively in many repeatable ways. Positive consciousness-enrichers or enhancers include the pursuit and attainment of happiness, knowledge, moral and religious virtues, love, interpersonal intimacy, beauty in art and nature, adventure, creativity in every constructive domain of human interest, and fulfillment of our beneficial or positive capacities, needs, desires, interests, and purposes.

Our worth as conscious individuals does not consist merely in our being instrumental receptacles for the realization of these universal and repeatable abstractions, as many philosophers traditionally assumed. Individuated consciousness is not just a worthless or merely useful bucket into which intrinsically valuable concrete universals may be poured. Individuated consciousness has its own inherent worth; and our fullest intrinsic worth consists of a synthesis of our individuated consciousness with the consciousness-enriching universals just mentioned.

In the existing cosmos, with a bit of luck and many wise choices, consciousness-enriching activities and goods can be and are available to individual human beings in great abundance, despite our spatiotemporal finitude. Embellished by such repeatable good-making properties, complex individual embodied consciousness is immensely and inherently good. It is immensely significant, meaningful, and valuable in and of itself right here and now, even with all its obvious faults and limitations.[37] The universe has a magnificent point simply because we are here, no matter what comes later. For each of us, our being here is a valuable end in itself. Our worth does not depend entirely on our being means to ends beyond ourselves, grand or otherwise.

Enduring Grand Objectives are hypothetical goods that arrive later and last for eons of time if not forever. If our universe and our individual lives have no Enduring Grand Objective, can we or our universe be anything but pointless, given the inevitable eventual extinction of our individual lives and of humankind itself? According to pessimists like Weinberg and Hawking, no Enduring Grand Objective will ever arrive, and they despair.

Barrow and Tipler, by contrast, situate the whole point of human existence in their own peculiar Enduring Grand Objective, the Omega Point. Omega will arrive (supposedly) at the end of our (supposedly) closed universe; and it will live happily and omnisciently ever after (subjectively).

Barrow and Tipler are not alone in locating the meaning of human life in some enduring Grand Objective; but grandiose objectives may be conceived in many different ways. Freeman J. Dyson finds our existence to be ultimately meaningful only if and because the existence of intelligent life can be prolonged infinitely in an open universe through radical biological adaptations;[38] but this just turns the infinite prolongation of intelligent biological life into another Enduring Grand Objective, another free beer tomorrow that never comes.

For traditional Christianity, the endless survival of individuals after death in Heaven or the Kingdom of God, however conceived, is the Enduring Grand Objective. Most Christians, following St. Paul, assume that life here and now is completely meaningless, an utterly pointless pilgrimage to nowhere, unless there is a resurrection and a Heaven. Heavenly survival after death is the whole point of human existence, without which everything is in vain. Traditional Christianity assumed and taught that life in this world has value and meaning only as a necessary condition for or means to a Grand Existence Beyond this World.

Suppose, however, that no Omega Point will ever arrive; our descendants will never space travel to far distant solar systems and galaxies; the existence of intelligent biological life will not be infinitely prolonged; and we as individuals will not survive after death. Would human life really be worthless, pointless?

All Enduring Grand Objective theories mistakenly locate the meaning and value of human life only or primarily in some distant future Grand Objective that will last indefinitely; but what gives these Enduring Grand Objectives their significance? Nothing more than exactly the same conditions–perhaps intensified, amplified, and prolonged–that make active, individuated, and enriched consciousness so precious here and now! Grand Objectives themselves have value and meaning only to the extent that they epitomize things that are presently valuable and meaningful. If life now is a purely instrumental pilgrimage, life in some glorious future can be only an instrumental means to some even Grander Objective that never comes. If no inherent meaning and goodness now exist, no enduring and glorified future version of it will have any.

Future meaning and goodness make no sense unless meaning and goodness are now available. Even without a "then," intrinsic worth exists now. Individuated conscious life is for living–for itself; and its intrinsic worth lies in living it. The meaning and value of our existence do not depend on some future Enduring Grand Objective, even if there is one. The significance of either

Heaven or the Final Anthropic Principle is greatly exaggerated, even if something like it is true.

What makes the Omega Point so valuable according to Barrow and Tipler? Things like power, presence, and knowledge (omnipotence, omnipresence, and omniscience)! But we have these and many other excellences now, not in their "omni" form, but sufficiently to make conscious life in this world a great gift, an immense if not an infinite good in, of, to, and for itself. Why are other Enduring Grand Objectives like immensely prolonged intelligent, affective, volitional consciousness in this world or in the next valuable? Isn't it because they contain the very same good-making properties that make our lives here and now so precious–active individuated consciousness enriched by happiness, knowledge, adventure, and their pursuit, by moral and religious virtues of many descriptions, by love, friendship, and other manifestations of interpersonal intimacy, by beauty in nature, society, and the arts, by creativity in every constructive domain of human interest, and by fulfillment of our beneficial capacities, needs, desire, and purposes? Value and significance-making properties and predicates may not be available in the same degree here and now as they would be in those idealized Enduring Grand Objectives in which we continue to survive after death, but they can be and usually are quite sufficient to warrant cherishing our present lives for their own sakes–not just as a means, a pilgrimage, to something beyond themselves.

v. The Meaning and Value of Infinitely Prolonged Existence

Hell would be infinitely prolonged individuated consciousness devoid of all value-enhancing enrichments including hope, together with the everlasting presence of their bad-making opposites. Its meaninglessness would consist largely in knowing that good-making properties are not and will never be supported by Hell's broad environment. Infinite duration (immortality or resurrection) and infinite complexity or richness in properties as such fail to differentiate Heaven and Hell beyond this life from heaven and hell on earth. Quality of existence matters immensely here and hereafter, not just quantity.

Both quality and quantity (intensity and duration) of consciousness-enhancing properties are important; but individuated consciousness can be "enriched" numerically with bad-making properties like misery, authoritarian animosity to knowledge, cowardly adventure-avoidance, innumerable moral and religious vices like hatred, resentment, selfishness, interpersonal insensitivity, philistine revulsion to beauty and the arts, and fulfillment of welfare-destructive interests, desires, and purposes. Quantitative "enrichment" of complex individual consciousness with bad-making properties results in intrinsic disvalue, not intrinsic goodness. Since bad-making properties may in principle be indefinitely prolonged, neither property-richness nor temporal endurance suffice for intrinsic goodness and meaningful existence. Intrinsically good lives involve both qualitative and quantitative enrichment of conscious experiences and activities with good-making properties, not bad-making properties; and meaningful lives involve conceptually understanding that the intrinsic, extrinsic, and systemic goodness of our lives is supported by our broad environments. Kind as well as number and complexity of properties enter into the constitution of inherent worth and meaningful existence, but vast spatiality or temporality are irrelevant, as are far distant Grand Objectives.

Sheer endurance is really not as valuable as supposed by those who compare our brief lives with the duration of the universe, or who yearn for everlasting survival after death. Sheer immortality or infinite endurance should not be confused with infinite goodness and meaning. According to traditional western religion, immortality–infinite endurance–could co-exist with and consist of nothing but endless evil and senselessness. Hell is forever. Nothing could be worse or more meaningless than everlasting survival, infinite duration, immortality, in a traditional Christian Hell.

Hell could be infinitely prolonged and infinitely complex or rich in properties, but Hell nonetheless! How so? As John Stuart Mill maintained, many different qualities of feeling are called "pleasure," and many others are called "pain." The agreeable feelings we derive from reading our favorite authors are qualitatively different from the pleasures of music, dining, sex, or sadism. Likewise, the disagreeable feelings that we get from reading Schopenhauer and Nietzsche (I wanted to say Kant!) are quite distinct from the pains of grief, guilt, loneliness, boredom, emphysema, a bee sting, and/or a severe injury or burn.[39]

Consider the relation between pains and endless duration. Mill gave little attention to the topic of qualitatively distinct pains, and he wasted no energy worrying about Hell. But, for the fun of it, let us ask how many different kinds or qualities of pain might exist and consider how a concept of Hell might be constructed from such information. The issues are partly empirical and partly logical. Logically, an infinite number of qualitatively distinct kinds of pain might exist. Hell, says traditional Christianity, is a place of infinite pain in multiple respects. First, Hell could be infinitely rich in distinct qualities of pain; next, each of these could be infinitely intense; finally, all of these combined could endure forever, unrelieved. Working out all the details would give us a modern version of Dante's *Divine Comedy* and John Milton's *Paradise Lost*!

So conceived, would the existence of conscious persons enduring infinitely complex and intense pain endlessly in Hell have any positive meaning or intrinsic worth? Surely not! Total extinction would be preferable by far, and part of Hell's misery supposedly lies in the realization, the conceptualization, that extinction is impossible. Hell would be so devoid of hope that it allows for no hope for extinction! A brief moment in such a Hell would be unthinkably horrible, and an infinite duration of it would be infinitely bad, endlessly meaningless. Obviously, immense or endless endurance (immortality) does not necessarily correlate with positive worth and meaning!

Both Heaven and Hell are at least logically possible as parallel or co-existing universes, but whether or not they actually exist we are not likely to know for sure in this life–unless we are confident that all possible universes are actualized. Still, nothing could be better, more meaningful, than infinite duration or immortality in a traditional Heaven, where individuated conscious existence is supposedly as rich in good-making properties as it can bear–within a totally supportive environment. Perhaps individuated consciousness and its excellencies do not endure forever; but Heaven might be nice anyway, since more of a good thing is generally better than less. To those who say that we would become too bored to want to exist after a vast period of heavenly survival, the proper reply is that we would be happy to try it for a few million years just to find out! Love, kindness, joy, creativity, curiosity, learning, and growth seem to be inexhaustible forever.

To return to the here and now, positive enrichments that make presently existing individuated consciousness so enormously valuable to, for, and in itself also make it instrumentally valuable to other persons (and to God, as explained in the next chapter). Our intrinsic and systemic worth here and now, and our extrinsic helpfulness and usefulness to others, are precious gifts that we contribute ultimately to God, who remembers and cherishes our intrinsic, extrinsic, and systemic worth forever. Others can recognize our intrinsic worth, as we can theirs, and relate respectfully to us as mutual members of a kingdom of ends, a kingdom of God. Our worth to others consists in what we mean to them here and now and the impact that we have on the quality and duration of their lives; but it has nothing to do with spatiotemporal vastness, or with an ultimate Omega Point, or with the interminable survival of intelligent biological or cybernetic beings, or with any other Enduring Grand Objective.

Conscious human existence here and now can become intrinsically disvaluable, something to be avoided or eliminated for its own nasty sake. For example, all of a terminally ill person's wakeful moments may be filled with overwhelming and unrelievable suffering, despite medicine's best efforts to provide pain relief, and he or she may beg for a merciful death. Along with many others, I believe that moral duty requires the expeditious and active elimination of such intolerable and immense intrinsic disvalue, especially when death is requested by hopelessly ill persons crushed by unrelievable bodily pain and/or mental distress. With good pain management, fortunately, most suffering is relievable today; but not all. Active voluntary mercy killing is sometimes a moral duty, although not yet legal in most countries. *Intrinsic disvalue* for individuated consciousness is not mere privation or deficiency in kind or number of properties. It involves the presence of properties that are undesirable and worth avoiding or eliminating for their own nasty sakes–like the excruciating and unrelievable sufferings that maliciously tortured victims, or terminally ill patients, themselves judge to be too horrible to endure.

If something is intrinsically good, its value does not depend on its consequences, even if they, too, are desirable. Life for us here and now, properly enriched, is an immense intrinsic good; and its inherent meaning and goodness are completely independent of any and all Enduring Grand Objectives or spatio-temporal amplitudes. With or without resettlement of biological or computerized human beings on other planets, or a traditional Heaven, or an Anthropic Omega Point, human and animal life can, and usually do, have immense and intrinsic significance and worth in themselves here and now. The goodness or worth of enriched, conscious, individual existence is inherent, in itself, and not merely instrumental. As Ralph Waldo Emerson said, "My life is for itself and not for a spectacle."[40]

More of a good thing is generally better than less; but more goodness requires more time–for ourselves and/or for posterity. It would be nice to know that our descendants will happily inhabit this good and beautiful earth, rich in life-forms, for many generations. We really are capable of caring deeply about future generations and for non-human species; but do we actually care enough? We tend to be very short-sighted and to consume wantonly, wastefully, and conspicuously the natural resources that future generations, human and non-human, will need for worthwhile lives; and we overpopulate the earth with other people who bear the same imperfections. We pollute our supportive environment so much with chemical and nuclear poisons that the earth may be uninhabitable in a few generations by almost everything except cockroaches. Stephen Hawking now believes that the greenhouse effect resulting from human endeavors that spew excessive carbon dioxide into the atmosphere will make the earth uninhabitable in less than a thousand years; and he pushes space travel and colonizing other planets so that a few of us will survive.[41] The trouble is, only a very few can survive this way; most human beings, most of our descendants, will perish with the dying earth. Does a species that befouls its own nest while knowing better really deserve to survive? Does our own species deserve to survive, given our enormous and largely unrestrained propensity to make the earth uninhabitable for our own kind and for all other forms of life? If we are stupid, greedy, and shortsighted enough to contaminate this fabulously beautiful planet earth to the point of uninhabitability, *Homo sapiens* does not deserve to survive. We must yet prove that we do.

To summarize, the Final Anthropic Principle affirms that infinitely many worlds, including our own, exist for the purpose of bringing about the Omega Point or God. Life and intelligence will gradually spread throughout our cosmos and all others. Biological life will be replaced eventually by computerized, robotized, android intelligence. In the far distant future, all universes, ours included, will merge into the Omega Point, an Ultimate Supercomputer that at the end becomes omnipotent, omnipresent, omniscient, and Divine. The Omega Point includes the full actuality of all possible histories, all possible universes, rolled up into one. Barrow and Tipler make many dubious assumptions concerning the survival and proliferation of intelligence. Most seriously, they are mistaken about the meaning and value of the lives we now enjoy. They presume that our lives have no immediate intrinsic meaning or worth, and that our true significance will arrive only with the Omega Point–trillions of years after our extinction.

Enduring Grand Objectives are attractive only because they amplify and prolong all the good things that make life here and now meaningful and worthwhile. Because our lives here and now are ends in themselves and not mere means to ends beyond themselves, their true significance is here and now; and the vastness of the universe is irrelevant. The value of life is in living and enriching it–intrinsically, extrinsically, and systemically. The meaning or worth of our existence depends in no way on some remote Enduring Grand Objective, even if one will eventually come to be; but meaning is, or would be, enhanced significantly if and when we know that our worth is supported by our ultimate metaphysical environment, God, in ways explained in the following chapters.

Ten

CONCEPTS OF GOD'S NATURE AND EXISTENCE

Theism is belief in God. Theistic Cosmology says that God caused the Big Bang, that God created the world. To understand and evaluate Theistic Cosmology, many questions must be considered: What does "God" mean? What does it mean to say that God caused the Big Bang? What evidences support Theistic Cosmology? Can Theism be defended against profound objections to it? Does order and design in the universe as disclosed by Anthropic Astrophysics indicate that God exists. Does cosmic contingency or dependence provide strong evidence for the reality of God?

Before examining the evidence, we must first consider the concept of God. Debating the existence or non-existence of flying warthogs would be inane until we first know what "flying warthogs" means. (They are the US Air Force's A10 fighter-bombers.) Debating the existence or non-existence of God is equally foolish unless we first know what "God" means. Most theists are also atheists; they are convinced that God, in many meanings of the term, does not exist. Early Christians, for example, were branded as atheists and persecuted by the Romans because they denied the existence of the finite, fickle, immoral, anthropomorphic gods of the Greco-Roman pantheon.

1. Two Concepts of God's Nature: Classical and Process Theology

"God" has innumerable meanings. We cannot discuss them all, but we will consider at least two very different but cosmologically relevant concepts of the nature of God–Classical Supernaturalism and a modified version of Process Theism. These concepts presently contend for acceptance by thoughtful theistic believers, and we must choose between them as intelligently as possible. All Theists believe that God exists, but not all Theists believe in the same God. Who is God? What is God like? How is God related to the world? Classical Theism offers one answer and Process Theism another. Prior to the twentieth century, Classical Theism was the dominant view of God among professional theologians in traditional Catholic, Protestant, Jewish, and Islamic monotheism.

Process Theism was anticipated by earlier thinkers, but it was developed primarily in the twentieth century by Alfred North Whitehead and Charles Hartshorne. Henri Bergson also made important contributions. Numerous philosophers and theologians such as Robert L. Calhoun, Nels F. S. Ferré, John B. Cobb, Jr., Daniel Day Williams, Schubert M. Ogden, David R. Griffin, John A. T. Robinson, Robert C. Neville, Marjorie H. Suchocki, Lewis S. Ford, Bowman L. Clarke, Sally McFague, Holmes Rolston, III, Arthur Peacocke, John Polkinghorn, Ian Barbour, Rem B. Edwards, and many others were deeply influenced by Whitehead and Hartshorne and made, or are still making, their own significant contributions to this theological perspective, even when they disagree with or about some of its features. Many contemporary religious thinkers who accept the essential point that God is in process do not consider themselves to be Process Theologians because they reject certain emphases of mainstream process thinkers like Alfred North Whitehead, Charles Hartshorne, John B. Cobb, Jr., and David R. Griffin. In this book, "Process Theology" is broadly understood to include all who ascribe to God a temporalistic or processing nature, and much room is allowed for honest disagreement on details among Process Theologians themselves. The version of Process Theism defended here is significantly modified to allow for God's ability to create our universe out of nothing, God's voluntary self-limitation in creating co-creative creatures, God's ability to experience passively the present self-creativity of temporal occasions, and God ability to influence events as they occur in time.

Process Theology is a rational or philosophical theology. As such, it aspires to satisfy credible criteria for a plausible rational theology. (1) It must be logically consistent, a chronic shortcoming of Classical Theology, say Process Theologians; and it must satisfy other criteria of rational explanatory adequacy like coherence, simplicity, comprehensiveness, clarity, and conformity with experience. The concept of "experience" is broadly construed; it includes sensory experience as well as religious experience, aesthetic experience, and moral, logical, religious, and mathematical intuitive experience.

(2) It must explicate the ideal of perfect excellence or supreme worshipfulness expressed in St. Anselm's concept of "that being than whom none greater can be conceived." "Divine perfection" is a valuational or axiological concept as well as a metaphysical notion. Developing a concept of God as a perfect being is not an empirical project. Rational theology allows plenty of room for honest intellectual doubt, disagreement, and growth; but a start must be made somewhere. Until thoughtful religious people reach agreement about what is ultimately admirable, they must simply agree to disagree about some, but not all, attributes of divine perfection. Faithfulness to the originators of Process Theology like Whitehead and Hartshorne must often yield to our most profound sensitivities about supreme worshipfulness.

(3) A rational theology must also be compatible with, that is, it must not be falsified by, the structure and contents of the world that natural science discloses. Science does not dictate rational theology, but it rules out many familiar religious beliefs, especially cosmological convictions, as unviable and untenable. For example, rational persons cannot accept a literal six days of creation in the face of astronomy and paleontology, cannot reject evolution in the face of biology, and cannot affirm rigid and universal determinism in the face of quantum physics. A rational theology cannot affirm that Adam and Eve once existed in an idyllic Garden of Eden, or that anything of any theological importance depends upon their having so existed. It cannot affirm that death originated as a consequence of human misbehavior, since organisms were dying for eons of

time before human beings evolved. Being compatible with natural science is not the same as being proved by natural science. Philosophical theology aspires to proof by reason or philosophy in a very broad sense, but not by empirical or sensory inquiry alone.

(4) A rational theology must also provide a plausible and coherent account of the immediate and the ultimate meaning and value of human life, indeed of all life. This account also must be firmly grounded in and compatible with scientific knowledge and critical rational reflection, as well as with our most profound religious, moral, and aesthetic sensitivities.

Classical and Process Theologians usually agree that they are trying their best to conceive of divine perfection and to answer the question of the meaning of human existence. Let us begin with divine perfection. What would God have to be like to be supremely worthy of human worship, love, service, and devotion? St. Anselm characterized God as "That Being than whom none greater can be conceived." By "greater" he meant "better." This understanding of divine perfection is at the heart of the Ontological Argument, through which Anselm hoped to persuade all rational persons of God's existence. Anselm argued that if we truly understand the meaning of the concept of God, we cannot deny God's existence without contradicting ourselves, that is, without affirming that a Divine Being who could not not exist might nevertheless not exist.[1] Both Classical and Process Theologians try to conceive of a God who exemplifies all desirable attributes or perfections, including necessary existence, and who is truly worthy of being loved with all our hearts, souls, minds and strength. To show that this can be done, we must do it; and doing it shows that it can be done!

Classical Theism began with Philo, a Jewish theologian living in Alexandria, Egypt in the first century A.D. The second through fifth-century Christian church fathers were profoundly influenced by Philo's project of combining the Greek philosophers' ideal of divine perfection with that of Biblical authors. From the time of the Eleatics, Greek philosophers conceived of God as simple, undifferentiated, unitary, passionless, and timeless Being. The attribute of rationality was often thrown into the bargain. Biblical writers, by contrast, conceived of God as both transcendent and immanent, as unitary but complex, as having real feelings and volitions as well as rationality, knowledge, or wisdom, as acting temporally and historically upon and within the world and its inhabitants, and as responding in time to events within nature and human history. Classical Theists fused Hellenic with Hebraic ideals of Divine perfection, thereby producing an unstable theological synthesis riddled with paradox and unintelligibility.

The Classical synthesis of incompatible ideals of Divine perfection lasted for almost two thousand years and is still going strong. Classical Theism is not identical with popular Judaism or Christianity, which are usually much closer to Biblical religion than to "big name" Classical Theologians like Augustine, Anselm, Aquinas, Moses Maimonides, Martin Luther, and John Calvin.[2]

The Classical understanding of God dominated early Christian and Medieval theology, and it still prevails in Roman Catholicism. It has been almost as influential among Protestants. To illustrate, consider the following contrast between Process Theology and Classical Theism as it was expressed by reformed Protestants in the highly influential "Westminster Confession" of 1647.[3] Though it is rational rather than revealed theology, Process Theism is actually much closer in many respects to Biblical and popular religion than is Classical Theism. "The Westminster Confession of Faith" identified the following central metaphysical attributes of God.

> There is but one only living and true God, who is infinite in being and perfection, a most pure spirit, invisible, without body, parts, or passions, immutable, immense, eternal,...almighty....[4]

Let us reflect upon this lists of divine attributes, though not exactly in this order.

A. Infinite in Being and Perfection

Plato's Principle of Plenitude dominated Classical Theology's ideal of divine perfection, but the fullness was in God, not in creation, as previously noted. God's perfection consists in infinite being, in knowing, but not necessarily in creating or actualizing all possibilities, an infinite number of them. In himself, God is pure being, pure actuality, an *actus purus*, in whom there is no becoming, and no unactualized potentialities. Just why this did not translate into the belief that God has actually created all possible worlds is unclear. Plenitude of Divine Being merged with plenitude of creation would imply that everything that God could possibly create was actually created from eternity, including all possible universes. Classical Theologians believed only that every mutually compatible niche in the great chain of being was filled in God's one universe. God's fullness and self-sufficiency in himself, not in creation, makes God perfect and supremely worthy of worship, service, and devotion, according to Classical Theology.

For many good reasons, Process Theology does not accept the Principle of Plenitude, either in God's Being or in God's Creating. The process understanding of divine perfection includes both Divine Being and Divine Becoming. It recognizes that choices must be made between things that could be created separately but not together, that infinity cannot be exhausted, used up, or fully actualized either timelessly all at once, or successively, and that divine creativity is interminable. It acknowledges that some possible worlds are too horrible, trivial, or boring to be created by a loving, morally upstanding, and aesthetically sensitive God. All Divine attributes in their integrated wholeness and togetherness are integral to God's perfection.

Process Theology completely rethinks all the old problems of theology in light of the hypothesis that aspects of both process or change and permanence or constancy are in God. Divinity includes both an unchanging timeless dimension, God's Primordial Nature, and a changing temporal dimension, God's Consequent Nature. In Classical Theology, God is only being; but in Process Theology, God is both being and becoming.

God's everlasting and changeless Primordial Nature includes God's necessary existence and God's eternal vision of possibilities for creation. Plato's ideal forms are relocated in the mind of God. Whitehead called them "eternal objects." Possibilities as such are empty systemic abstractions that have no definiteness and significance to and for themselves and are deficient in definiteness even for others. Eternal objects are not the ultimate, independently existing, causally productive, concrete realities that Plato thought them to be. They lack the definiteness, concreteness, and power of existing actualities in physical space and time; and they are devoid of intrinsic subjectivity–the immediacy of self-awareness, self-creation, and self-enjoyment possessed by concretely existing individual centers of conscious experience, activity, and valuation like God, ourselves, and most if not all non-human animals and living things. Process thinkers disagree about the full extent of God's knowledge of possibilities; but we need not settle that question here.

Whitehead probably included little more than God's envisionment of possibilities, eternal objects, and their relevance to possible worlds, in God's Primordial Nature; but for many good reasons other process thinkers, Hartshorne and Cobb especially, have considerably enriched the notion. The Primordial Nature also includes God's necessary existence, necessary creativity, and God's enduring and essential general capacities for knowing everything that actually exists and that might possibly exist (omniscience), for loving all concrete actualities (omniloving), and for creatively influencing (omnipotence) and being influenced by (omnipresence and omnisensitivity) every actual entity. God's unchanging and ever dependable love, compassion, and all-around moral virtue or righteousness belong to the Primordial Nature. The abstract essence of God, the Primordial Nature, is deficient in actuality; but the fullness of God includes both a Primordial and a Consequent nature. The two are separable in thought, but not in reality.

The Primordial Nature of God transcends every particular cosmic epoch or created universe while existing necessarily in relation to all particular epochs and universes and what transpires within them. Critics of Process Theology like Mark W. Worthing[5] who regard the God of Process Theology as completely immanent in our finite cosmos totally ignore the everlasting, necessary, and transcendent Primordial Nature of God. Without a Primordial Nature, a contingent and purely immanent God would die either the heat death or the Big Crunch death of our universe;[6] but Process Theology affirms that God's Primordial essence transcends and endures before, through, and after all created worlds. So does God's Consequent Nature as such if God has been infinitely creative in innumerable worlds.

In the Consequent Nature, God's transcendent and everlasting abstract capacities are concretized in relation to actual worlds. The Consequent Nature is God's actual creation of and interaction with individuals in space, time, and history. God comes to know, love, influence, and be influenced by all actual entities *as they concretely exist and become* within spacetime. God's experiences of successive events in time are themselves successive and thus temporal.

God's experience of value is enriched by every concrete value realized in all actual worlds; all values created by and within all creatures are taken into God's Consequent Nature and saved there forever. God is the supreme valuer of every intrinsic value actualized within every world, including our own. God experiences all values (and disvalues) realized by concretely existing spatio-temporal individuals; and after they have perished to themselves, God remembers and profoundly cherishes (or deplores) them forever. Existing events possess their own present moment of relative independence and immediacy of self-enjoyment and self-creativity. As temporal occasions perish to themselves, they achieve permanent being or "objective immortality" in God's infallible memory. God's Consequent Nature continuously assimilates and treasures the ongoing order and concreteness of all spatiotemporal actualities.

God is constantly being created, says Process Theology, but not the necessary *existence* or the abstract *essence* of God, both of which belong to the Primordial Nature. The Consequent Nature consists of the full *actuality* of God's decisions about, experiences of, interactions with, and responses to concretely existing creatures as they are constantly being created. God exists necessarily, but the full actuality of Divinity is contingent, depending in part upon God's own freely creative acts and in part upon free decisions made by God's creatures. God affects the world, but the world also affects God, for better or for worse. God is affected by values and disvalues realized in all of creation.

An unsolved problem remains after we realize that our lives are intrinsic ends and that their significance does not depend upon their contributions to some far distant future Grand Enduring Objective. The problem is that all temporal goods, including those that are intrinsic, are transient, fleeting, and ephemeral. We ourselves, and the very best moments of our lives, perish in time. Through memory, we can recover traces of our most precious moments; but eventually we die, and our memories die with us. Traces of the concrete values realized in our lives may remain in the memories and even in the genes of others; but eventually they also die with their memories and their genetic endowment. Some day the human race will perish–after thousands of years perhaps if we are lucky and are good stewards of the earth, which we do not seem to be. Will no trace of our worth ultimately remain? Bertrand Russell thought not, and the idea filled him with profound pessimism. In "A Free Man's Worship," he wrote:

> All the labor of the ages, all the devotion, all the inspiration, all the noonday brightness of human genius is destined to extinction in the vast death of the solar system, and that the whole temple of Man's achievement must inevitably be buried beneath the debris of a universe in ruins–all these things, if not quite beyond dispute, are yet so nearly certain that no philosophy which rejects them can hope to stand.[7]

Process Theology offers an attractive solution to the problem of the transience of all created goods–the Doctrine of Contributionism. All created values are ultimately contributed to God, who remembers and cherishes them everlastingly. All created events and values become objectively immortal in God.[8] Some, but not all, Process Theologians also affirm subjective immortality–the survival after death of individual subjects in the alternate spacetime system of Heaven.

Many theologians assume otherwise, but a traditional Christian Heaven does not solve the problem of the transience of created values. As my friend Tom Dicken indicates, "Immortality offers a continuing *creation* of value, not a conservation of value."[9] We can make little or no sense of an utterly spaceless and timeless Heaven where everyone is disembodied and nothing ever happens; but if spatiotemporal events transpire in Heaven, they too perish to themselves.

Time, whether Heavenly or worldly, is indeed a perpetual creation and rebirth of events; but it is also a "perpetual perishing," as John Locke put it. Unless the concrete values realized in both worldly and Heavenly events are known fully and preserved forever by God, ultimately they are lost forever; but in God's Consequent Nature there is no ultimate loss. God gives the intrinsic worth of all creatures objective immortality and intrinsically valuates them without end. God is the supreme intrinsic valuer of all intrinsic, extrinsic, and systemic values in all actual worlds.

The objective immortality of all creatures and created values in God are not just additional remote Enduring Grand Objectives. Enduring Grand Objectives are ends in themselves and have intrinsic worth, supposedly; but the means to them do not; and we are the intrinsically worthless means! But, says Process Theology, God values creatures like us as ends, not merely as means. According to the Final Anthropic Principle, the Omega Point has intrinsic worth; and we are significant only as means to that terminal condition. In traditional Christian theology, this life is but a miserable pilgrimage to what is truly worthwhile–pie in the sky by and by. Enduring Grand Objective theories degrade our lives here and now into extrinsic or systemic goods; but Process Theology affirms that the *intrinsic* worth of our lives here and now is ultimately contributed to God, who cherishes us and every living creature forever.

God's love is not limited to humanity; it extends to every experiencing subject that ever exists. All animals, not just human ones, and not just members of contemporary species, are included. Panexperientialism or panpsychism, to which some process thinkers subscribe, says that all existing individuals including sub-atomic particles (but not aggregates like rocks) are experiencing subjects having some degree of intrinsic worth. We do not have to decide that issue here. The essential thing is that intrinsic values actualized in all presently living creatures, in our hominid ancestors, and in every extinct individual and species that ever lived, are not lost. They exist forever in God. Past values are lost to themselves and to our present selves for the most part, but no achieved goodness is ever lost to God.

God does not timelessly actualize all possibilities, that is, all possible individuals, their qualities, and relations, according to Process Theology. God alone could know it, but an infinite number of possibilities may have been actualized in an infinite number of universes within God's supercosmic past or present if God is infinitely creative. Superspacetime may be an attribute of God! Yet, infinite possibilities for further creativity always remain for the supercosmic future of an infinitely creative God. Neither God nor man can use up infinity, not even in an infinite amount of time.

B. A Most Pure Spirit, Invisible, Without Body

Erroneously, Classical Theologians often identified the biblical notion of "spirit" with the Platonic/Neoplatonic/Cartesian notion of an immaterial and potentially disembodied mind or soul. According to Plato, immaterial, nonspatial human minds are temporary prisoners in their spatially extended bodies. Platonic immortality encompassed both existence before birth and survival after death for disembodied, nonspatial, immaterial souls.

By contrast, in the Biblical tradition, body and spirit are inseparably unified; and survival after death takes the form of the resurrection of a dramatically transformed body, relocated ultimately in an alternate spacetime system, but never completely non-spatial or immaterial. In the earliest centuries of the Christian era, Platonic mind/matter dualism spilled over into Classical Theology. God himself was understood to be incorporeal or "without body" as the "Westminster Confession" put it. Since incorporeal things are imperceptible, God's invisibility indicated incorporeality to Classical Theologians.

In Biblical, creedal, and popular religion, God, the invisible spirit, is embodied, at least metaphorically. God has hands, feet, a face, right and left sides, a backside, and such. God is pictured as a large, humanoid, white or tawny skinned, blue-eyed, gray-bearded male who sits on a white throne, wears a jeweled crown, and has a string of lieutenants on each hand, the right hand (where Jesus sits) being most favored.

Neither Classical Theology nor Process Theology takes this physical humanoid imagery literally, however, and even this has a Biblical basis. The Second Commandment in the *Old Testament* prohibits making graven images or likenesses for religious purposes of anything (including humankind) that is in heaven above, the earth beneath, or the water under the earth (*Exodus* 20:4). Implicit in the Second Commandment is a profound metaphysical insight: God's form, whether physical or not, cannot be represented by any familiar finite physical forms. God is infinite being, not physically finite and humanoid being. Judaism and Islam allow no representative art as aids to worship and thus generally take the Second Commandment much more seriously than Christianity, except for the early Christian iconoclasts.

For Classical Theologians, human beings are made in the spiritual, not the physical, image of God because an incorporeal being has no physical image. Process Theologians repudiate mind/matter dualism and insist that the mental and spiritual aspects of all actual entities, including God, are always embodied. God does not literally have humanoid hands, feet, eyes, face, and so forth; but, contrary to Classical Theology, Process Theologians think that God has a body and is not totally incorporeal. God's body is the universe, or at least *some* world or worlds, perhaps even infinite Superspacetime itself. God may be embodied in infinitely many actual worlds for all we know. God's body or bodies belong to the Consequent Nature; but God's Primordial Nature transcends and persists through all embodiments in all spatiotemporal phases of all the worlds God's "hands" have made.

God's spirituality and mentality are related to the world as our own spiritual and mental dimensions are related to our bodies. If we really understood that, the analogy would be much more illuminating! In our universe, organic wholes have both an internal and an external unity plus properties that are influenced by but not reducible to the sums of their component parts. Protons are influenced by but not reducible to their constituent quarks; and the same is true of atoms in relation to their protons, neutrons, and electrons, of molecules in relation to their component atoms, of living cells in relation to their molecules, of brain-consciousness as related to brain cells, and of Divine consciousness as related to the whole of creation. In some mysterious way, our cells, organs, and bodily processes affect our conscious experiences and activities. Our conscious experiences and activities as embodied in our brains also affect our other bodily processes, organs, and cells without violating any natural laws.

We are to God as our cells are to us, and God is to the world as our consciousness is to our bodies, but with important differences as well as similarities. The well-or-ill-being of our cells, especially our brain cells and brain waves, affect and are affected by our consciousness;[10] and God affects and is affected by the well or ill being of our conscious experiences and activities. The important differences are: (1) human consciousness has only a limited sensitivity to bodily events, but God is completely attuned and responsive to all worldly events, and (2) we are largely unaware of events external to ourselves, but everything is internal to God, who misses nothing. Both we and God are embodied; and just as our stream of consciousness, the dominant society of events in our bodies, can affect the rest of our bodies without violating any natural laws, so can God affect the world. No actual beings are purely spiritual, invisible, incorporeal, and disembodied. God is literally the all-inclusive spatiotemporal reality in whom we live and move and have our being–and our becoming.

Religiously, God's ability to affect the world, to influence and communicate with us creatures without overwhelming our freedom, is just as important as God's ability to be affected by the world. Just *how*, in technical metaphysical terms, this is possible may be and is debated extensively;[11] but *that* it is so is indispensable to our concept of that being than whom none greater can be conceived. The details of just how this is possible need not be resolved here.

C. Immense

To have a body is to be spatially extended. As Descartes noted, all bodies are extended; this is what being a body or being physical means. Spatiality is the defining characteristic of embodiment. Paradoxically, despite its claim that God is without body, Classical Theology affirmed that God is immense or omnipresent. These words intimate spatiality in a Being who supposedly exemplifies no spatiality or corporeality at all. "The Westminster Confession" affirms the immensity of God based on *I. Kings* 8:27, which says that Heaven and the highest Heaven cannot hold God, and on *Jeremiah* 23:23-24, which says that God is not far off but fills Heaven and earth. Other Biblical passages also affirm the omnipresence of God. *Psalms* 139:7-10 presents God as an inescapable presence who cannot be evaded in Heaven, Hell, or the uttermost parts of the sea; and St. Paul affirmed, according to *Acts* 17:28, that God is the being in whom we live and move and have our being.

Classical Theology embraced the paradox of God's spatiality with one breath (immense, omnipresent) and denied it with the next (pure spirit, incorporeal); but how can a being who is nowhere be everywhere? How can a being who is everywhere be nowhere? With no evasiveness, Process Theology attributes both spatiality and temporality to God. God's Consequent Nature is that most inclusive spatiotemporal reality within which or whom we live and move and have our becoming. Localized moral agents like ourselves are only finite parts of our local spacetime system; but God is the all-inclusive ultimate reality.

A temporally ordered looseness of fit obtains between God, the whole of all inclusive Superspacetime, and our spacetime. This looseness allows room for creaturely freedom and creativity. Individual events within the whole of God's reality, for example, those composing human streams of consciousness, enjoy a fleeting moment of relative independence, originality, and creative self-synthesis (to which God is passively sensitive) before they perish to themselves and gain objective immortality within God. This slight departure from process orthodoxy will be better explained in what follows.

God always and necessarily has a body, some body, because God is always and necessarily a creative, loving, social being who creates others to love.

If God is embodied, can we see God? Yes, in a sense; but we cannot see all of God's body, the whole of our spacetime, or the whole of God's Superspacetime; and we cannot see the transcendent Divine Primordial Nature or the privacy of God's own consciousness. But every time we look into another persons eyes, or behold the good earth, or gaze at the wondrous starry Heavens above, we see some of the components of God's immanent Consequent Nature. All things are divine, even the mundane, though most of us do not realize this. Appropriately qualified, as in the preceding discussion, God is all in all.

D. Without...Parts

Classical Theology affirms that in himself God is pure, undifferentiated unity and simplicity, or "without...parts," as the "Westminster Confession" put it. In our thinking about God there is complexity, says the classical theory; but no counterpart for this complexity exists in God himself. We think of God as having a plurality of desirable attributes or predicates; and we have many names for God's diverse parts–omniscience, omnipresence, omnipotence, omnicompassion, and so on; but in God's own nature, these multiple attributes are so thoroughly integrated that the many are simply one. We think that God performs many acts, knows many things, is present with many individuals, causes many events, loves and is compassionate toward many creatures; but in God's own reality, all this apparent diversity exists as pure and undifferentiated Parmenidean unity and simplicity of Being. God is the simplest of all beings, Being Itself, though we think of God as the most complex.

By contrast, Process Theology rejects the classical unbridgeable gap between the way we think about God and the way God really is. As the simplest possible being, God could only be that being than whom none poorer in properties can be conceived. Process Theology conceives of God as the supreme, self-conscious, unitary individual who is richest, not poorest, in good-making properties, and who is capable of endless further enrichment through infinite creativity. No other being surpasses God in complexity or any other desirable attribute, but an endlessly creative God is constantly self-surpassing. God's experience of value is enriched continuously through ongoing interactions with created worlds. Instead of being without parts, an infinite number of real parts exist, not just in our thoughts, but in God's own concrete actuality. In the final chapter of this book, we will return to the topic of God's simplicity and complexity.

E. Without...Passions

The most dramatic difference is that Classical Theology refuses to attribute any feelings to God; whereas Process Theology sees feelings as the very essence of God's love, mercy, and compassion. Most Greek philosophers depreciated the affective, appetitive parts of the human soul; feelings and desires were deemed greatly inferior to rationality, thus altogether unworthy of Divinity. Aristotle's Unmoved Mover was "impassible," meaning "without any feelings whatsoever–totally devoid of all affections and desires." His God is emotionally as well as physically unmoved and thinks only about thinking, never about our world and its denizens. Neither omniscient nor omnicompassionate, Aristotle's Unmoved Mover knows and values only logic, thinks without feeling about thinking–nothing else, and does not know or care about us or anything in our world.

Regrettably, the earliest Christian theologians accepted Greek philosophical prejudices against feelings. The lavish emotional, affective, and appetitive language applied to God in Biblical religion was then regularly dismissed as woefully inadequate metaphorical speech; and all spatial and temporal imagery was branded as totally misleading and impious metaphor. In Classical Theology, human beings were made in the rational, not the affective, emotional, or physical image of God, who is literally impassive, empty of all feelings whatsoever, thus not literally loving, merciful, or compassionate.

The God of Classical Theology is not literally or physically male because he has no body at all; yet, psychologically and behaviorally, this God is stereotypically masculine. Like big boys who don't cry, the classical God has absolutely no feelings, emotions, or desires whatsoever about anything. Reflecting a tradition that dates back at least to St. Athanasius, St. Anselm emphatically denied that the Divine part of Jesus suffered on the cross; only his human nature suffered because "The Divine nature is beyond doubt impassible."[12] In Classical Theology, God entirely lacks not just undesirable feelings but all feelings whatsoever. St. Thomas Aquinas said that God "loves without passion" and that "Mercy is especially to be attributed to God, provided it be considered in its effect, but not as an affection of passion. To sorrow, therefore, over the misery of others does not belong to God."[13]

Anselm, Aquinas, and other Classical Theologians held that we experience God as if he has feelings, but in God himself no feelings exist. This appalling compromise came about when Classical Theologians, led by Philo in the first century A.D., combined two fundamentally incompatible ideals of divine perfection–that of the Greek philosophers, and that of Biblical religion. When forced to choose which religious language to take literally and which to construe metaphorically, the Greek philosophers always won. Their outlook was literal truth, so most of the language of the *Bible* and of ordinary believers was dismissed as impious and misleading metaphor. Nowhere is the conflict between these two incompatible ideals of divine perfection more obvious than in St. Anselm's description of God as compassionate in terms of our experience, but not in His own being and experience.

> Truly thou art so in terms of our experience, but thou art not so in terms of thine own. For, when thou beholdest us in our wretchedness, we experi-

> ence the effect of compassion, but thou dost not experience the feeling. Therefore, thou art both compassionate, because thou dost save the wretched, and spare those who sin against thee; and not compassionate because thou art affected by no sympathy for wretchedness.[14]

Thus, in Classical Theology, God is literally impassive, only metaphorically and inaccurately compassionate; and Greek ideals of Divine perfection always trump Biblical values and ideals.

Ministers dare not talk like that to their congregations! They could not get away with it! Ordinary believers are led, or misled, to believe that God is in himself all the good things that we experience him to be. The "Westminster Confession" says that God is "most loving, gracious, merciful, long-suffering...,"[15] then takes it all away with the qualification "without passions." Classical Theologians drive an infinite wedge between the reality and the appearance of God. God is really not loving but appears to be, not compassionate but appears to be, not merciful but appears to be. Divine appearance corresponds in no way with divine reality!

Process Theologians reject such duplicity and hold that God is quite literally all of these immensely good things. They think that the Greeks were totally wrong in devaluing all feelings and affections and in regarding them as greatly inferior to reason if not completely worthless. Yes, love, mercy, and compassion always involve cognitive elements; but without their affective components, their intrinsic significance is lost. Yes, many desires and emotions are bad and unworthy of God; but, Process Theology insists, many feelings and desires are exceptionally good and very worthy of Divinity. The good ones belong to God.

Impassivity, total incapacitation for every feeling, is an imperfection, not a perfection. Feelings are stereotypically feminine not masculine attributes, but having just the right feelings and desires is one of the most majestic features of both human and Divine persons, male or female. God literally suffers with us in our sufferings and rejoices with us in our joys. God literally preserves and cherishes forever the goodness of our unique lives, activities, experiences, and values. The created goodness of the world is ultimately contributed to God, who does *not* respond to it "without passion."

The authors of the "Westminster Confession" asserted that the chief end of "man" is to "Glorify God and enjoy him forever,"[16] but the classical notion of divine impassivity implies that nothing in the universe contributes anything whatsoever to God. What then is the point of loving and glorifying God, asks Charles Hartshorne, if God's experience and happiness are not enriched one tiny bit by our glorification, love, adoration, and devotion? *We* benefit from religious devotion, we are told; but we and our world mean nothing to God.[17]

In Classical Theology, since there is no passivity or receptivity in God, nothing that happens within the world ever affects God. God is pure act, pure causation; the world is totally passive, pure effect. God is in no way passive, sensitive, or receptive in relation to the world; in no way is the world active in relation to God. In no conceivable way is God the effect of anything that happens in our spacetime. The "Westminster Confession" affirms,

> God hath all life, glory, goodness, blessedness, in and of himself; and is alone in and unto himself all-sufficient, not standing in need of any creatures which he hath made, nor deriving any glory from them, but only manifesting his own glory in, by, unto, and upon them: he is the alone fountain of all being, of whom, through whom, and to whom, are all things....[18]

Process Theology replaces divine impassivity with the Doctrine of Contributionism, according to which all created value is ultimately contributed to God, who is enriched by it, feels deeply about it, and preserves and cherishes it forever. God's happiness is enhanced by our happiness, and God's sadness is increased by our sadness and woe. All the good we create is ultimately created for God; all the evil we inflict on any sentient creature is also ultimately inflicted on God. God literally rejoices with all who rejoice, and suffers with all who suffer.

F. Immutable...Eternal

Negatively, "eternity" means "having no beginning or end." The biblical God exists from everlasting to everlasting. When Classical Theologians accommodated biblical everlastingness to Greek ideas of timelessness, they redefined "eternity" positively (without a negation) to mean "all time all at once." In Classical Theology, eternity is the simultaneity in God of the past, present, and future of all creation–a *totum simul*. For God, everything happens timelessly, all at once. God comprehends all change and mutability in a changeless and immutable way. In no conceivable respect does God change. God is so "perfect" that any change would be for the worse, as Plato and Aristotle, but not the *Bible*, decreed.

In Process Theology, God has both an immutable Primordial Nature and a changing or developing Consequent Nature. Change with respect to God's necessary existence or desirable ethical attributes like love, compassion, and moral goodness would indeed be for the worse and would make God unworthy of supreme devotion, service, and adoration. But many kinds of change and many feelings are extremely desirable and valuable, despite what the Greek philosophers believed. Not every change is for the worse; not every change alters fundamental goodness. Some types of change are undesirable but others are desirable. God's experiences and choices change as the Divine Individual constantly creates and interacts with living creatures in spacetime; and this type of change is not undesirable at all. In fact, it is just what we would expect of a perfect being if our religious ideals and intuitions have not been perverted by Greek philosophers!

Introducing temporality into the concept of God requires a small modification of the classical notion of divine omniscience. Classically, God knows absolutely everything, the past, present, and future of the entire universe, changelessly, immutably, simultaneously, and infallibly. In Process Theology, God infallibly knows the past and present of any actual universe, plus all general tendencies toward the future. But God can not know future free decisions that have not yet been made because they simply are not there to be known.

Both theologies agree that God knows everything that is there to be known. Unlike Classical Theologians, Process Theologians deny that future free decisions already exist somehow to be known before they are made. God may (or may not) know all possible decisions as possibilities, depending on just how detailed a knowledge of possibilities can be; but God learns which free and creative decisions the creatures actually make only when they are made, not timelessly in advance. God cannot know things as actual until they become actual. God knows all things according to their appropriate modes of being–actualities as actualities, and possibilities as possibilities. All of history–natural, human, and Divine–is a partly unpredictable adventure in creativity for both God and God's creatures.

G. Almighty

Classical Theology followed Plato's suggestion that God could not be changed or affected either by himself or by anything other than himself, and it regarded causal relations between God and the world as completely one-sided. God is almighty or omnipotent, the sole originative causal agent in all of reality, who determines everything. Indeed, in thinking that anything might exist, God thereby creates its existence; so all possibilities are actual if God knows all possibilities. Through programmed or predestined chains of secondary or worldly causation as we experience them, God ultimately causes everything that happens within the universe; but nothing that occurs within the world affects God or brings about effects within God.

Divine omnipotence, understood classically as ultimate and total causality, implies that no freedom or originative causality exists within the universe or even in God, if all Divine decisions follow inevitably from the Divine Nature. Clearly, creatures have no free will and are not co-creators with God.

Christendom accepted predestination with relatively little protest up to the twentieth century, and those who dissented were condemned as Pelagian heretics. Many Christians now realize that Biblical predestination can be interpreted as applying only to *classes* of individuals (for example, all who come to believe or to love) that were chosen by God from eternity, but *not* to specific individuals elected eternally. Each individual must freely decide whether to become a member of such classes. In our era, free will is popular in the churches, including those that once championed predestination.

Many religious people now acknowledge the obvious–that without free and genuinely creative creatures, God is responsible just as much for all sin and evil as for all righteousness and goodness within the world. Something essential to human dignity would be lacking if we merely act out a pre-existing script and contribute nothing original and personal to the drama of creation. Classical Theologians squirmed and double-talked endlessly but unsuccessfully to avoid these implications.

In Process Theology, God influences all worldly events, including all occasions of human consciousness; but God does not absolutely determine them. God presents us with possibilities for creative self-development and endeavors gently to persuade us to make the right choices; but there is no compelling, no omni-causation. We are co-creators with God. We originate our own free and creative choices for better or for worse. We are responsible, not God, for our choices of good or evil. Hitler and his cronies and collaborators, not God, caused the Holocaust. Process Theologians refuse to "pass the buck." With Harry Truman, they affirm that "The buck stops here!"

In many ways, Process Theology is much more intelligible and attractive than Classical Theology, so we will hereafter construe the question of God's existence in process terms. The Primordial Nature of God is the locus of transcendence, of necessary existence, and of all other desirable general Divine attributes; so we really want to know if we can and do have good reasons for thinking that God's Primordial Nature and the full actuality of God's consciousness are real. The content of God's Consequent Nature is the world, which undoubtedly exists, so its reality is not in question. What we want to know is this: Is the observable world all that there is, or does it have an enduring holy mind of its own? Does a Divine, Holy World-Soul really exist?

2. Conceiving of God's Existence

Does God exist? What do we want to know when we ask? Aristotle said that "There are many senses in which a thing may be said to be," or as President Bill Clinton might put it, "It all depends on what the meaning of the word 'is' is." For Aristotle, possibilities do not exist in the same way as actualities; forms do not exist in the same way as matter; substances do not exist in the same way as attributes. Does God exist in the same way that things ordinarily exist?

A. Ordinary Existence

Theologians usually insist that the being of God is very different from the being of ordinary everyday things. How does Divine existence differ from ordinary existence? Ordinarily, we think that something exists if it (1) is entirely located within our familiar public world of spacetime, as opposed to the spacetime of dreams or hallucinations, (2) occupies a finite region of the everyday public world of spacetime, (3) can be publicly detected in spacetime through sensory perception (directly, or with the aid of magnifying instruments), (4) is the effect of perceptible causes located in common spacetime, (5) is the efficient cause of perceptible effects that exist in common spacetime, and (6) the denial of its existence is not logically self-contradictory. This list applies in past, present, and future tenses. To this list, Charles Hartshorne would add that (7) contingent existence is always competitive and excludes other contingent existence.

Degenerate or marginal senses of "exist" also function in ordinary language. Ideas and fictional entities may be said to exist in our minds, thoughts, imaginations, dreams, or in myths or stories, even though they fulfill only the 6th criterion: Denying their existence is not self-contradictory. Santa Claus, elves, fairies, the present King of France, numbers, logical self-contradictions, and Captains James Kirk, Jean Luc Picard, Kathryn Janeway, and Benjamin Sisko exist only in this degenerate "intramental" or "intrafictional" sense. Fictional roles exclude no other actual beings from the domain of existence, even though the actors playing the roles do. If pressed to say whether they "really" exist, our usual answer is appropriately negative. They are too far removed from our paradigms of ordinary existence. Of course, we can always change our minds if and when convinced that some suspect item really fulfills our most essential criteria.

Many entities in the annals of contemporary science-minded cosmology exist only marginally by ordinary standards. Did an initial singularity exist? Obviously not by the first four criteria. It was not located within our spacetime and did not occupy a finite region of it. Because it was infinitely small, it was not perceivable; and its cause (a collapsing antecedent universe?) was not located within our system of spacetime. Perhaps it did not exist by the fifth criterion: if it is only a theoretical fiction or construct, it had no effects within common spacetime; but if it actually initiated our universe, everything observable is its effect. No self-contradiction results from denying its existence; but since it occupied no spacetime, it is difficult to see how it could exclude the existence of anything else.

Do antecedent, co-existent, or parallel universes exist? Not by the first five criteria, so far as we really know: they are not located within our spacetime, do not occupy a finite region of it, cannot be perceived, are not effects of perceptible (to us) causes, and have no perceptible effects as far as we really know. Perhaps they exist in an expanded, marginal, or metaphorical sense: although they are not in ours, they and their components may exist in *some* system of spacetime or Superspacetime, as may also the Heaven and Hell of traditional theology, or the many worlds of Big Fizz Cosmology. Denying their existence is logically possible (criterion 6), but mere possibilities are not actualities or even probabilities. They do not compete for existence with anything else that we really know to exist, but they may compete for locus in Superspacetime with other worlds.

Does our universe as a whole exist? Not by criteria one, two, four, and five, and only partly by three. It is not located within itself, not a mere part of itself, can be perceived only in part, and is not the effect of causes within itself. Naturalists, who contend that the universe as a whole is eternal, uncreated, and uncaused, deny that it exists in sense four. Because it is not the effect of anything, the whole universe did not result from causes within itself, even if all of its parts are effects of perceptible causes in common spacetime. Theists, Oscillationists, and Big Fizz and Big Divide plenitudists deny the application of criterion four to our universe as a whole because its cause is not a perceptible object within our system of spacetime. Can we deny its existence without self-contradiction? This may be something like denying our own existence without self-contradiction. At any rate, since it includes everything whose existence we could ever verify, it does not compete for existence with anything verifiable.

Did inflation exist? We do not know, but its partisans affirm that it somehow "in principle" existed in senses four, five, and six: its in principle perceptible causes were there in the earliest fractions of a second of our spacetime system; cosmic isotropy and the distribution of galaxies and stars are its perceptible effects; but denying inflation is not self-contradictory. Inflation would compete with a more leisurely pace of cosmic expansion during the fraction of a second that it supposedly existed.

Do the infinitely condensed singularities at the cores of black holes exist? Perhaps, at least in senses one and three through six. These singularities have a position in our spacetime, but since they have no magnitude (like Cartesian minds) they do not occupy a finite region of it. With powerful telescopes we can look toward their black cores, determine that they are caused by gravitational collapse, and see some of the stuff they consume. Their effects are perceptible, even though they are not. We can think consistently about their possible non-existence; and since they have a definite locus, their existence might exclude the presence of other black holes in that locus.

Did the Big Bang exist? Probably, at least in senses five and six. Its effects surround us and are us; thoughts about its non-existence are logically coherent. It does not seem to satisfy the first, second, third, and fourth criteria. It was not located *within* our system of spacetime because it *is* our system of spacetime in its earliest stages; for the same reason, it does not occupy a finite part of familiar spacetime because it encompasses all such regions; its very early stages may have been in principle perceptible but were definitely not so in practice. Its cause was neither perceptible nor natural, but its existence excludes the existence of other universes that might have been created instead.

Do non-extended Euclidean points exist? When conceived of as mere points,[19] do particles like quarks, photons, electrons, and monopoles exist?

Some physicists regard this as just a careless way of speaking about "point-like" entities that are very, very small;[20] others suggest that the notion "has reached its limits of validity and usefulness."[21] By criterion one, if they are anything more than purely conceptual constructs, they have position but no magnitude or volume within real spacetime. By criteria two and three, they clearly do not exist: they are too small to be regional; and they are in principle imperceptible. They do not exist in any obvious way by criteria four and five: if they have perceptible causes or effects, the process by which this happens is very mysterious indeed; only entities with magnitude or spatial extension are perceptible and have perceptible effects. Incidentally, most philosophers today reject Cartesian dualism with its non-extended minds and extended matter mainly because no one can figure out how non-extended entities can act on extended entities; but if real matter is also non-extended, the mind/matter problem is a whole new ball game! So, too, if real minds are extended! By criterion six, non-extended points are not logically necessary beings. In no clear sense could their existence be competitive, unless one point excludes another; but since they are all in some sense identical, counting them might be difficult.

Do non-extended Cartesian minds exist? Here philosophers disagree about the answer but not about the meaning of the question. Dualistic mind/matter interactionists think that non-extended minds exist only in senses four and five: they are affected by bodily events, and they cause and are affected by bodily events. By criterion six, the existence of other minds can be denied without self-contradiction, but not our own. I contradict myself if I deny that I exist. Descartes was right about one thing; unless I exist I cannot deny that I exist. Materialists affirm that minds exist as brains in public spacetime, and they can be perceived if our skulls are cracked open; but Dualists and Idealists deny that minds exist in senses one through three. Clearly, our direct access to our own conscious awareness is introspective, not sensory. Since minds have their own unique identity, they exclude other minds.

Now for the really important question: Does God exist? Popular religion conceives of God as an old, gray bearded, white skinned, blue eyed, male humanoid, dressed in a white robe, sitting on a throne. So conceived, God would exist in something remotely resembling the first three and the fifth and sixth criteria, except that the relevant spacetime system is that of Heaven, or both Heaven and earth, but not this world alone. If these criteria are strictly applied only to their natural home–the everyday spacetime of this world–naturalists are correct: God no more exists than fairies, elves, unicorns, and winged horses. Clearly God does not exist in the ordinary way, but then neither do most if not all of the other things or realities just discussed. Does God exist in some extraordinary way? Is God's existence at all intelligible?

Paul Tillich insisted that it is atheistic to say that God exists, meaning in part that thinking that God's being is like the being of ordinary things is totally wrong-headed.[22] Amazingly, Naturalists, Classical Theists, and Process Theologians fully agree that God does not exist–by most ordinary criteria. They agree that God does not fulfill any of the first four criteria. Neither does the universe as a whole, for that matter. God is not a finite physical, perceptible object or being located within a finite region of our system of spacetime. We could never literally see God sitting in a chair, standing by a waterfall, or walking through a garden.

Naturalists insist that God does not exist by *any* of the seven criteria, and this settles the question. Theologians, both classical and process, think that God fulfills the fifth criterion by being the efficient cause of the Big Bang that created our universe, and perhaps by causing miraculous events within the universe. Process thinkers, who view the universe as the body of God, think that God fulfills criterion 3 in principle, though in practice we cannot perceive the whole universe, the whole Consequent Nature of God. Classical Theologians would not agree because they think that God is completely incorporeal or disembodied, that God occupies no space, includes no space, and has no position in space; for that reason an immaterial God could not be perceived.

Naturalists contend that God's existence is a contingent matter and thus fulfills criterion 6, but many theologians in both camps think not, agreeing with St. Anselm that necessary existence is an essential divine perfection and that it is logically self-contradictory to deny the existence of a perfect Being whose non-existence is not possible. Theologians also insist that nothing could compete with God's existence, and in that sense God's existence is not falsifiable.

Unlike Naturalists, theologians in both camps insist that the question of God's existence or reality is not definitively resolved just because God does not fulfill the first four criteria. The same must be said for singularities, as well as for the universe as a whole, for non-extended Euclidean points, quarks, electrons, and for Cartesian minds. This means only that God's existence is radically different from that of ordinary everyday things and that having a unique mode of being is integral to Divine perfection. Not fulfilling criteria four, six, and seven means that God exists necessarily; and, since every other ordinarily existing thing fulfills criteria four, six, and seven, ordinary existence is contingent existence. We must now give more attention to this distinction between necessary and contingent existence.

The Teleological Argument for God developed in the next chapter reasons from the observed order or design of the world to the existence of an intelligent and benevolent Designer. The Cosmological Argument developed in the final chapter reasons from our experience of the contingent existence of the world to the necessary existence of God. Whether this reasoning is sound must be and will be considered in these later chapters. For now, we must try to comprehend the difference between contingent and necessary existence, as understood in philosophical theology. Much that is theologically significant hangs upon this distinction.

B. Necessary and Contingent Existence

Formal (or modal) as well as factual (or ontological) dimensions of meaning belong to both necessary and contingent existence. Both logic and ontology inform us about their differences and similarities.

Necessary Existence

Formal meaning: Existence that could not not be; non-existence is not possible. Denial of necessary existence in this formal sense is logically self-contradictory.

Factual meaning: Existence that is self-sufficient, eternal or everlasting, uncreated, and indestructible. Denial of necessary existence in this factual sense is not self-contradictory.

Contingent Existence

Formal meaning: Existence that might or might not be; non-existence is possible. Possible or actual non-existence.

Factual meaning: Existence that is causally derived from or dependent on something other than itself; if something exists contingently, it has a cause. Contingent existence is created and destructible. Denial of contingent existence in either the above formal or the present factual sense is not self-contradictory, though it may be factually false when contingent entities really do exist.

Usually, contingent existence both comes into being and perishes in time; but it is logically possible for an existing thing, for example, an everlasting universe that does not come into being and perish in time, to be causally dependent on God in some ways throughout endless time. Big Bang Cosmology shows that our world is not like this, however, for it came into being within the finite past.

In Classical Theology, necessary and contingent existence mark the difference between a supernatural God and the natural world. In Process Theology, they mark the difference between the Primordial Nature or general abstract essence of God and the contents (some world or other) of the Consequent Nature. World-events come into being and perish to themselves in time; then they are assimilated into God and never perish in, to, and for God's contingent Consequent Nature. God remembers, preserves, and cherishes created actualities everlastingly. They are ultimately integrated into the contingencies of his Consequent Nature, where they are never lost. (Masculine pronouns are applied to God only as a last resort in Process Theology for simplicity of expression and should not be taken literally.)

Classical Theology treated every general aspect and every particular detail of God's reality as necessary; yet it insisted, inconsistently, upon the contingency of creation itself and of God's will or choice to create. Process Theology assigns necessity only to the abstract essence or Primordial Nature of God that persists unchanged through all other changes. The Primordial Nature includes not only God's envisionment of possibilities for creation, but also God's necessary existence, necessary creativeness, and necessary and flawless capacities to know, love, influence, and be influenced by all creation. God's Necessary Primordial Nature cannot possibly not exist; and it is self-sufficient, everlasting, uncreated, and indestructible. The Primordial and Consequent Natures are logically or conceptually distinct, but they are never separated ontologically. They are not two different Gods, but are inseparable ontological aspects of the full reality of one God who exists necessarily but creates and interacts contingently with particulars.

3. Critique of Process Theology

On many issues, Process Theologians disagree among themselves and not just with atheists and Classical Theologians. To fully understand and appreciate disagreement within the process camp, read Lewis Ford's *Transforming Process Theism*, where Ford even disagrees with himself![23] Almost all Process Theologians agree that Whitehead's original formulation of Process Theology requires many additions and revisions. Many objections to this novel temporalistic way of thinking about God are raised by those who have no stomach for theology at all. These will be faced in the following chapters. Other objections are raised by those who are greatly attracted to many features of process thought but who believe that a few amendments are needed to bring it in line with certain religious and rational requirements. We will next look at a few ways in which I and other sympathetic critics think that Process Theology needs to be amended.

A. God's Influencing and Being Influenced by the World

Mainstream Process Theology has serious problems about God's ability to know, value, and retain the values inherent in the subjective immediacy of moments of created time, and about God's ability to influence individuals within the world. Its fine-grained analysis of "time," taken from Whitehead, seems to be incompatible with its religiously appealing claims that all created values are ultimately contributed to God and that God's responsiveness to individuals and occurrences within the world adds value and direction to them.

Except for the data and subjective aims that nascent occasions receive initially from their predecessors, the free and self-creative moments of subjective immediacy that make up our ongoing streams of consciousness are understood by most process thinkers to be closed to all other happenings. They can neither influence nor be influenced by, neither experience nor be experienced by, other entities during their own brief "duration" or "specious present" of self-creative independence. Supposedly, this causal shut-up-ness prevents the independence, freedom, and self-creativity of temporal occasions from being overwhelmed by the presence and power of God or by past events within the world. Once they get underway, temporal occasions are closed even to God, Whitehead thought, Most process thinkers agree. This implies that God can know and value temporal occasions only after they achieve their final "satisfaction" or unified definiteness, but not in or during their self-creative subjective immediacy. About that, God can know nothing whatsoever. Except at the very beginning and end of their becoming, God can neither experience and influence nor be experienced and influenced by created actual occasions.

For preserving creaturely freedom, the standard process fine-grained analysis of temporality may be overkill. From a deep religious perspective (about which there is ample room for honest disagreement), the God of Process Theology may be deficient or less than perfect because he cannot be omnipresent to, omniloving of, or omniscient with respect to all actualities, specifically, those that are still becoming. That they can be present to God is also doubtful because God is also still becoming. At best, the standard process God can be present with, love, and know our subjective immediacy only as it was, perhaps as we would know it through very short term memory, but never as it is in its vibrant subjective immediacy of becoming where its primary value resides.

Robert C. Neville, whose own concept of God is very weak, argued effectively that the most important values of existence are located in the immediacy of subjectivity and that the God of mainstream process thought is gravely defective in being unable to know, love, and evaluate the immediate subjectivity of entities in becoming. In being unable to know immediate subjectivity directly, God also cannot remember it and thus cannot give it objective immortality.[24] The chief culprit here is the prevailing process analysis of the fine-grained features of temporal becoming, which for many reasons needs to be modified or abandoned, as explained later. Process thought's greatest contribution to theology is its emphasis on God's temporality and on God's sensitive and receptive responsiveness to events in the world, but not its fine-grained analysis of the nature of time itself as it applies to God and to us.

Contributionism, the view that all created values are ultimately contributed to God, requires a deeper understanding of divine time and causation. The self-creative independence of creaturely events would still be intact if God's own temporally ordered subjectivity is reconceived to be continuously but passively present to, sensitive to, and receptive and appreciative of, developing occasions within the world in their immediacy of becoming. As Neville suggests, God might even be able actively to influence the internal concrescence of created events,[25] but this would have to be through very modest spiritual promptings that do not overwhelm our freedom.

Traditional Whiteheadian thought distinguishes two kinds of time or process, the transitional succession of two or more atomized actual occasions, and their singular internal self-development or concrescence. Human consciousness and all other enduring realities within the world are composed of prolonged societies of successive momentary occasions. Charles Hartshorne modeled God after this pattern; God is an infinitely prolonged society of divine actual occasions. Whitehead, by contrast, thought that God is a single, continuous, everlasting concrescence; and many process thinkers side with Whitehead on this issue, as do I. My view is that all of time, God's and ours, is much more like concrescence (properly reconceived) than like atomized actual occasions. Concrescence occurs continuously; atomization occurs discretely. Discrete actual occasions supposedly reach a stage of final unified definiteness or "satisfaction" when their internal processing is completed. This achieved definiteness is then hurled at or infused into succeeding occasions as "data" to be assimilated, then further processed by them in light of their own emerging objectives, aims, feelings, and choices.

Most process thinkers accept the theory of discrete or atomized actual occasions within the world, even if not for God; but I, for one, just cannot find any totally discrete atoms of time anywhere; and I am convinced that the conventional process account of the very nature of time must be challenged. We experience becoming, I believe, as *continuously* (that is, without sharp atomization) exhibiting degrees of unified definiteness, receiving and transmitting data, manifesting ongoing and usually extended-range purposiveness, exercising selection periodically, interweaving and refining feelings, being receptive to novelty and open possibilities, and perishing at some indefinite point. Past moments flow into present moments more like streams than like squirts. Yes, there is perpetual perishing; but past moments penetrate into present moments, and present moments penetrate into future moments, without abrupt atomization. At a *very* fine-grained level (tenths of a second or less), our conceptual distinctions between past and present, and present and future, are arbitrary. At no absolutely precise experiential point does the past lapse into the present, which then in turn becomes abruptly past to its successor. Receptivity, definiteness, continuity, duration, purposiveness, synthesis, unification, achievement, self-enjoyment, and self-creativity are ongoing features of temporal becoming, the stuff we and God are made of; but time is nowhere sharply atomized into discrete epochs that endure only for a jiffy.

Except for never forgetting the past, God's experience of becoming is not radically different from ours, I believe, with respect to continuous concrescence, persisting creativity, ongoing receptivity and synthesis, constant unity and definiteness, and enduring responsiveness. God constantly assimilates data and value from the world's past *and present* (God's Consequent Nature) and gives relevant novelty, purpose, and direction back to developing events within the world (God's Superjective Nature). We lose much of the past because our capacity for memory is very imperfect; but God's memory, by contrast, is flawless. God assimilates the achieved value of the world without losing the value of its subjective immediacy because God directly experiences our subjective immediacy–something else that orthodox process thought cannot affirm. Our subjective immediacy perishes to us but not to God, who knows it forever as directly as we do fleetingly, and who cherishes it everlastingly.[26] Things become settled and definite for God as they do for us, not abruptly but continuously. At some *indefinite* point, experiences become definite and past for God, as they do for us. Neither God's nor our own temporality is sharply atomized into completely discrete occasions. God continuously experiences and grows in value-satisfaction in being acted upon by and in acting upon the world; but neither God nor we experience the terminal technical "satisfaction" of the *abrupt* termination of temporal occasions accepted by most process thinkers. Time just isn't that atomized or abrupt.

As Lewis Ford heavily emphasizes,[27] Whitehead's position, widely held, was that God is a single, present, everlasting, active experience (or concrescence, to use the technical word for it) that *never* reaches final completion, unification, and definiteness (or technically, "satisfaction"). This implies that God's consequent nature cannot influence particulars in the world in any way, for only entities that have achieved final definiteness can exert efficient causal influence on ("objectify" themselves within) other entities. God cannot be prehended or experienced. Ford's latest position, proclaimed in his *Transforming Process Theism*, retains God's influencing the world primarily if not entirely by persuasion, but Ford drops divine influencing through efficient causation from past definiteness into present becoming. Ford's God, like Whitehead's, never achieves "satisfaction" and thus can never be prehended or experienced by us or anything else. Religiously, this is a fatal flaw.

Ford recognizes that the traditional process account of temporal concrescence must be revised; and he proposes that "concrescence" be extended to cover the future. In some mysterious way, God exists in, concresces in, acts in, and dispenses creativity from the future. God provides initial aims and creativity to worldly events through some spooky form of causation from the future.

I am convinced that process philosophy's unresolved problems can be best remedied by drastically revising the very notion of temporal concrescence, not in Ford's way, but much more radically than he proposes. Process thinkers, I believe, must go back to square one, reinterpret the very nature of time itself, and give it a radically non-Whiteheadian analysis. The outcome must not deny that process is metaphysically fundamental, and it should save all that can be saved of Whitehead and Hartshorne. Here are some preliminary and admittedly incomplete suggestions about how to do it.

Time is clearly the stuff that reality is made of, but what is time made of? Following Whitehead, most process thinkers identify two distinct types of temporality: (1) *transition*, the succession of atomized temporal occasions after the self-creative becoming of each one ends, its subjective immediacy ceases, it achieves the unity, definiteness, and permanence of being called "satisfaction," and it objectifies what it can of itself for its successor(s) as it perishes; and (2) *concrescence*, the internal process of becoming of each single atomized occasion, its subjective immediacy, self-creativity, and self-enjoyment in the present moment. Once initiated, the becoming of an immediately present occasion cannot be influenced or experienced by anything outside itself, said Whitehead. Becoming, this second mode of temporality, also involves temporal elements–both duration or temporal thickness and succession. Human level occasions or concrescences endure from a twentieth to a tenth of a second; gluon concrescences that hold quarks together endure for only a trillion trillionth of a second;[28] but no occasions are infinitesimally thin.

The elements of concrescence that endure briefly but successively together in their internal becoming are so organically interrelated that they cannot be separated or sub-divided into distinct atomized units like those involved in *transition*. They are not separate occasions, only components of a single occasion. These inseparable elements, most of which are classified as successive "phases" of becoming or concrescence, can be differentiated conceptually; but they cannot be sub-divided physically or metaphysically.

The successive *phases* were analyzed "genetically" by Whitehead in slightly different ways in different writings,[29] but they involve at least (a) an initial phase in which a nascent occasion both derives data from (prehends) its predecessor(s) and derives its initial aim from God. The received initial aim consists of an awareness of the possibilities or "eternal objects," some quite novel, that it might actualize, as well as a slightly weighted aim, purposiveness, or lure towards the best of these possibilities; (b) an intermediate phase of self-creativity in which it decides for itself what it will be, what real or immediately relevant possibilities it will actualize, best or not; and (c) a final phase, called the "satisfaction," in which its definiteness and unity are finalized, after which it perishes and infuses what it can of itself into its successors; (d) the whole becoming of concrescence is characterized by a "subjective form," which is *how* the occasion feels, processes, integrates, and values its constitutive elements and by (e) creativity, the presumably ubiquitous category by which every occasion is partly self-creative.

The subjective form persisting throughout each occasion provides it with a kind of unity from the outset to the end, but *how* an occasion reacts to its components gradually develops, grows, and becomes more unified and definite. Subjective forms are thus odd mixtures of unity in variety, of beauty as classically understood. Note carefully that all the elements that define present becoming are *modeled on psychology*–experiencing data (receiving information), having aims, purposes, attractions, revulsions, feelings, and values, awareness of indefiniteness, alternatives, and choosing among them, as well as discerned duration, unity, and definiteness. In most occasions, Whitehead thought, these psychological constituents of time itself merely exist unconsciously; only rarely do they rise to consciousness.

This Whiteheadian model of temporality is wrought with perplexities. For one thing, although the elements composing becoming are drawn from psychological self-awareness, the theory, which pays lip service to faithfulness to experience, does not order these elements accordingly. In lived experience all these elements are like subjective forms–they occur throughout becoming, not sequentially. They pervade our present moments of becoming; they do not follow one another successively in experience.

Another serious difficulty springs from Whitehead's insistence that only at its outset can an actual occasion experience anything outside itself, and only after it perishes can anything outside itself, and including itself, experience it. Only after it achieves the fully unified definiteness of terminal "satisfaction" can anything outside of itself prehend it. Applied to God, a serious theological quandary results. If God is a single everlasting concrescence with no beginning and no end (no satisfaction), it follows logically that God never experiences anything, and nothing else ever experiences God. God never has either an aboriginal beginning or temporal beginnings where data can be received; God also never has endings or "satisfactions" where he can objectify himself for others. Thus, God cannot know us and we cannot know God! What an embarrassment for a theology that revels in God's availability to us and our availability to God! Something has to give.

The Hartshorne/Cobb view that God is an infinitely prolonged society of actual occasions rather than a single everlasting concrescence is initially attractive because from this perspective God has temporal beginnings and endings (satisfactions); thus God can receive data from creation and provide feedback to it. Before you embrace this view, please consider just one of many difficulties with the society of occasions theology. How many divine occasions must occur per second for God to be present at the outset and terminus of a succession of gluon occasions, each of which endures for a trillion trillionth of a second or less? Like most other enduring entities composed of societies of successive occasions, gluon occasions do not all occur in sync; they are not like those rare fireflies that flash simultaneously in perfect harmony. So how densely packed would God's occasions have to be just to cope with all the out-of-sync gluons in existence? How densely packed per second must God's occasions be just to cope with gluons? Now consider all other kinds, durations, and unsynchronized varieties of occasions composing our incredibly complex universe. How dense must a Divine society of actual occasions be in order to be there at the beginning and end of each occasion in the world? The obvious answer is that God's occasions would have to be infinitely dense! God *always* has to be there doing his job; God can't just flash in and out of existence; God must exist continuously; God has to be an infinitely dense continuum just to cope; God has to be an everlasting concrescence continuously interacting with all creation. God must be an infinitely prolonged, infinitely dense continuum; no matter where we cut, God is there. But if God is an everlasting continuous concrescence who acts upon and is acted upon by the world, the very notion of concrescence must be drastically revised.

Applied to ourselves, the requirement that only terminal satisfactions can be experienced means that, despite the vaunted value of subjective immediacy, we can never experience or know ourselves directly and immediately. That is, the occasions that constitute our stream of consciousness cannot know themselves directly and immediately. Process thinkers generally accept Hartshorne's view that all introspection is really retrospection on immediately past self-occasions. This means that we never really know or experience ourselves in the full subjective enjoyment and self-creation of the present moment. We know how we were, but never how we are. Self-enjoyment and self-creativity are gone by the time they are experienced, so these vital values are never really known as such. This is another very good reason for revising the process analysis of the becoming of the present moment. How can anything be remembered (retrospected) if it was never experienced or known in the first place?

"Satisfaction" is the most vicious culprit in the standard process analysis of temporal concrescence. As it terminates, every occasion supposedly resolves all its indefiniteness and achieves absolute unity and definiteness. This presumably happens every fraction of a second in human experience, but this is totally at odds both with quantum physics and with our own immediate experiences of temporal becoming.

Quantum physics says that real indefiniteness is a persisting feature of quantum-level physical occurrences, and quantum non-locality involves immediate perceiving and knowing, not temporally ordered retrospection. Indefiniteness of position or locus and velocity are only occasionally and never simultaneously resolved. Wavicles, which cannot be rigidly atomized, may persist through extended periods of time as unresolved sets or superpositions of pure potentialities until their wave functions collapse, according to some quantum physicists. Instead of classical definiteness of "simple location," wavicles are smeared out over small but indefinite regions of the spatiotemporal continuum. Particles are embedded in waves, and waves are embedded in even broader quantum fields. But none of this could be true if every occurrence begins in disarray and ends in completely unified definiteness. Certainly, our own temporally ordered subjective experiences of becoming do not begin or end that way.

If all indefiniteness is totally resolved every tenth or so of a second in terminal satisfactions, none could ever be transmitted from one occasion to the next, so no indefiniteness should ever persist in our lives. In reality, our temporal streams of awareness are riddled with all kinds and degrees of *enduring* disunity and indefiniteness, mixed with persisting unity and definiteness. If Whitehead's theory of becoming-terminating-in-satisfaction were true, we should never experience any lingering perceptual, conceptual, volitional, motivational, or affective confusion, unclarity, ambivalence, ambiguity, or indecisiveness whatsoever. But we do! So Whitehead's theory of becoming/concrescence is wrong. To account for *the transmission of indefiniteness*, we could say that our "satisfactions" end in definite indefiniteness, but Whitehead's account of concrescence cannot be saved by that kind of double talk.

Traditional Whiteheadian models or metaphors for temporality were drawn from two primary sources, our own subjective experience of temporal selfhood and quantum physics. These are the right models, but I doubt that Whiteheadian analysis does real justice to either. My recommendations for reform, especially for rethinking the whole notion of concrescence, will doubtless require refinement, but here is a beginning. (1) The elements of concrescence identified by Whitehead do not follow one another progressively and successively; they are present continuously through every occasion. (2) Occasions within our own streams of consciousness are not analyzable correctly into discrete atoms or epochs of experience; they begin and end indefinitely. (3) The epochal or atomistic model of time that Whitehead abstracted from quantum physics as he knew it is an incomplete and misleading model; he neglected waves and quantum indefiniteness; and non-local immediacy was unknown to him.

(1) In my (our) own immediate experiences of present moments, admittedly durational, the elements that Whitehead adduced to be successive do not occur successively but continuously. They are simultaneously present throughout concrescence, just like subjective forms. Moments of subjective immediacy are continuously and constantly being flooded with data or derived information; these data, feelings about them, valuations of them, and aims with respect to them are continuously present throughout every present moment of self-becoming, along with high degrees of unity and definiteness. These elements don't just occur only at the beginning, in the middle, or at the end respectively of present moments. Temporalistic consciousness continuously synthesizes data, possibilities (thoughts), and feelings into high degrees of unity and definiteness.

Furthermore, practically all of this happens almost unconsciously, automatically, and irresistibly; and very little of it can be accurately described as involving conscious choices or voluntary effortfulness. Creative synthesis does not involve intentional creativity. We cannot choose to turn on or off our continuous synthesizing of multiplicity into unity. We cannot control it; it transpires not because of us; it happens to us and maybe even in spite of us. Decisions, properly so described, occur only intermittently, not in every present moment, certainly not every tenth of a second or so. Active synthesis of plurality into unity is always there in consciousness, but not active choosing or voluntary effortfulness. Freedom is intermittent, not ubiquitous. Arriving at a real decision about anything takes a while, and the preliminaries run through many present moments during which deliberations but not decisions occur. At times we actively deliberate, but even that takes a lot of time, not just fractions of a second. Our subjective aims, our plans of life, cover extended periods of time, not just the present moment. They do not achieve determinate "satisfaction" every tenth of a second. They exist continuously, not discretely.

High degrees of definiteness and unity run throughout our present moments of subjective immediacy, self-enjoyment, and self-awareness. Decisions, occurring only intermittently but not constantly, do cut things off into memorable definiteness; but unity and definiteness are constantly there to a significant degree, and I am always "satisfied" in that technical sense of the term. Clearly, unity and definiteness do not happen just at the "end" of each present moment. Time is not experienced as a series of oscillations that begin with indefiniteness and end with definiteness. Experienced temporality melds indefiniteness with definiteness throughout; it does not vacillate identifiably between indeterminate beginnings and determinate endings. Present moments don't even have discrete and clearly identifiable beginnings or endings. That brings us to atomization.

(2) Process thinkers should just drop the claims that the elements of concrescence exist successively rather than together, and that temporality, whether divine or in the world, is sharply atomized into discrete occasions. This conceptual theoretical construct, out of touch with all experience, creates all the problems. Our experiences of subjectivity are definitely not so synchronized or atomized. A quantum-like indefiniteness characterizes every present moment of temporal concrescence, experientially. Over the *very short term*, past, present, and future are not sharply divided or atomized. Instead, they interpenetrate and flow continuously, not abruptly and discretely, into one another. Experientially, we cannot tell with infinitesimal precision where the present ends, the past begins, or the future arrives. Perhaps something like this is what Lewis Ford means by the presence and activity of the future; but over the very short run it makes just as much sense to speak of the presence and activity of the past.

Our philosophical theories, including our theories of time, should be grounded in both subjective and scientific experience. Real time, the basic stuff of reality itself, has to resemble experienced time, which is more like a continuum than like atomized squirts, bursts, flashes, pulsations, or epochs. This does not mean that it is infinitely divisible into real parts capable of existing independently of durations; but the boundaries of its real parts, concrescences, are quantum-fuzzy; their durations flow into one another and are not sharply differentiated. Over fractions of a second, indefiniteness separates immediate past, present, and future. This does not mean that experiences and events are not at some point definitively past, over and done with; but we can't tell *precisely* when because no absolutely precise "when" separates the present from the past or from the future. Rejecting temporal atomism does not mean that no intrinsic causal connectedness exists between events; rather, it makes this intelligible. The energy of one event infuses and flows imperceptibly and directly into the energy of another, and we can tell *exactly* where one (the temporal cause) ends and another (the temporal effect) begins. The cause is, in part, in the effect.

Few process thinkers have questioned Whitehead's sharp temporalistic atomism; but his analysis of becoming or internal concrescence confuses our theoretical constructs, Whitehead's constructs, with both temporally ordered experience and reality and thus commits Whitehead's own "fallacy of misplaced concreteness." Theory should coincide with experience. When Hartshorne confronted the substantial discrepancy between our *experience* of time and the mainstream process *theory* of time, he conceded, in these words, far too much.

> We here confront one of the subtlest problems which event pluralism has to face, that of the apparent continuity of process, its apparent lack of discrete units. Dewey, Bergson, Peirce, all three careful thinkers much interested in the analysis of experience as such (and to them Husserl and Heidegger could, so far as I know, be added), found no definite discreteness in the becoming of human experience. And no process directly exhibited in human experience seems to come in clearly discrete units. Here is a splendid example of a seemingly strong (empirical) case for a philosophical view, a case which is nevertheless inconclusive, and indeed can be opposed by perhaps a still stronger though non-empirical case.[30]

That the Whiteheadian theory of the very essence of time is radically non-empirical is a devastating admission! The theory fails to agree both with self-awareness or subjective immediacy and with physics, quantum or otherwise.

(3) Whitehead's epochal, atomistic theory of time was at best only incompletely abstracted from quantum physics as he knew it in the 1920s, and, as suggested already, it is definitely at odds with quantum physics as we know it today. Its physical as well as its psychological models for temporal atomization are inadequate and misleading. Contemporary readers familiar with quantum physics understand that it blends indefiniteness with definiteness, particles with waves. The standard Whiteheadian view of time was modeled microscopically on particles alone, not on wavicles. Macroscopically, it was based on successive and discrete bursts and flashes, while ignoring their underlying wavelike constitutions and origins. Actual occasions are atoms or particles, not of space, but of time; each is discrete; and each has a definite beginning, middle, and end. So the theory says, but the theory is wrong.

What would happen if our theory of time gives at least equal weight to the wave model of differentiation? Definiteness (atomization or particularization of sorts) would still be there, but it would be smeared out, have fuzzy edges, and lack simple temporal or spatial location and infinitesimally precise discreteness. Waves have definite peaks and troughs, but they flow almost seamlessly into one another. Like circles, determining *exactly* where one ends and another begins is quite arbitrary. So it is with actual occasions. We cannot leap from "All energy is *transferred* only in discrete packets or quanta" to the conclusion that "All energy *exists* only in discrete particles or quanta," for some energy exists as indiscrete waves and encompassing but indefinitely bounded fields, not as discrete particles; and time is like that. We should go back to discussing "events," as in Whitehead's pre-systematic metaphysics, rather than "actual occasions," technically conceived.

Since no concrescences, not just God's, but ours as well, have absolutely discrete beginnings and endings, they cannot be the absolutely closed units that Whitehead took them to be; and the absolutely definite beginnings and endings he required are not the insuperable obstacles to mutual and self experiencing that he took them to be. Past events can penetrate present events sufficiently for efficient causation. Inherited definiteness of data, energy, purposes, and feelings are always and continuously mixed with degrees of openness and indefiniteness. No strictures pertaining to terminal "satisfactions" preclude one event's direct awareness of its own or another's concrescent subjective immediacy because significant degrees of unity and definiteness are *always* present, and because counting concrescences as "first" then "second" is arbitrary over the very short run. Because something like instantaneous non-locality obtains within events, they can know themselves directly, not merely retrospectively.

All of this should be just as true of divine as of human level concrescences. God's ongoing concrescence never begins or ends absolutely; yet, incredibly significant degrees of unity and definiteness are always present to God. This includes the complete definiteness, the full burden, the total joy and worth of the world that God takes into himself and saves forever without losing the subjective immediacy that we creatures lose to ourselves in the perpetual perishing of time. It is not true that God never experiences anything, and nothing else ever experiences God because God is a single everlasting concrescence with no beginning, no end, and thus no satisfaction. If unity and definiteness define "satisfaction," then *God's concrescence is always satisfied*, continuously satisfied, to a very high degree in this technical sense; otherwise, if achieved only in terminal satisfactions, God's everlasting concrescence has no unity or definiteness at all. God and nothingness would be indistinguishable!

Nothing, especially God, has the absolute beginnings and endings required by the epochal theory of time. All concrescences, including God's, are highly definite and unified throughout. All concrescences, including God's, can receive information without having absolute beginnings; they can transmit data without having absolute endings. Nothing has to await absolutely terminal satisfactions to experience itself or something else, or to be experienced by something else. God and the world are indeed available and open to one another–continuously.

Most process thinkers probably agree with Whitehead that the closedness of concrescences to one another is essential for freedom and self-creation; but this cannot be true if the traditional account of concrescence is flawed from the very outset. (1) Temporal atomism and (2) sequential phases of concrescence just don't exist! For many reasons, our freedom is not overwhelmed by God's presence with and to us in the present moment, contrary to orthodox Whiteheadianism. God can be sensitively, passively, receptively present as well as actively present; our capacities for including, recognizing, and assimilating God are exceptionally limited even when God is fully available. God values our conscious individuality with its independence and freedom sufficiently to insure that his immediate presence with and to us does not overwhelm us.

Repudiating temporalistic atomism allows God to know and experience us, not just as we were, not merely after our robust present moments perish, but as we are in the full vibrancy and value of our unique subjective immediacy, self-enjoyment, and self-creativity. This is where our intrinsic value primary resides. To that, God gives objective immortality. Ultimately, we contribute what we are immediately, not must what we were retrospectively, to God. Unhindered by the strictures of technical "satisfactions," which don't exist anyway, God can respond lovingly, providentially, and temporally or historically to particular events within the world as they occur, not in a timeless eternity, not from a non-existent future, and without having to wait around forever for a Whiteheadian "satisfaction." By reconceiving the present moment of time, all of the religious advantages of process thought can be preserved and can flourish unfettered by untenable and unempirical theoretical distinctions and strictures.

B. Our Freedom and God's Self-Limitation

God is not the sole creative agent operating in the universe; we creatures are free to choose between good and evil. This is an important part of the process resolution of the problem of theodicy–reconciling the reality of a good and powerful God with the brute fact of evil in the world. This theodicy is developed much more fully in the next chapter, but a difficulty that arises in connection with it must be considered here.

If God is not the sole creative agent in the universe, is this because God freely chose to limit his freedom? Or is it because God is impotent to create unfree creatures or to interfere with creaturely freedom? Process thinkers agree that existing realities are partly self-creative, but they may disagree about whether this is so by metaphysical necessity or by divine choice. Mainstream Process Theologians seem to believe that God is merely the final but never the efficient cause of events within the world; but these terms need to be defined carefully. Some might disagree verbally because they define "efficient" and "final" causation differently. As here understood, Divine efficient causation is God's power to create, infuse, or reorder energy. Divine final causation is God's purposiveness, which includes God's power to present developing spatiotemporal occasions with "initial aims," the set of viable possibilities open to them for both limited and long-term self-creation and choice. (But there is nothing "initial" about them!) Mainstream Process Theologians hold that God influences individuals within the world primarily if not entirely by gently persuading or luring them toward the best, by presenting them with aims that are slightly but not overwhelmingly weighted toward the good, but never by directly causing or forcing them to conform to the Divine will. Since miracles, by definition, involve Divine efficient causation, and since God never exercises efficient causation, no miracles or acts of God could inject energy directly into the world or redirect it, either at the beginning or subsequently. God never causes or prevents evil because God totally lacks the power to do so. Since God never causes anything, God never causes evil, if that is much consolation!

Although he would not formulate his position in terms of efficient versus final causation as just defined, David Griffin, among others, vigorously defends this dominant process solution to the problem of theodicy.[31] Griffin identifies efficient causation with transition between occasions and final causation with the internal concrescence of occasions, (which unfortunately precludes purposiveness or final causation between occasions); but he also recognizes that final causation involves efficient causation.[32] Mainstream process thinkers hold that God simply lacks the power to bring anything about unilaterally, whether good or evil. This implies, fortunately, that we cannot blame God for evil; it also means, unfortunately, that God could never create a universe out of nothing, or work any miracles, or do anything except nag! God works persuasively, at best, but not efficiently, in all natural causation; but since God never reaches "satisfaction," (on non-Hartshornean interpretations), it is difficult to see how God could even work persuasively.

As exemplified in David Griffin's books on theodicy and in various writings by John B. Cobb, Jr., Lewis Ford, Bowman Clarke, and many others, process orthodoxy says that God's power over the world is *primarily if not entirely* that of final causation, not efficient causation as just defined. They might word the issue differently, but these thinkers clearly hold that God lacks the power to do much of anything except persuade. God influences the world and its denizens by presenting attractive ideals, by "luring" toward the best; but beyond that God seems to *do* very little, if anything.

Process Theologians say that God shares creativity with the creatures,[33] but this is usually tempered if not contradicted by their insistence that God has no choice in the matter because creativity is a universal metaphysical category that necessarily characterizes all actual entities in all possible worlds.[34] This implies, says process orthodoxy, that God absolutely lacks the power to create the deterministic universes in which predestinationists, Newtonians, mechanistic materialists, and others believed. God has no choice but to create co-creative creatures; God simply does not have the power to do otherwise, to create absolutely predetermined events, individuals, or worlds.

I would like to see the options open to Process Theology expanded in many ways. To illustrate, let me flesh out briefly a notion of a supremely worshipful being, a being than whom none better or more worshipful can be conceived, that is much closer in some respects to the non-process tradition. It seems to me, and I recognize enormous room for honest disagreement about this, that God would be religiously deficient if he really lacks the power to influence or modify events except through persuasion or final causation, and if he absolutely could not create deterministic universes, however repulsive they might be morally and aesthetically, because he lacks the power to do so and not because they are repulsive.

I regard freedom/creativity as highly desirable, but not as a metaphysically necessary feature of the kind of universe that a truly worshipful Divine being would create. This implies, contra Whitehead and Hartshorne, that creativity/freedom is not a metaphysical category that is universally and necessary present in all possible temporal occasions and universes. Universality and necessity can be separated; creativity could be cosmologically universal in our world but not metaphysically necessary for all possible or conceivable universes. Even if creativity exists contingently in *all* actual universes, assuming more than one, this is by God's choice, not because of metaphysical necessity. If creativity is ubiquitous in our universe, this is because God freely and voluntarily made it so. Ideally, divine omnipotence involves *having* the power to create both deterministic and non-deterministic universes, but *choosing* from goodness rather than from metaphysical necessity to create free creatures.

Creativity may not be ubiquitous even within our own universe. It may characterize mainly higher or more complex actual occasions like those in living beings, particularly conscious animals and human beings. Lewis Ford allows that although persisting elemental physical particles like quarks, protons, atoms, and molecules are novel in their individual actuality, they do not exhibit novelty of form. Except for somewhat rare sensitive and creative occasions in the material world, most persisting physical particles just do not have conceptual aims, and do not creatively modify their aims.[35]

But this means that most elemental physical particles manifest no creativity whatsoever. This may also be true of many of the dull moments of human experience. I suspect that quantum-level wavicles are uncreative most of the time, but periodically they may manifest it. As indicated in Chapter Six, neither we nor an ace predictor like God can tell or predict when individual particles will change orbits, exactly where they will appear next within broad-band orbital shells, which slits they will go through, the directions they will fly when scattered, or which ones will decay and produce atomic radiation. Quantum uncertainty, indeterminateness, and spontaneity widely pervade physical reality; but this does not imply their universal presence. Creativity may be neither metaphysically necessary nor cosmologically universal. Either way, when present it is a precious gift of Divine grace.

As usually conceived by process thinkers, God lacks a kind of power that Kierkegaard and so many others believed to be of immense worth–the power to share power freely and voluntarily rather than necessarily. Wouldn't a being than whom none better can be conceived have that kind of power? Wouldn't a truly worshipful God have efficient as well as final causation at his disposal?

Shouldn't God be able to *do* as well as *lure*? Shouldn't God have the power to create life-supporting universes from nothing, or to work an occasional miracle or two? Ultimately, each of us must answer these questions for ourselves; but many devout and thoughtful people answer affirmatively.

Miracles are another can of worms, I know; and the view I will now express is process heresy. But the "omnipotence as only persuasion" God of process orthodoxy seems to me to resemble too much a celestial George Bush, or some other conspicuous wimp, who espouses noble ideals like kindness and gentleness but never budgets for them or does anything about them except cajole and nag. George Bush was not that President than whom none greater can be conceived, and the God of process orthodoxy is not that Ultimate Reality than whom none greater can be conceived! (This was originally written about "Big Bush," as Rev. Jesse Jackson calls him; but it will probably also apply to "Little Bush." Only time will tell.)

Many traditional theologians and believers think that God has the power to create both free and unfree creatures but actually does the second (predestinationists); others think that he has that power but actually does the first (freewillists); others try incoherently to have it both ways. Process orthodoxy says that God just doesn't have it, partly because of a power deficit, and partly because all creatures are necessarily free (ubiquitous creativity).

Without wanting to have it both ways, I and many others think that God has all consistently conceivable power at his disposal, but God uses it freely and benevolently to create free creatures and the kind of universe required to sustain them. Adopting this view would help immensely in healing the alienation between Process Theologians and more traditional believers; but standard brand process thinkers won't like the suggestion.

The alternative process view of God's power, espoused here, concurs with many critics that the divine power-deficit at the heart of the dominant process outlook is religiously intolerable. Only a God who is not a supercelestial wimp is supremely worthy of human worship, service, and devotion, according to this minority perspective. On this voluntary self-limitation view, God voluntarily limits his own power and chooses to share power/creativity with free creatures. God exercises a *desirable balance* of both persuasion and efficient causation in relation to events within the world. Values, not metaphysical necessities, determine the balance. God is good, not impotent. Events within the world are partly self-creative because God, who could have created another kind of universe, chose instead from the outset to share creativity with us creatures and thus to limit his own power over and knowledge of the future.

Although God's omnipotence was often equated with omnicausality by Classical Theologians, it can mean instead that although God could predestine all, God actually chooses to influence all without determining all. God voluntarily uses his power to do the best that an omni-influential agent could do, which may include creating universes from scratch and performing infrequent miracles, as well as delegating creative power to others. Freedom within creation is a deliberate, gracious, and voluntary gift of divine self-limitation, not an impersonal metaphysical necessity about which God has no choice. Process thinkers can hold the minority view that creativity is general if not universal in our cosmic epoch, perhaps in all others, because God cherishes it and chooses to make it so, not because God had no alternative.[36]

David Griffin critically discusses and rejects the view that co-creative creatures are free because God voluntarily self-limits his own power. He recognizes that this "hybrid free will defense," as he calls it, is currently supported by such prominent theologians as John Hick, Emil Brunner, and L. Harold DeWolf.[37] To this list must be added Arthur Peacocke,[38] John Polkinghorne,[39] Nancey Murphy,[40] Ian Barbour, Diogenes Allen, and many others who accept Process Theology's emphasis on God's inclusiveness, temporality, sensitivity, affective capacities, persuasiveness, and causal efficaciousness, but who are reluctant to call themselves Process Theologians because they think that Process Theology can make no place for creaturely freedom bestowed by voluntary divine self-limitation. Arthur Peacocke explicitly affirms panentheism, the all-inclusiveness of God, and divine self-limitation, while deploring their historical association in the twentieth century with Process Theology, which he repudiates.[41] Ian Barbour says that Process Theology is the best model of God available, while preferring God's voluntary self-limitation in creating co-creative creatures.[42] I also wish to see God's voluntary self-limitation recognized as a legitimate minority perspective for Process Theologians, even if this implies that many of the technical distinctions in process orthodoxy must be extensively revised or abandoned, especially Whitehead's detailed analysis of the nature of temporal concrescence and its application to God, which few have previously questioned seriously.

C. How Process Theology Can Affirm Creation *Ex Nihilo*

Many temporalistic theists object to mainstream Process Theology's clear repudiation of the traditional Christian view that God created our universe out of nothing, *ex nihilo*, at some point in the finite past. They affirm instead that God created our universe out of the chaotic remains of some prior universe or cosmic epoch, which in turn was also created out the chaotic remains of some still earlier universe, and so on to infinity, because *every finite actuality was partly created by and out of some prior actuality*.

As David Griffin put it, "Creation of our particular world was not initiated by a creation *ex nihilo*, in the sense of a total absence of finite forms of actuality, but was a creation out of chaos, out of a less ordered realm of finitude."[43] Integrated into Process Theology, the claim that every reality is created partly by and out of antecedent temporal realities (and partly by God) implies that our universe or cosmic epoch is just the latest member of an infinite sequence of antecedent universes that God created necessarily because God is necessarily creative, social, loving, and embodied in *some* universe–*ad infinitum*.

I wish to show that and how we can retain valuable process insights such as that God is necessarily creative, social, loving, and embodied in some actual universe and still affirm creation *ex nihilo* for our universe. Without relating his metaphysics to recent developments in scientific cosmology, Robert Neville, both a friend and a severe critic of process theology, has previously championed creation *ex nihilo*.[44] However, most philosophical-minded Process Theologians have not been able to conceive of a way to get around the principle that all realities are partly created out of prior actualities and still preserve God's necessary creativity, sociality, love, and embodiment. I will show that and how it can be done quite successfully employing concepts that are quite readily available in contemporary Big Bang astrophysics and cosmology, and that reasons given by process thinkers for repudiating creation *ex nihilo* can be bypassed. In developing these points, I also hope to show how process thought can relate its insights to contemporary scientific Big Bang Cosmology, and that traditional process thought contains elements out of which a process understanding of creation *ex nihilo* can be constructed.

i. A New Framework for Understanding Creation *Ex Nihilo*

In answering the question "Is God Creator *Ex Nihilo*?" on the *Process and Faith* website, John B. Cobb, Jr. replies that "Whitehead knew nothing of the 'Big Bang' and thought instead of cosmic epochs evolving out of earlier cosmic epochs with no singularities involved. Process theology followed him."[45] Process thinkers have indeed followed Whitehead in affirming that our universe, our cosmic epoch, was created out of the ashes of some temporally antecedent universe, and that both universes belong within an infinitely prolonged series of created universes that collectively fulfill the necessity of divine creativity, sociality, love, and embodiment. Charles Hartshorne affirmed, admittedly with some hesitation,

> That actuality is finite in space I readily believe. It is certainly finite in some respects; for to say otherwise would be to say that everything thinkable was also actual, and this is absurd. But the serious question concerns the past of the creative process. Is there an actually infinite regress of past stages–if nowhere else, then at least in the divine becoming? If not, how can a first stage be either avoided or made intelligible, if every experience must have antecedent objects...? So Kant's first antinomy, his most potent argument, stares us in the face. All I can see to do is to reject his disproof of the possibility of an actual infinity....This question I cannot at present answer to my own complete satisfaction.[46]

Hartshorne elsewhere fleshes out his cosmology of finite space and infinite time by linking it to Whitehead's doctrine of cosmic epochs, telling us that "I incline to Whitehead's view of cosmic epochs, each with its own laws."[47] Presumably all of this means that a series of spatially finite cosmic epochs extends infinitely into the past, and that our universe was created out of the remains of the preceding epoch. The same interpretation must also be placed upon Lewis Ford's "Alternative to *Creatio Ex Nihilo*," which affirms: "For if the world is not created from nothing, it can possibly have an infinite past. If every creative act creates itself out of past acts, *ad infinitum*, the world must have an infinite past"[48] and upon Cobb and Griffin's "Process theology rejects the notion of *creatio ex nihilo*, if that means creation out of *absolute* nothingness. Process theology affirms instead a doctrine of creation out of chaos."[49] Griffin positions this chaos within a temporally ordered set of oscillating universes, explaining that "There was no beginning. The chaos from which our world began can be considered the final state of a previous world. Creation is the gradual bringing of order out of chaos."[50]

How does all of this relate to what is going on in contemporary scientific cosmology? Today, for the most part, cosmology is being done by astrophysicists rather than by philosophers or theologians. As seen in earlier chapters, most of these scientific cosmologists do not believe in God and seem to know little or nothing about process philosophy. They wish to leave the impression that their atheistic cosmological speculations are somehow "scientific," although this is far from being the case, as later explained. Still, for convenience, let us call cosmological speculation being done by astrophysicists and other professional scientists "scientific cosmology."

Contemporary scientific cosmology is very diverse. The variety that best correlates with the views of mainstream Process Theologians is Oscillationism, even though process thinkers have not explicitly affirmed it by using the word "Oscillationism." As explained in Chapter Four, contemporary scientific Oscillationists usually affirm that our universe is but the most recent in a temporally infinite series of cosmic epochs, that it was created entirely, not by God, but by an influx of energy from an antecedently existing universe, that this prior universe originated from its own Big Bang, enlarged to the maximum allowed by the tension between the expansive kinetic energy of its Bang and the constrictive force of its gravity, began to contract after gravity ultimately prevailed, and finally ended in a Big Crunch, from the ashes of which our own Big Bang rebounded.

Most scientific Oscillationists also affirm that the series or set of Bang-to-Crunch epochs extends infinitely into the past. They do so primarily because they think that this is a way of avoiding God. As Alan M. MacRobert recognized in *Sky and Telescope* in 1983, "The idea of an oscillating universe, in which the Big Bang resulted from the recollapse of a previous phase of the universe, gained currency merely because it avoided the issue of creation, not because there was the slightest evidence in favor of it."[51]

The naiveté of the view that an infinitely prolonged natural or spatiotemporal order of things needs no God would be readily apparent to philosophers, from Aquinas to Whitehead and beyond, who understand that an infinitely prolonged universe or set of successive universes would likely lack the complete self-sufficiency essential for naturalistic atheism and would be contingent upon God in many respects. For instance, God could and most likely would be required by each cosmic epoch to squeeze out any residual entropy or chaos inherited from an antecedent epoch, to select desirable laws (especially life-supporting ones) for each new universe, and to choose its initial conditions (like the quantity of stuff, energy, or mass in the universe, the strength of the basic physical forces, and the asymmetry of matter over antimatter–or vice versa). Process thought would add that God is essential to provide each spatiotemporal occasion in every epoch with an "initial aim" that includes novel possibilities to be creatively actualized by the choice or initiative of every creature, and that God preserves and cherishes forever in his faultless memory the values created by existing individuals in each cosmic epoch and gives them "objective immortality."

Pure Oscillationism, which affirms a single infinitely prolonged strand of successive universes, has some stiff competition in contemporary scientific cosmology. The main competition comes from the many worlds view, or what I call "Big Fizz Cosmology," according to which both time *and space* are infinitely extended and creative. Space in today's astrophysics is not just nothingness or an empty Newtonian or Kantian form that separates physical objects and processes. As Whitehead recognized, a lot is going on in so called "empty space"[52] Actual occasions constantly occur there, but they do not consolidate into persisting societies.

As documented in earlier chapters, today's cosmologists are convinced that space itself is a kind of physical something, a field with its own physical properties, its own actualized mass/energy and density. It has a fine-grained foamy texture, best described by the laws of quantum physics; and it can be bent, stretched, shrunk, warped, vibrated, and knotted. The seemingly emptiest spatial regions are seething or bubbling with "virtual particles" awaiting birth or actualization. Scientific-minded cosmologists think that quantum indefiniteness allows these virtual or real potential particles to be converted briefly into actual particles, so long as they promptly cease to exist so as not to violate–for more than an instant–the principle of the conservation or constancy of energy

Matter and antimatter particles are constantly being created in empty space; usually they annihilate one another almost immediately, but perhaps not always. The cosmology proposed by highly influential Inflation Theory says that effervescent virtual particles occasionally escape from "empty space" into more enduring actuality, as allowed by the random fluctuations recognized by quantum theory, and then they inflate into entire universes. This happens more than once; most Inflation Theorists think that it happens an infinite number of times to actualize every possible world. Process thinkers should agree here with Hartshorne (and Leibniz) that the notion of actualizing every possibility is absurd since there are incompossible possibilities within and between every conceivable world. Quantum Cosmologists seem to think that every possibility is actualized, even if it takes an entirely new universe to accommodate each one. Process thinkers dissent, however, on the grounds that for moral and aesthetic reasons, God would not create the innumerable horrible, trivial, or boring worlds that are logically possible.

Our spacetime system, the only one we can observe directly (at least in part), the one whose origins we can trace back to a chaotic Big Bang, originated around 15 billion years ago. All events that compose our spacetime system are causally connected with other events within that system, which is in principle traceable back to the Big Bang. The cause of the Bang itself lies outside our spacetime system; it is transcendent; but it may or may not have been God.

Most Quantum Cosmologists, those who apply quantum theory to quantum questions, hold that our universe is but one of infinitely many universes spawned, not by God, but by and from the near-nothingness of quantum-foamy empty space. According to this many worlds Big Fizz inflationary scenario, the relevant infinitely fertile "empty space" is not a part of, does not belong to, our cosmic epoch. Big Fizz Cosmology postulates a transcendent quantum-fizzy Motherspacetime or Superspacetime within which infinitely many child-worlds or universes co-exist in infinitely extended space throughout infinite time. After child worlds are thus spawned, they may or may not then begin to oscillate

Let us consider the "many worlds" notion of infinite Superspace that supposedly accommodates an endless number of independently co-existing and spontaneously conceived child universes to see if it can help us to conceive of creation *ex nihilo*. According to cosmological theories widely accepted today, since infinite Superspace has always existed, it co-exists with infinite Super time. When a spatiotemporally finite universe like ours expands, it pushes into pre-existing Superspacetime, not into absolute nothingness. We have seen that many scientific-minded cosmologists take all of this stuff very seriously!

Developments in contemporary cosmology outlined thus far may strike you as utterly wild speculation, having little or nothing to do with empirical natural science, even if it originates with professional astrophysicists. Indeed, it is just that! All postulated antecedent and contemporary universes, and the infinite Supertime and quantum-foamy Superspace within which they are located, transcend our cosmic epoch and are totally inaccessible to human experience. They exist before and beyond our spacetime system in a time prior to the beginning of our time and in a space beyond and outside of our space, so we can never observe them. They are supernatural realities, if real at all, that transcend our system of nature or spacetime. If they exist, they are supernatural other worlds. Even science, if this is science, cannot get along without the supernatural. As philosophical postulates or explanatory hypotheses, their reality (or lack thereof) can be considered and debated–as was done in Chapters Three through Nine. Obviously, any explanatory appeal to realities that transcend our spatio-temporal natural order of things always leaves empirical natural science far behind. Hereafter, "scientific" cosmology will appear in quotes.

So, what does this have to do with creation *ex nihilo*? The concept of transcendent Superspacetime developed by Big Fizz Cosmologists is purely theoretical and has nothing to do with verifiable natural science, but it may nevertheless be extremely useful to theologians! I began by saying that Process Theologians have been unable to conceive of how to make sense out of creation *ex nihilo* and still affirm infinite Divine creativity, love, sociality, and embodiment. This is largely because they assumed that finite space is the only possible complement to infinite time. Hartshorne, for instance, says that "The divine actuality so far as I can grasp the relevant concepts, must involve a numerically infinite number of past creatures, but the creation need not, and I think must not, be spatially infinite";[53] and he repeatedly asserts the finitude of space while affirming the infinity of time.[54] By default, if in no other way, other Process Theologians seem to agree.

What would happen if, contra Hartshorne, the conceptual framework of process theology were expanded to include not only Hartshorne's infinite Supertime, but also the infinite Superspace postulated by so many contemporary "scientific" cosmologists? Here, our objective is simply to extend our way of *conceiving* of the arena of infinite Divine creativity, love, sociability, and embodiment; and this has nothing to do with *verifying* propositions about other transcendent worlds, which we mortals could never do. Neither infinite Supertime (previously assumed or affirmed by Process Theology) nor infinite Superspace (hitherto denied by Process Theology) are verifiable by us. Only God could do the job.

Within infinite Divine Superspacetime, God could be infinitely loving, social, embodied, and creative without being tied to a single temporal strand of spatially finite antecedent-and-successive universes. Within infinite Superspace and throughout infinite Supertime, God could create many co-existing universes out of nothing, or nothing more than "empty" Superspace itself; and God could be infinitely creative, social, loving, and embodied in relation to them. No co-existing universes would *have* to be created out of antecedent universes, although some might be. As God wills, some or all co-existing universes could be completely independent causally of all the others, so the crucial barrier between mainstream process theology and traditional Christian theology would no longer exist.

Divine creation of universes *ex nihilo*, thus understood, always *presupposes* other actualities, that is, God's embodiment *somewhere* in Superspacetime, but actual universes or Divine bodies *need not be created out of* other actualities, such as temporally antecedent universes. Process Theologians can consistently affirm that throughout everlasting Supertime, God may create, as willed, many independently existing universes out of nothing, or the near-nothingness of "empty" Superspace; and if, once initiated, some universes form an oscillating series, this is not true of our universe, which God could have created *ex nihilo*.

In infinite Superspacetime, all child universes could be so far removed from every other–infinitely far apart if necessary–that they could never contact or causally influence one another or be derived causally from preceding universes. Or, if God wills, some might have tangential contacts with others, being connected perhaps by wormholes or creative acts of God. Some of these co-existing child universes might even be Heaven, Purgatory, or Hell; and God might be able to figure out how to get us from one to the other! "Beam us over, God!" After we die, God could just reconstitute us in transformed and much improved resurrected bodies (as John Hick suggests) in the spacetime of another world that co-exists with our universe in infinite Superspacetime. Again, the point is just to conceive of such things, to make them intelligible, not to verify or confirm any beliefs we may have about them.

The concept of infinite Superspacetime is neither the Newtonian notion of absolute space and time, nor Einsteinian relativity spacetime. It derives not from classical or relativity physics but from quantum physics applied imaginatively to cosmology. My suggestion that God might recreate an improved edition of us in another co-existing spacetime system is not as un-Whiteheadian as it may seem. If order as we know it is usually a complex emergent achievement from pre-existing order, this could not be true of creation *ex nihilo*; and even if true, in light of what quantum physicists have discovered about non-local causality, we can no longer assume that all causal influence requires spatiotemporal contiguity or proximity. The telepathy in which Whitehead believed[55] does not presuppose that. According to quantum physics, what Einstein called "spooky action at a distance" is a reality; and within Superspacetime, that action could transcend local universes. Whether it actually does or not, we do not know.

If God is actualized in both infinite Supertime and infinite Superspace, the everlastingness of divine sociality, love, and creativity would not be subverted if a finite universe like ours was created out of nothing about 15 billion years ago. Why should God's everlasting creativity be tied to a single temporal strand of spatially finite universes, of which ours is the most recent member? God could be everlastingly creative in Superspace as well as in Supertime, where particular universes need not emerge from antecedent universes. To reconcile Process Theology with the creation of our universe *ex nihilo*, we need a concept of Divine Superspacetime as God's sensorium and arena for infinite creativity, as further explained in the following discussion. If my analysis is successful, Process Theology should adopt the view that God's potential embodiment is coextensive with infinite Superspacetime; and God's actual embodiment is coextensive with all the worlds God has chosen to make within Superspacetime. God's present body is not confined to our finite Big Bang spacetime epoch, which may or may not have antecedents, depending on the plausibility of Oscillationism.

Taking the general concept of Motherspacetime or Superspacetime from contemporary many worlds Big Fizz Cosmology does not and should not commit Process Theology to much of the unwieldy baggage often attached to it. Process thinkers will want to reject the Principle of Plenitude, so popular with today's "scientific" cosmologists, according to which all possible worlds are actual worlds. Instead, in infinite Superspacetime, God creates all the worlds that he chooses, but not all possible worlds. For many good reasons, God is not driven by the ideal of Plenitude, which requires the creation of *all* possible worlds. God may have created an infinite number of worlds in Superspacetime, but God understands that infinity cannot be used up and that an infinite number will always remain to be created. God also realizes that many possible worlds are too horrible, or too trivial and boring, to be created at all. As Whitehead noted, "It is not true that God is in all respects infinite. If He were, He would be evil as well as good."[56] Divine Superspacetime need not be conceived as resembling the quantum-foamy spacetime of our universe, in which actual particle-occasions are constantly emerging spontaneously but briefly from virtuality. Instead, Superspacetime is God's arena for deliberate but selective creativity; and it has all the properties that God wants to give to it, even though we may not know what they are.

Mainstream Process Theologians were unable to conceive of creation *ex nihilo* because they were wedded, implicitly if not explicitly, to the model of a single strand of spatially finite oscillating universes extending infinitely into the past, each member of which arises causally from both God and from its immediate predecessor. Hartshorne affirmed "an infinity of earlier universes, each produced out of its predecessor, more or less catastrophically or gradually;" but God created them all, including our universe, out of their predecessors.[57] This cosmological model precludes the possibility that a universe could arise causally only from God at some point in the finite past–the essence of creation *ex nihilo*. It assumes that God's infinite creativity was only temporally ordered; but it may also be spatially ordered as Divine Superspacetime, where God might be everlastingly creative of multiple universes that have no causal relations with *our* system of spacetime; and our system of spacetime could arise directly from God's Superspacetime and creative will alone, without being preceded by antecedent universes. Other universes or cosmic epochs could be "beyond" ours spatially, to use Whitehead's word for it, without being "before" ours temporally, as mainstream Process Theology has assumed.

ii. Elements of Superspacetime in Process Thought

Elements out of which a theory of Divine Superspacetime can be constructed already exist in Process Theology. In discussing the possible existence of many independent worlds in an essay in *Science and Philosophy*, Whitehead proposed that and how we might conceive of independently existing universes that have no causal, temporal, or even spatial relations with one another.

> We can imagine that, in the realm of existence, there may be an alternative space-time process other than that of nature; but nature and the alternative process do not conjoin to make one process. In fact we are aware of such alternative processes in dreams, where we apprehend a process of events which in respect to nature are nowhere and at no time.[58]

Despite any philosophical problems we might have with Whitehead's dream world analogy, this shows that historically the idea of multiple independent worlds is not entirely alien to process thought. The most effective and trouble-free way to conceive of independent worlds and to relate process theology to contemporary Big Bang Cosmology is to think of other worlds as co-existing, not in dreams, but within Divine Superspacetime, within which some worlds (like ours) could be created deliberately out of nothing, that is, out of the real potentiality and virtuality of genuinely "empty space."

Whitehead was unaware of Big Bang cosmology, as Cobb indicates. Hartshorne, by contrast, was well aware of it; but he neither made a serious and detailed attempt to relate his cosmological commitments to it nor verbally affirmed Oscillationism. However, he clearly had a concept of Divine Supertime, that is, of God's time before (and after) our time, the time of our fifteen billion year old universe. He wrote that:

> Certainly someone ought to correlate metaphysics and physics. For instance, even if the supreme reality is a kind of becoming, then it seems there must be a sort of divine time (even Barth says something like this), and the correlation of this with worldly time, as construed by relativity physics, is a neglected and apparently extremely formidable task. Perhaps this is rather a problem in cosmology than in pure metaphysics, cosmology being the application of metaphysical principles to what science reveals as the structure of our "cosmic epoch." Yet unless either physicists or metaphysicians have erred, there must be an at least *possible* way of harmonizing what the physicists say is true of our cosmic epoch and what metaphysicians say is true of all possible epochs.[59]

As we have seen, today's "scientific" cosmologists do not restrict themselves only to our epoch, but this just makes them metaphysicians in disguise.

I suggest that today's metaphysical (and only pseudo-scientific) cosmologists have done Process Theologians a great service in providing us with a concept of Superspace to complement the Supertime that Hartshorne and mainstream Process Theologians postulate to accommodate antecedent cosmic epochs.

Superspacetime is the proverbial transcendent space beyond our space and time before our time. Although the concept of Superspacetime originated with infinitely many worlds atheism, it can be united with the process concept of God to form the notion of a Divine Superspacetime, within which both infinite divine creativity and universes created out of nothing are possible and conceivable. If time and space are inseparable, as process thought and contemporary physics both suggest, then divine Supertime, affirmed by Hartshorne, also implies divine Superspace. Divine Superspace can be more inclusive than the finite space of our own and preceding oscillating epochs; it can embrace other co-existing universes.

Hartshorne likely had only the spacetime of our cosmic epoch (or similar antecedent oscillating epochs) in mind in insisting upon the finitude of space. If so, his insistence on the spatial finitude of our cosmos in no way conflicts with affirming infinite Superspacetime as the ultimate arena for divine creativity. As far as I have been able to determine, Hartshorne does not give a good argument for his insistence that space must be finite. He just affirms spatial finitude without argument, as if it were intuitively certain or obvious, which it clearly is not to contemporary "scientific" cosmologists; but his writings were never informed by the concept of Superspacetime that they have developed.

A good argument for the finitude of our space can be given, namely, at or immediately after the onset of the Big Bang, the space of our universe began as finite (slightly larger than a singularity); it has since expanded at a finite rate (the Hubble constant or cosmic expansion rate, plus perhaps a brief exponential but still finite inflation rate); and the expansion has endured for only a finite amount of time (about 15 billion years). From these premises we can conclude that our space is finite. A parallel argument shows that our time is finite and has a "first moment"; but this is perfectly compatible with the idea that our finite spacetime exists within and is expanding into the "empty" quantum-foamy virtuality of infinite Superspacetime, which has no "first moment."

We might conjecture, as suggested to me by Lewis Ford, that Hartshorne would argue for the finitude of space by appealing to the premise that there can be no actualized infinities at all, that such things are unintelligible, from which we could conclude that there can be no actualized infinity of space, that space is finite. Yes, but when reflecting on the far distant past, Hartshorne bites the bullet and reluctantly admits that process thinkers must affirm an actual infinity if they hold that each creaturely event is created out of some other creaturely event–*ad infinitum*; otherwise one must affirm creation *ex nihilo*![60] In these passages, Hartshorne clearly affirms an actual, not just a potential, infinity of past events for our world and for God. Anyone who wants to avoid creation *ex nihilo* is logically committed to an actualized infinity and thus must repudiate the above argument that space is finite.

As quoted earlier, Cobb says that Process Theologians accept Whitehead's notion of distinct "cosmic epochs." Whitehead invented this terminology, though he was not very specific about its scope. Under the influence of early quantum theory in the 1920s, Whitehead thought that our own cosmic epoch is dominated by electromagnetic energy that exists only in discrete quanta. He defined a "cosmic epoch" as "the widest society of actual entities whose immediate relevance to ourselves is traceable."[61] Our present cosmic epoch can be traced "to an aboriginal disorder, chaotic according to our ideals,"[62] Whitehead believed; but there are other cosmic epochs "far beyond our immediate cosmic epoch" that are ordered very differently from our own.[63] He knew nothing about Big Bang Cosmology, which was still in its infancy when these words appeared in *Process and Reality* in 1929; and he did not explain whether his "beyond" is to be construed spatially, temporally, or both. Mainstream process theology has interpreted Whitehead's wording temporally; but "widest" and "beyond" are actually spatial words, not temporal words; he did not say "oldest" or "before."

Perhaps Whitehead spoke better than he knew! Or perhaps he knew about Superspace as well as Supertime! Isn't it just his "extensive continuum" construed not simply as the realm of "real potentiality" for our own cosmic epoch, but "in its full generality beyond the present epoch"?[64] Notice especially his emphasis on potentiality. The in-depth explication of Whitehead's concept of "extensive continuum" by Jorge Luis Nobo is almost perfectly compatible with the understanding presupposed here.[65] Whitehead distinguishes this more general extensive continuum from that of our own epoch, which is dominated by societies of electromagnetic occasions.[66] He describes it as

> ...a vast nexus extending far beyond our immediate cosmic epoch. It contains in itself other epochs with more particular characteristics incompatible with each other.... We cannot discriminate its other epochs of vigorous order in our own epoch. This ultimate, vast society constitutes the whole environment within which our epoch is set.[67]

Whitehead uses the spatial word "beyond" rather than the temporal word "before" to refer to alternate cosmic epochs. He certainly does not say that our epoch's "whole environment" is merely temporal, as pure Oscillationism would have it. Co-existing universes in infinite Superspace are no more "traceable" by us than antecedent universes in infinite Supertime.

iii. Process Objections to Creation *Ex Nihilo*

As documented earlier, mainstream Process Theologians have clearly repudiated the traditional Christian belief in creation *ex nihilo*, and they have given a number of reasons for rejecting this belief. With one such reason I wholeheartedly agree, namely, the (Protestant) *Bible* teaches only that our universe was ordered out of chaos, but not unequivocally that it was created out of nothing.[68] Let us begin with the reasons that John Cobb, Jr. gives in his *Process and Faith* website discussion of "Is God Creator *Ex Nihilo*?"

First, Cobb explains, the traditional theology of creation out of nothing reserves the word "'creation'...for a single act, the one in which the world is brought into being out of nothing." To this he opposes the process view that "God is creatively at work at all times and places." But these positions are not really opposed. Whether Cobb intends to make a historical point or a logical point here is unclear, but much of the hostility of mainstream Process Theologians toward creation out of nothing may issue from confusing historical associations with logical connections. It is true historically that traditional Christian theology tended to reserve the word "creation" for God's origination of our universe from nothing, but it did not deny that God is creatively at work at all times and places. It just used other words for God's ongoing creativity, words like "sustaining" the universe and exercising general and special "providence" over and within it.

Traditional concepts of God's sustaining and providential activities were usually qualified by the deterministic or predestinationistic assumption that everything that happens is implicit in creation itself from the very outset, or from the immutable vantage point of God's changeless eternity. Perhaps something like this is what Cobb has in mind. In their *Process Theology: An Introductory Exposition*, Cobb and Griffin raise this more subtle metaphysical objection. They tell us that the doctrine of creation out of absolute nothingness "is part and parcel of the doctrine of God as absolute controller."[69]

Viewed logically rather than historically, creation out of nothing, ongoing creation, and the creation of co-creative creatures are in no way incompatible with one another. Creation out of nothing is logically contradicted by the mainstream process assumption of creation out of something, but not by the notion of God's ongoing creative activity within our world; and God's creating co-creative creatures is logically contradicted by the traditional notion of creating totally programmed non-creative creatures, but not by the notion of God's creating the universe out of nothing. No *logical* obstacles exist to combining creation *ex nihilo* with ongoing divine creativity and divine creation of co-creative creatures.

Cobb clearly wants to make a logical point when he says in "Is God Creator *Ex Nihilo*?" that "...the implication of the doctrine of creation is that God is quite external to the world and the world quite external to God." Closely related is Cobb's charge that creation *ex nihilo* encouraged "exclusive emphasis on divine transcendence."

Historically, Classical Theologians consistently affirmed God's immanence as omnipresence and made some solemn efforts to take this seriously; so it is not entirely true that Classical Theology made God and the world to be totally external to one another. The real difficulty is that what the Classical Theologians gave with one hand, they usually took away with the other. They did indeed characterize the contrast between God and the world so severely (pure being/pure becoming, pure cause/pure effect, spatially extended/incorporeal, and so forth) that the two were "quite external" to and mutually exclusive of one another.[70]

Does creation out of nothing inevitably involve such catastrophic contrasts? I can't see that it does. The opposition here is between our universe or epoch as caused by both God plus a series of antecedent worlds extending infinitely into the past, and as caused solely by God at the beginning of its own finite past. Both have God as a causal factor; the latter has only God. Necessary and everlasting Divine creativity, sociality, love, and embodiment presuppose the everlasting actualization of other universes *somewhere* in Superspacetime, but God's creative actualization need not be confined to a single line of temporally ordered and spatially finite cosmic epochs in Supertime, of which ours is the latest member. If, through either metaphysical necessity or God's voluntary self-limitation, the laws of quantum physics apply throughout Superspacetime and its products, and are not limited just to our spacetime and its antecedents, then every actualized universe is grounded in indeterminateness, spontaneity, and creativity, just as process metaphysics affirms. However, there is no *logical* necessity that "empty" Superspace be quantum-fizzy. A purely Newtonian Superspace is at least *logically conceivable* and thus possible.

In "Is God Creator *Ex Nihilo*?" Cobb himself recognizes that "the event in which our universe arose certainly seems to be markedly different from all the subsequent events"; and process metaphysics has its own ways of differentiating between God, the world, and occasions within the world without implying that God, the world, and finite occasions are "quite external" to one another.

As we have seen, some Process Theologians believe that God lacks the power to prevent evil, to work miracles, to create a universe out of nothing, or to bring about any effects where "persuasion" is not involved.[71] But must process thinkers presume that persuasive final causation applies absolutely everywhere? Might there not be some "markedly different" situations, for example, originating universes–creating the mass/energy out of which partly self-creative actual occasions emerge–in which God acts only as an efficient cause without being a final cause in the sense of giving initial aims to occasions that issue from pre-existing societies? Insisting that God, who has his own aims for newly created universes, must be able to persuade everything by imparting initial aims to successive occasions could not apply before the first moment of creation *ex nihilo*. Before that, nothing exists to be persuaded; the first moment of creation out of nothing succeeds nothing. Beginning with the very first moment, however, something may exist to be persuaded. The absolutely original grandly unified and undifferentiated mass/energy presumed to exist at the very beginning of our Big Bang might not be susceptible to persuasion; but as soon as it is sufficiently unwound, expanded, and diversified to generate actual occasions, it would. We cannot simply equate physical energy with persuasive creativity; the basic physical conditions that make partly self-creative entities possible must come first. Dictating that persuasion must be exercised even on the non-existence that preceded our Big Bang is an irrational demand, like insisting that circles must be squared. Non-existence cannot be a co-creator with God; but from or very near the outset, a new universe created *ex nihilo* could be.

In his website discussion, Cobb relates the process view of infinitely prolonged ongoing creation to Big Bang Cosmology by indicating that the latter calls for an initial "singularity" from which our universe emerged, and by doubting that this means strictly "out of nothing." About this, at least four points need to be made.

First, singularities are defined as being infinitely small, dense, compressed, hot, and curved; an initial singularity has no magnitude or locus in our spacetime since that is what emerged from the initial singularity. Some versions of Big Bang cosmology really do affirm that our universe emerged from a singularity. Clearly, something infinitely small *is* absolutely nothing empirically and physically. Not even God could perceive something infinitely small, and nothing can be physical that is absolutely devoid of all spatial properties, having no size at all, because spatial extension is the very definition of the physical. As all modern philosophers agree, "All bodies are extended."

As noted earlier, initial singularities have many problems that make them cosmologically unattractive. In brief, being absolutely nothing empirically and physically is surely one of the most serious difficulties; another is that non-physical things cannot be physical causes, so an initial singularity does not provide a physical explanation for the origin of our universe. Closely related is the problem that no one knows what would make a singularity explode because no known laws of physics apply to them. Again, cosmic epochs separated by singularities could not belong to a single, continuous, spatiotemporal, causal sequence because space, time, physical causation, and all natural laws break down completely and do not exist in or apply to singularities. Yet again, we could not reason back to singularities separating cosmic epochs, or to earlier epochs themselves, by extrapolating from the natural laws that we know because these laws presuppose spacetime for their application and terminate absolutely at singularities.

Second, Oscillation Cosmology is not bound inextricably to the idea that successive universes arise from and are separated by singularities. Many contemporary Oscillationists agree with Stephen Hawking that quantum effects would prevent a prior universe undergoing gravitational collapse from shrinking to a singularity. According to Big Bounce Oscillationists, a universe or cosmic epoch being terminated by a Big Crunch would rebound from a small finite state of intense compaction into a subsequent cosmic epoch initiated by a Big Bang without first proceeding all the way to total collapse into an infinitely condensed singularity.

As Cobb indicates, Whitehead thought "of cosmic epochs evolving out of earlier cosmic epochs with no singularities involved." Process Oscillationists would presumably find Big Bounce Oscillationism very congenial, for it requires no singularities between crunch/bounces. It has its own problems, as indicated in earlier chapters, but by appealing to it Process Oscillationism could affirm a Big Bang that rebounds from an antecedent universe without having to embrace troublesome singularities.

Third, our universe may not be derived from a singularity or a crunched-up antecedent cosmic epoch at all. If and when singularities form at the end of a Big Crunch, why don't they just stay there forever? No one knows what would cause a singularity to explode. No physical laws that we know could account for it, for all of them break down in singularities. The quantum fluctuations to which Inflationary Cosmology appeals would not do the job because they presuppose the laws of quantum physics, which, along with all other natural laws, would also break down in singularities. Inflationary Cosmology does not derive its many worlds from singularities or from crunched-up antecedent universes. Inflation requires just the right kind of quantum-foamy "empty space" in Superspacetime; and singularities and crunches just aren't the right stuff.

For the reasons just given, with or without singularities, Process Cosmology need not and should not give an oscillationistic account of the origin of our universe. The most plausible view is that our world or cosmic epoch was not created out of a preceding universe. Instead it was created out of nothing (without other-world antecedents) within divine Superspacetime. If our low mass universe is open, as it now appears to be, especially in light of the very recent revolutionary discovery that the rate of Hubble expansion is increasing, not decreasing as previously assumed,[72] then our universe does not belong within any kind of an oscillating series because all members of such a series must be closed to sustain infinite oscillations.

Fourth, Cobb doubts that the nothingness to which contemporary cosmologist appeal is really nothing. Although singularities are empirically nothing and have many other problems, what about the "empty space" of Superspacetime? Well, it is not a full-fledged antecedent universe, so we are at least that close to creation out of nothing. Superspacetime may but need not have the actualized quantum-foamy physical mass/density that contemporary cosmologists assign to "empty space" *within* our existing spacetime system; on no empirical or scientific grounds can we infer that Superspacetime is like our universe's quantum-fizzy spacetime "vacuum." It could be closer to a realm of real potentialities than to an actualized energy field. Aside from the co-existing universes that God has created, Superspacetime could consist mainly of potential rather than actual occasions; and nothing is to potentiality as something is to actuality.

As the everlasting arena for creativity, Divine Superspacetime is God's infinitely extensive potency for creativity and social sensitivity. Its "spontaneity" is God's well-considered selectivity and creativity.

The actualized regions of Divine Superspacetime would contain, not all possible worlds, but only those universes deemed desirable by an infinitely loving Creator. Just how many co-existing worlds there are, if any, only God knows; but at least one universe must exist in perpetuity to satisfy God's loving, social, creative nature and the plausible requirement that all minds are embodied. Any number of successive and/or co-existing universes could come and go, given an infinite amount of time to play with them. Unlike us, God doesn't have to rush to do anything. Presumably, as many universes would co-exist as God freely chooses to be involved with; but only God knows how many.

The view proposed here does not locate God entirely outside of our cosmos. It allows for all the divine immanence that metaphysics and religion find desirable; but it recognizes, as do most Process Theologians, that God's Primordial Nature, comprised of the everlasting and omnipresent features of divinity, transcends our cosmic epoch. It also does not violate Whitehead's "ontological principle," according to which explanatory reasons are always located in actual entities, but not necessarily in actual occasions.[73] God is not located in Superspacetime; rather, it is located in God, the ultimate all-inclusive actual entity, without whom there would be no space, no time, no actuality, no potentiality.

Finally, Hartshorne maintained very explicitly that the finitude of past time is inconceivable. After conceding that if we conceive of the past as infinite, what we could know of it is "negligibly small," he then argued,

> Conceive of it as finite, and then it seems fairly clear that we never grasp what is meant by a first stage of creation, a process preceded by no process. All our thinking seems to break down at that point. We would have either an effect of an inconceivable cause, or something which simply transcended the causal idea, and hence our concept for explaining concrete things.[74]

In response, we must distinguish the finitude of our spacetime, which is conceivable, from the infinitude of Superspacetime. Creation of our universe *ex nihilo* does not presume an absolute "process preceded by no process." It presupposes the everlasting processing of Divine creativity, which need not be located solely in oscillationist Supertime but could be expressed in many worlds that either co-exist within and/or are created successively within Divine Superspacetime. If so, God's occasions or experiences of created worlds would always be preceded by other divine occasions or experiences, even if the series of occasions that constitute our world originated *ex nihilo* around fifteen billion years ago. The God of Process Theology can be both the final (purposive), efficient (creating *ex nihilo*), and formal (the Divine vision of eternal objects) cause of a universe created out of nothing. If efficient causation acts from the past to the present, God's creative act of bringing our world into being out of nothing could be in God's past without being in our world's past. Creation *ex nihilo* is possible and conceivable without violating the "no process preceded by no process" principle from God's perspective, though it might seem so from a non-process-theism human perspective. If the "all creation" refers to Superspacetime, God could still be "not *before* all creation but *with* all creation"[75] while definitely and necessarily existing "before" the creation of our spacetime system fifteen billion or so years ago.

Is it God as transcendent cause, or the world as an *ex nihilo* effect, that Hartshorne regards as inconceivable?

If God is everlastingly creative in Superspacetime, God's creation of our universe out of nothing would not be an inconceivable effect of "an inconceivable cause" because God, the cause, really is conceivable, at least in the abstract. Hartshorne has argued extensively and persuasively that we can and do have an abstract concept of God (the cause) without knowing God's full concreteness. The crucial issue is whether a universe caused by God alone is any less conceivable than a universe produced by God out of an antecedent universe. If God is conceivable at all, then a universe caused by God alone would not result from an "inconceivable cause." Perhaps it is inconceivable that a necessarily creative, loving, social, and embodied Supreme Becoming should exist without having created anything to love, but other universes in Superspacetime having no causal relations with our own epoch could fill that bill.

Hartshorne's main point could be that a universe created out of nothing would be an inconceivable effect. I contend, and I believe Hartshorne would agree, that the notion of causation as such is broader than that of physical, that is, spatiotemporal, causation. It is the notion of conditions that are either necessary and/or sufficient for producing an effect. Even if, contrary to the absolute incorporeality and timelessness of the classical God, all efficient causal conditions must be in some sense spatiotemporal, then the relevant spatiotemporality for creation *ex nihilo* could just be transcendent Divine Superspacetime; it need not be the spacetime of an antecedent universe from which our universe was causally derived. Our Big Bang could have been created out of nothing within God's Superspacetime without violating any defensible presupposition of Process Theology.

Thus, subtle and not so subtle replies can be given to the central objections that mainstream Process Theologians have raised against the traditional doctrine of creation *ex nihilo*. The preceding account of how Process Theology can accommodate creation *ex nihilo* may need a bit more tweaking and development here and there; but its affirmation would permit Process Theology to avoid alienating those more conventional Christians who are convinced that in the beginning, God created our universe out of nothing.

In sum, with a few minor revisions, Process Theology is the most religiously viable and rationally intelligible option available to us today. It can be saved from some of its own mistakes like its contention that God cannot know us in our subjective immediacy, that God creates free creatures because creaturely creativity is a metaphysical necessity not a Divine voluntary self-limitation or choice, and that God could not and did not create our universe out of nothing.

Armed now with a better understanding of what it means to exist and a more viable concept of God, whether God exists and whether God caused the Big Bang can now be addresses more intelligibly.

Eleven

THE BIOPIC TELEOLOGICAL ARGUMENT

No appeal to infinitely many worlds, either antecedent or contemporary, can explain the existence and order of our life-supporting world unless infinitely many worlds actually exist. Likewise, God cannot explain our life-supporting world unless there actually is a God. Theistic Cosmology affirms that God caused the Big Bang, that God is THE necessary condition for its occurrence; but this claim is true only if God exists. What reasons support the belief that God exists?

Rational evidence for God was expressed traditionally in philosophical arguments for God's existence. Two traditional arguments for God, the Teleological and the Cosmological, are of special interest to us because they directly invoke what we know about the cosmos.

When he considered evidences for God drawn from our knowledge of the world, the philosophical theologian John Hick concluded that "The universe, as presently accessible to us, is religiously ambiguous in that it is capable of being interpreted intellectually and experientially in both religious and naturalistic ways."[1] Since all phenomena can be interpreted in both Theistic and Naturalistic terms, neither position can win a clear victory over the other, Hick contends. Yet, the crucial issue is not whether all observations can be interpreted in a certain way. Rather, it is whether one interpretation is more defensible rationally than another. Hick is too generous and kind toward Naturalism. We saw in Chapter Two that no strong case can be made for Naturalism, especially when measured by its own appeal to scientific method alone; and in the ensuing chapters we saw that contemporary atheistic cosmologists fail to explain adequately the origin, order, and existence of our universe without God.

Can a good case be made for God's existence based partly on the order or design of the universe disclosed to and through contemporary Anthropic or Biopic Cosmology? Granted that they will not be absolutely certain, can our inquiry produce theistic results that are rationally warranted and compelling?

1. God's Purpose for the Universe and Cosmic Teleology

The Teleological Argument or Argument from Design affirms that the observed order of the world provides powerful evidence both for the existence of God and for divine attributes like power, intelligence, and benevolence or good intentions. A well-designed cosmos implies not only *that* God exists but also something about *what* God is like. The Argument from Design expresses the deep religious intuition that the ultimate cause of the universe knew what it was doing and did it well from commendable motives. Perhaps no one ever really believes in God without that intuition, but reflection may make it plausible.

From the time of Plato, innumerable versions of the Argument from Design have been offered, criticized, defended, and attacked. Today's Anthropic Cosmology presents us with dazzling indications that our world was deliberately designed by an intelligent and benevolent being of Divine proportions. Amazingly, most Anthropic Cosmologists reject a theistic explanation of the data and favor some version of an infinite world-ensemble metaphysics, according to which a life-supporting world like ours occasionally happens accidentally in a infinite number of tries. Barrow and Tipler acknowledge the possibility of a Theistic Anthropic Cosmology, but they and most other Anthropic Cosmologists reject it. Should we follow their lead?

A Biopic Teleological Argument for God contends that all of life, not just intelligent life, requires God. Is the Biopic Teleological Argument for God's existence defensible? Consider the following premises of a strong Biopic Teleological Argument for God.

Premise 1: Our universe is exceptionally fine-tuned for the production of an immense variety of intrinsically valuable complex forms of life.

Premise 2: This fine-tuning was caused by either by the existence of: A. infinitely many universes, or B. the Principle of Plenitude, or C. nothingness, or D. God–a transcendent, benevolent, Supercosmic Intellect.

Premise 3: It was not caused by the existence of: A. infinitely many universes, or B. the Principle of Plenitude, or C. nothingness.

Premise 4: Probabilities favor God.

Therefore: Our universe's fine-tuning for life was caused by an existing God, a transcendent, benevolent, Supercosmic Intellect.

Both Teleological and Cosmological Arguments for God reason from something that is known to be true of the world through sense experience to the existence of a transcendent Ultimate Reality who best accounts for that something. The Teleological Argument reasons from the presence of order, design, and purpose in the universe to the existence of a supreme, intelligent, skillful orderer or designer, and benevolent purposer–God. The Cosmological Argument, examined in Chapter Twelve, reasons from contingency or dependence in and of the world to the existence of a Divine ground for all contingency.

Teleological and Cosmological Arguments for God contain some empirical premises that are known to be true on the basis of observation and inductive inference; but scientists are usually not interested in these kinds of experiential truths. Other premises in the arguments are philosophical and must be defended philosophically, with no pretense of doing natural science. Philosophy can operate at a level of generality that goes beyond the natural sciences, even though the line separating them is not exact. Although not done in the Teleological Argument, philosophy can even appeal to other types of experience–intuitive, introspective, religious, mystical–to which natural science, confined to sense experience, cannot appeal. Philosophy's repertoire of arguments, concepts, and analysis transcends the natural sciences. Absolute certainty is not available to us anywhere, especially in philosophy and natural science. A fallibilistic approach to philosophy concedes from the outset that absolute certainties are unreachable, but it also eschews absolute skepticism. Without giving us absolute certainty, philosophy can at least give us rationally warranted assertions and an enlightened faith–something far superior to blind faith.

Now let us examine and defend the premises of our Biopic Teleological Argument more carefully.

A. Extraordinary Cosmic Coincidences that Favor Life

The first premise of the Biopic Teleological Argument points out that our universe is exceptionally well-designed for the production of an immense variety of intrinsically valuable complex forms of life, human and otherwise. Atheistic versions of the Anthropic Principle, discussed previously, agree that scientific cosmology has discovered impressive empirical evidence that our universe is extraordinarily fine-tuned to engender and sustain our existence. If it were not, we would not be here. But we are here! Why must we regard our universe as fine-tuned for life, and what best explains that?

Our universe is exceptionally suitable for the production of complex intrinsically valuable life during certain prolonged cosmic periods like the one in which we exist. Exceedingly small changes in the most basic physical features of our universe would make complex and valuable life impossible. About that, very little disagreement exists. Tiny changes, usually much less than one percent, in the numerical values of fundamental physical features of nature like gravity, electromagnetism, the strong nuclear force, and the weak force, would have resulted in a lifeless universe. So would minute changes in initial conditions, natural laws, and many other primary components of the cosmos. All the fundamental components of our cosmos, as well as their harmonious interrelationships, must be calibrated with incredible exactitude to produce a life-sustaining universe.

In what follows, most of the numbers are omitted; but physicists attach precise numbers to all the incredible life-supporting "coincidences" that make up the basic physics and chemistry of a life-supporting universe like our own. To the finite range of every numerical value essential for the production of a life-sustaining universe corresponds an infinity of numerical values that would insure lifelessness. For example, the speed of light at roughly 186,000 miles per second is a very fundamental constant in a relativity universe; but, as with all numerical values essential for life, a World-designer could get it wrong in an infinite number of ways. If the speed of light were 184,000, 188,000, or 189,000 miles per second, life would probably be impossible. An infinite number of ways to go wrong correlate with the very small range of permissible numbers for every physical condition that supports life. An infinite number of whole numbers above 186,000 miles per second, and an infinity of fractions between all these whole numbers, would yield lifeless universes.

Every possible lifeless universe would be an actual universe if all possible universes are actualized, as dictated by the Principle of Plenitude. This includes infinite variations on the numbers for all physical conditions that could enter into the constitution of possible universes. According to Stephen Hawking,

> It seems clear that there are relatively few ranges of values for the numbers that would allow the development of any form of intelligent life. Most sets of values would give rise to universes that, although they might be very beautiful, would contain no one able to wonder at that beauty.[2]

Getting all the numbers right for life is a task for Infinite Intelligence. All the physical fundamentals of our universe and their collective harmony were precision-tuned for life production, and this is overwhelming evidence that our universe was deliberately contrived for life by a Divine Being who loves life. Consider just a few cosmic fundamentals that are specially designed for life.

i. Matter/Antimatter Asymmetry

Either broken matter/antimatter symmetry, or primordial asymmetry, is necessary for the emergence of valuable life. We really do not know whether symmetry prevailed in the beginning, or whether asymmetry was an original feature of our universe. Assuming broken symmetry, as do most cosmologists, the processes that upset the balance of matter and antimatter in the earliest universe are very mysterious; but if symmetry had not been broken, the universe would not have evolved beyond a primordial fireball fueled by endless matter-antimatter collisions and explosions. Astrophysicists believe that when symmetry was broken, only one material particle survived for every billion matter-antimatter pairs that were annihilated. Without these exceptionally rare survivors, there would be no stable baryonic matter (protons and neutrons) and leptonic matter (electrons and neutrinos). No stable physical world would exist at all, and if no stable physics, then no life. A universe of stable antimatter would serve just as well for life, but it too would require either broken symmetry or primordial asymmetry.

Symmetry might have snapped in an infinite variety of ways, very few of which would be life-supporting. It looks as if the symmetry/asymmetry deck was stacked deliberately; so also were all the other most basic features of our cosmos. Of course, for all we really know, the universe was created from the outset with a great preponderance of matter over antimatter as an initial condition; but this primordial asymmetry would also be a stacked deck. For us to exist, our universe must contain just the right amount of free matter.

ii. A Weaker or Stronger Force of Gravity

If the force of *gravity* were slightly *weaker*, the universe would have expanded too rapidly for galaxies, stars, and planets to form. Gravity would have been too weak to pull local clouds of gaseous mass/energy together into heterogeneous clumps. With weaker gravity, the universe would have undergone a rapid heat death. Mass/energy would have dissipated too fast for life to form. The evolution of valuable life is impossible without just the right tug from gravity to form planets with suitable physical and chemical conditions for life's emergence and development. To be life-supporting, hospitable planets also must have stars (suns) produced by gravity over billions of years of time that furnish them with just enough sustained energy, not too little, not too much.

If the force of *gravity* had been slightly *greater*, suns would suck in their planets or fail to release enough energy to sustain life. On planets with slightly increased gravity, living things would have to be extremely small and light in weight to avoid breaking apart when they fall. If gravity had been somewhat more powerful, no suns and planets at all would have formed; all regions of the universe in which they actually formed would have undergone rapid gravitational collapse. More powerful gravity would yield a universe full of black holes but no galaxies, stars, and planets. If a universe with greater gravity were to begin with a Big Bang, it would end shortly in total gravitational collapse.

iii. More or Less Mass/Energy

If the force of gravity were unchanged, but the initial quantity or density of *mass/energy* or physical particles in the universe were slightly greater or less, the results would resemble those from variations in the force of gravity. With either too much mass/energy or too much gravity, a universe would not expand rapidly enough. It would be too hot for life, and its total duration would be too short for life to evolve. With too little mass/energy or gravity, the universe would expand too rapidly and quickly become too cold for life. Stephen Hawking points out that "If the rate of expansion one second after the Big Bang had been smaller by even one part in a hundred thousand million million it would have recollapsed before it reached its present size."[3]

iv. The Size and Age of the Universe

Both the *size* and the *age* of the universe are closely related to its mass/energy; and as B. J. Carr and M. J. Rees put it, "The Universe must be as big and diffuse as it is to last long enough to give rise to life."[4] A closed universe no bigger than our Milky Way would pass through its entire expansion/contraction cycle in about a year of our time; and valuable life could not evolve. In a small universe of short duration, nucleosynthesis could not produce heavy elements like carbon, oxygen, and nitrogen, so essential to life as we know it. Even non-carbonaceous life forms, if there are any, would require some heavy elements, though we do not know exactly which ones or in what proportions.

No enduring life forms could exist within universes consisting solely of volatile hydrogen and/or helium gasses–not even in infinite quantities and configurations. Life requires a diversity of cohesive chemical elements; it cannot be composed solely of hydrogen and helium. No heavy elements were produced in the Big Bang. They require billions of years of cooking time in stellar furnaces. The earliest stars contained no chemical elements heavier than hydrogen, helium, and traces of deuterium and lithium. No life evolved until after the first generation of stars synthesized heavy elements and then exploded as supernovas to scatter their contents through spacetime. Afterwards, gravity slowly reassembled this heavy stardust into later generations of stars with planets. At our stage of cosmic development, this has happened many times; but if it never happened, no life would exist. Anthropic Cosmologists firmly believe that life could not have evolved much faster anywhere than it actually did on earth. Barrow and Tipler recognize that their arguments "use evolutionary time-scales as a crucial step."[5] This assumption seems fair enough.

v. Variations in the Electromagnetic Force

If the *electromagnetic* force that binds electrons to atomic nuclei were much weaker, no atoms would form, not even hydrogen, much less the heavier elements. If it were only slightly weaker, all stars would be inhospitable blue giants. If it were slightly stronger, no long-lasting hydrogen-burning stars would exist. All stars would be red dwarfs that could not explode as supernovas to distribute heavy elements throughout the universe. Whether slightly weaker or stronger, no long-lasting main sequence stars like our sun would exist to provide the duration and stability of life-supporting conditions essential for the evolution of life.

vi. Alterations of the Strong Force in Atomic Nuclei

If the *strong force* that binds protons and neutrons in atomic nuclei were slightly greater, there would be no atomic nuclei, no protons, and thus no atoms at all. If atoms were to form, all hydrogen would burn quickly into helium. The physics is complex, but if the relative strengths of the electromagnetic and strong nuclear forces had been ever so slightly different, either three helium atoms would not have fused to form carbon, or carbon would have been so unstable that it would have fused quickly with a fourth helium atom to form oxygen.[6] Either way, the cosmos would contain no stable carbon. The very existence of carbon in the universe for the construction of carbon-based life is a striking cosmic coincidence. If the strong force had been slightly weaker than it is, the universe would be composed entirely of hydrogen gas, with no heavy elements and no living things. No deuterium, heavy hydrogen composed of one proton and one neutron in its nucleus, would have formed because the nucleus of deuterium is too easily torn apart. No hot stars would exist in an all-hydrogen universe, for even if proto-stars had formed, nuclear fusion could never commence without deuterium to ignite it.

vii. Variations in the Weak Force Controlling Nuclear Decay

The weakness of the *weak force* that controls nuclear decay must be very precise to produce biochemical life. A precisely calibrated weak force is essential for fusing protons into elements heavier than hydrogen. If the weak force were slightly weaker, all the hydrogen in the universe would fuse quickly into helium; and there would be no slow-hydrogen-burning main-line stars like our sun. Heavier elements like carbon, oxygen, nitrogen, and the chemical compounds upon which life depends would not exist. No supernovas could explode to distribute heavy elements through the cosmos if the weak force were any stronger. If the weak force had been only slightly stronger than it is, everything would have turned immediately into iron. What a dead universe that would be!

viii. Different Spatial Dimensions

Innumerable features of our world are exceedingly fine-tuned for the production of intrinsically valuable lives. With only two instead of *three spatial dimensions*, no life could exist. Without three spatial dimensions, there would be no sub-atomic particles, no atoms, no molecules, and no organic chemistry to form biochemical life. Edwin A. Abbott's two dimensional "Flatlanders"[7] exist only in fiction, not in fact. Stephen Hawking explains how awkward, indeed how impossible, two-dimensional life would be. Two dimensional animals "would have to climb over each other to get past each other," and a digestive tract passing all the way through them would cut them in half![8] More than three spatial dimensions would also be incompatible with life; so, Hawking concludes, "Life, at least as we know it, can exist only in regions of space-time in which one time and three space dimensions are not curled up small."[9]

ix. Additional Fine-Tuned Features

Contemporary astrophysicists have uncovered a vast array of additional fine-tuned physical features of the universe like large number coincidences and exacting initial conditions that are necessary for the appearance and development of all life, especially complex valuable life. Very detailed and thorough discussion of the incredibly fine-tuned cosmic coincidences presupposed by the existence of life were published in 1989 by John Leslie, who interprets the data

Platonically,[10] and in 1993 by M. A. Corey, who interprets the data theistically.[11] For more details, read their books!

We have seen enough at this point to understand why Anthropic Cosmologists accept the Anthropic Principle, which affirms that the production of valuable lives is the purpose of our universe. For very good reasons, Atheistic Anthropic Cosmologists accept the Anthropic Principle, the first premise of our Biopic Teleological Argument for the existence of God. Our universe is indeed exceptionally fine-tuned for the production of an immense variety of intrinsically valuable complex forms of life. But they reject the second premise, that God did it; so we must scrutinize their reasons for doing so.

B. Inadequate Non-Theistic Explanations

No matter how impressive the cosmic coincidences are that sustain our existence, the Biopic Teleological Argument for the existence of God does not succeed if it can be shown that our universe's fine-tuning for life results from something other than a divine transcendent benevolent Supercosmic Intellect; but this cannot be done.

In earlier chapters, we saw that most Anthropic Cosmologists reject the theistic solution and try to account for the life-sustaining order of our world with the hypotheses that it was caused by: i. Infinitely Many Universes, or ii. the Principle of Plenitude, or iii. Nothingness. All these non-theistic solutions presuppose that our favorably ordered world was produced by blind chance. *Earlier chapters demonstrated that none of these alternatives can be defended.* This does not show with absolute certainty that God is the best explanation because other unanticipated explanations might arise (always a problem with disjunctive arguments); but these will just have to be confronted if and when they appear. Without repeating every detail, consider these reminders of why atheistic explanations of cosmic teleology fail.

i. Infinitely Many Worlds

Most Atheistic Anthropic Cosmologists think that a divine Supercosmic Intellect can be avoided by appealing to the unverified and unverifiable existence of infinitely many worlds. They claim that there are or have been an infinite number of transcendent antecedent and/or contemporary universes, that this infinity of universes in Supertime or Superspace actualizes all possible individuals, universal properties, and relations, and that it occasionally includes a life-supporting universe like ours purely by accident. If these metaphysical assumptions are indefensible, infinite worlds metaphysics fails to provide a plausible alternative to a Divine Supercosmic Intellect. Earlier, we determined that these assumptions do fail–for the following reasons.

No good scientific or empirical reasons or evidences show that infinitely many worlds exist, even if prominent scientists say so. No infinitely transcendent Supertime or Superspace containing an infinite number of particular spacetime systems or universes can be verified directly or inferred inductively or indirectly from available evidence. The creation of our world by an infinity of antecedent worlds, by infinite Mother Spacetime, or by infinite world divisions is not an empirically confirmed belief. "Scientifically" postulating infinitely many universes flagrantly, infinitely, violates Ockham's razor. No antecedent oscillating universes can end in singularities because, by definition, time, space, and physical causation are non-existent in singularities. Singularities provide no spatiotemporal or causal corridors for linking universes. Yet, without singularities, increasing entropy would carry through an infinity of antecedent oscillating universes and make our world infinitely chaotic–anything but life-supporting. But here we are!

Roger Penrose maintains that although singularities are infinitely small, dense, and curved, *they can nevertheless be structured with low entropy*. The initial singularity that originated our universe was structured with low entropy or disorder, and it has been losing it ever since. Recall that since singularities are not empirical entities, any claims about them can be translated without loss of empirical content into the language of creation *ex nihilo*. Thus, Penrose's claim about the low-entropy singularity that initiated our universe is empirically indistinguishable from the theistic claim that the original stuff that God created from nothing (or from a "singularity") at the beginning was divinely ordered with low entropy.

That God created the low-entropy grandly-unified mass of energy that began our universe is much more plausible than that it was caused by an antecedent universe's Big Crunch. Most cosmologists today agree that an antecedent universe's final singularity, culminating its Big Crunch, would contain extremely high entropy or disorder.[12] Even if the initial singularity of our universe were really nothing, as in "created out of nothing," our cosmic epoch was ordered from the outset with low entropy. No high entropy Big Crunches could ever produce the low entropy Big Rebounds required by Oscillation Cosmology because great chaos is never the direct and immediate cause of great order–another well confirmed empirical truth. More technically, no state of affairs being torn apart by rippling Weyl curvature is ever followed immediately by its total elimination.

Atheistic Anthropic Cosmology's other worlds are just as transcendent, non-empirical, and Other Worldly as Heaven and Hell in traditional Christian theology. Atheists may believe in them by a heroic leap of faith, but no rationally defensible evidence supports their existence. Naturalists should reject all cosmological other-world-ensembles for the same reasons that they repudiate religiously based Other Worlds. Yet, Naturalism can be defended only by appeal to Other Worlds! We know, because of the Big Bang, that *our* world has not always existed.

Even if infinitely many worlds were actual, they would not explain the existence of our life-supporting world. An infinity of universes in time or space will *not* actualize *all* possible individuals, universal properties, and relations. Infinity cannot be "used up," not even by infinity. Infinity minus infinity equals infinity. Numerical spatiotemporal infinity alone does not guarantee that a life-supporting universe will come along after a sufficient number of lifeless ones have been used up. Lifeless universes could easily be repeated inexhaustibly in time and/or space, either by causal necessity, or by accidental quantum fluctuations. Atheistic explanations commit the inverse gambler's fallacy.

David Hume maintained that all possible configurations of matter would be actualized in an infinite amount of time, but the very idea is incoherent, for many configurations are incompatible with others. It is logically possible that a monkey typing the letter "a" would type only that one letter from infinity to infinity, to the exclusion of all other possible worlds. Nothing in the concept of infinity requires any diversification at all. A monkey banging on a typewriter could just hit one key forever–in Eternal Recurrence. The assumption that our life-supporting universe is made more probable by the prior existence of an infinite number of lifeless worlds commits the inverse gambler's fallacy.

A numerical spatiotemporal infinity should not be confused with infinite individual, qualitative, and relational diversity. Bare infinity lacks a mechanism for insuring *any* diversity, much less the *right kind* of physical diversity for life. Unless some Divine Agent of Limitation and Selection chooses diversity, neither spatial nor temporal infinity as such is compelled to manifest any diversity at all among actual universes, much less infinite diversity. Some diversity among bubble universes, if any others exist, may happen accidentally through quantum fluctuations; but an infinite number of accidental universes cannot use up an inexhaustible class of lifeless universes to guarantee the actual appearance of a life-supporting one like ours. Given an infinite number of shoes, it is possible that *no* shoe fits! Trash universes could easily be repeated endlessly, even if caused by quantum fluctuations. Without God, they probably would be. Mere infinity gives no assurance of any diversity at all; inane sameness can be reiterated forever. Either the Principle of Plenitude or God might account for diversification among universes, but God alone really explains it because plenitude will not work. Why not?

ii. The Principle of Plenitude

The Principle of Plenitude affirms that all possible worlds are actual. It is a theological principle, not an empirical or scientific fact. It is empirically unverified and unverifiable; and it violates the principle of parsimony, Ockham's razor, in every conceivable way.

The Principle of Plenitude is derived from Platonic Theism, where it defines the Greek ideal of divine perfection. There are other arguments for its truth, but none that can really be defended philosophically.[13] Theists need not accept it unless they agree that Divine perfection necessitates the actualization of every possibility–including horrible as well as beneficent possible worlds; but actualizing all possibilities is not possible; the very idea is incoherent; and infinity cannot be exhausted. World-ensemble metaphysics is not natural science. Blind faith may postulate infinitely many diversified worlds; but that is all it is–blind faith. With or without singularities, no evidence available to us indicates that infinitely many worlds exist, diversified or not. Consider the following argument.

> If infinitely many diversified worlds exist, a life-supporting world like ours will come along occasionally purely by accident.
> Therefore, our life-supporting world came along purely by accident.

Deductively, this conclusion does not follow. From "If p, then q" alone, we cannot conclude "q". Only if we also know "p"–that infinitely many diversified worlds exist–can we conclude "q"–that a life-supporting world like ours will come along occasionally purely by accident. We do not know that "p" is true, and we have many good reasons for thinking that it is false. The existence of an infinity of diversified worlds–plenitude of creation–may be advanced as an explanatory hypothesis, but it does not explain a life-supporting world, and it cannot be successfully defended.

Admitting that other worlds are logically possible concedes absolutely nothing about their actuality. For every contingent "possibly so," there is a "possibly not." Possibly, many other worlds exist; possibly, no other worlds exist. These possibilities are mutually exclusive. Possibly, you are extremely wealthy; but that puts no money into your bank account. We should not confuse abstruse possibility with high probability, credibility, actuality, or reality. Atheistic Anthropic Cosmologists regularly make such confusions.

Plenitude, the actualization of all possible worlds, is logically incoherent. The non-existence of every possible world is also a possible world. Mutually exclusive whole universes and sets of universes are logically possible, and so are mutually exclusive alternatives or variations within any one universe. Our own world (or *all* worlds) could have been either life-supporting or non-life-supporting, quantum or non-quantum, oscillating or non-oscillating, Divinely created or not so.

Plenitude's requirement that all possibilities be actualized in an infinite number of universes is not logically coherent. As Leibniz knew, there are logically incompatible, mutually exclusive, or "incompossible" possibilities. It is logically possible that an infinite number of worlds are all life-supporting, and that none of them are; but these possibilities are incompossible, that is, they are possible separately but not possible together. Possibly, nothing exists; possibly, something exists; but these possibilities cannot both be realized. Possibly, every strand of past oscillating universes ended in nothingness, in which case there would be nothing now; possibly they all ended in openness, in which case they could not explain our Big Bang. The realization of all possibilities implies that nothing now exists, which is obviously false! It also implies that everything now exists, which is equally false! Possibly, a single lifeless universe recurs eternally in every tiny detail; but this would rule out all possible life-supporting universes, including our own. The very idea of realizing all possibilities makes no sense. We cannot infer from an incoherent concept that our kind of universe is bound to occur accidentally from time to time. It is logically impossible, logically false, that everything possible is actual. Incompossibles are possible separately but not together. Choices between possible worlds have to be made; actual existence is competitive and selective. Who chose our life-supporting cosmos? Not plenitude!

Plenitude is an abstract, disembodied, normative principle. At best it can only be a formal cause. Nothing totally abstract, disembodied, and normative can make actual choices or be the efficient cause of anything. Abstract, disembodied, normative principles bring about results only when contained within and intentionally acted upon by concretely existing individuals who understand them. God is required for the causal efficacy of plenitude, or more selective creation. A good God would not act upon it, would not create every possible evil or trivial world.

Atheistic appeals to plenitude to avoid God are self defeating. If all possibilities are actual, then a God who selects among possible worlds exists since such a God is possible! Indeed, all conceivable gods exist, assuming that the idea of such is coherent. And no God exists, since that too is possible.

iii. Nothingness

Nothing is ever caused by pure nothingness. Out of nothing, nothing comes. All experience and theory grounded in experience, including quantum physics, is against the Big Accident hypothesis that nothing causes something. Quantum physics does not abandon causality altogether; it always retains necessary causal conditions and relinquishes only sufficient causal conditions. Abstract disembodied physical principles like quantum laws cannot exist and act within nothingness, for then nothing would be something after all. No good empirical scientific evidence shows that infinitely transcendent Superspacetime exists beyond or before our spacetime, or that it has a quantum bubbly structure that produces spontaneous fluctuations. Theology may require Superspacetime, but not science. Non-quantum worlds devoid of quantum perturbations are logically possible in infinite numbers. Pure nothingness cannot guarantee anything, especially that endless universes are governed by quantum laws, or that an occasional quantum universe is accidentally life-supporting. Since non-quantum worlds are possible, universes capable of quantum fluctuations require their own explanation and do not ultimately explain anything. What selects for quantum instead of for non-quantum worlds? Nothingness selects nothing.

B. Probabilities Favor Divinity

The best explanation for our universe's fine-tuning for life is that an existing God, a wise and benevolent being of supercosmic proportions, chose and created it knowingly and intentionally. Various strategies tend to show that this conclusion is true. Up to this point my strategy has been to demonstrate that alternative explanations of the cause of the Big Bang are indefensible. Non-theistic explanations really do not work. Often, the winner of a contest is the last one left standing on the field.

Yet, it would be nice to have more positive or constructive evidence for thinking that God chose the life-supporting features of our world. Probabilities positively favor Divinity, but this needs some explaining.

In one sense, we cannot determine the probability of the occurrence of a life-supporting universe. Probabilities are normally assessed statistically by comparing an *actual* instance with other actual members of a known class; but no existential probabilities are available for unique entities that belong to classes having only one member. Charles Sanders Peirce remarked that universes are not as plentiful as blackberries. As far as we really know, our 15 billion-year-old universe is unique, the only one of its kind. We have access to no other actual universes with which we can compare it.

Recognizing that we experience only one world, David Hume argued that we are entitled to proclaim the truth of our theory of the origin of the universe only if innumerable universes have been formed before our eyes. Obviously, this has not happened. Neither theists nor atheists have ever seen worlds formed before their eyes, so they are both in the same boat in this respect. Atheistic world-ensemble metaphysicians seem to be unacquainted with David Hume, or they read him very selectively! Hume's own supposition of innumerable worlds existing in infinite time is inconsistent with his own skepticism!

In another sense, however, probabilities about ordered universes can be determined. A unique actual entity can be placed within a class of *possible* entities, and its absolute probability can be calculated. To determine the probability of a life-supporting world order, our actual ordered-for-life universe can be compared with all possible members of the class of universes that might exist. Appeal to absolute probability is commonplace in quantum mechanics, where determining the probability of a quantum event involves summing over all logical possibilities for that event.

Recall that physicists can assign numbers to every basic physical dimension of the universe–its density, expansion rate, entropy, the force of gravity, the strong nuclear force, the weak and electromagnetic forces, the speed of light, Hubble's constant, Planck's constant, and so on. For life-supporting universes, permissible numbers for *each* such basic component of life-supporting physical reality fall within a very limited range. This leaves an infinite range of numbers for each component that would result in lifeless universes, all of which are possible worlds. *For every measurable component of life-supporting universes, the range of right numbers is finite; the range of wrong numbers is infinite.* This is also true of the way in which all the relevant numbers must be integrated harmoniously in order to make a life-supporting universe.

Each wrong number correlates with a possible lifeless universe. Within the class of possible designs for universes, life-supporting ones are infinitely improbable. Recall Stephen Hawking's comment that

> There are relatively few ranges of values for the numbers that would allow the development of any form of intelligent life. Most sets of values would give rise to universes that, although they might be very beautiful, would contain no one able to wonder at that beauty.[14]

Steven Weinberg conjectures that "The existence of some form of life will turn out not to require any very impressive fine-tuning of the laws of nature,"[15] and he doubts that we will find "any sign of the workings of an interested God, in a final theory."[16] Yet, even if conditions favoring the existence of life are fairly broad and do not require extensive fine-tuning, the range of permissible physical numbers is still very finite, and the range of impermissible numbers is infinite. Our own world order is very impressive because the number of ways to go wrong with any two basic physical components is infinity multiplied by infinity–which is still just infinity, but putting it this way makes the point more impressive! If all possibilities are actual, no life-supporting universe exists now anyway, since it is possible that no life-supporting universes exist at all. An infinity of life-defeating aberrations corresponds to every finite range of cosmic conditions that would support life. Incomprehensible (to us) intelligence and skill are necessary to get single as well as conjoint conditions right for a life-supporting cosmos.

The order of our universe most resembles the intricate order of the most complex products of purposive human intelligence, not the order of chaos or chance. Anthropic Cosmologists are well aware of this. Roger Penrose, for example, points out that a life-supporting universe must begin in a state of extremely low entropy or disorder, and that the possibilities and probabilities for high entropy universes are immensely greater than for low entropy universes. To understand how enormous the odds are against getting a low entropy universe, Penrose suggests, we should picture the Creator poised before a system of space as voluminous as our physical universe, each point of which represents a distinctive universe. The Creator's task is to stick a pin into this vast system of space at exactly the one right point that will produce an initially low entropy life-supporting world. The rest of the volume corresponds to lifeless high entropy universes, so they would be exceptionally easy to hit; but the chances of hitting the tiny volume that represents a life-supporting low entropy universe are exceedingly minute.[17] The Creator's aim must be precise to an accuracy of "one part in $10^{10^{123}}$" if the intended target is a low entropy life-supporting universe. This number is so enormous, says Penrose, that we could not write it down in ordinary mathematical notation even if we wrote a "0" on every particle of matter in the universe![18]

Penrose actually underestimates the odds because his huge number is still finite; and the odds against life are really infinite. The number of non-life points on the Creator's target is infinite; life-points are finite indeed! Arranging a universe for life is no task for mere chance. The probability is overwhelming, indeed infinite, that a numerical infinity of universes in space and/or time would be wrong for life. Only a Divine Supercosmic Intellect could successfully select for and create a low entropy universe.

In 1980, Fred Hoyle, traditionally no friend of Theism, was overwhelmed by the realization that the chances are astronomically small that any universe would accidentally produce two things essential for life, enzymes and carbon. Hoyle says he was "plagued by the thought that the number of ways in which even a single enzyme could be wrongly constructed was greater than the number of all the atoms in the universe," and he found the conclusion irresistible that enzymes are produced

> by thought, not by random processes. Rather than accept the fantastically small probability of life having arisen through the blind forces of nature, it seemed better to suppose that the origin of life was a deliberate intellectual act. By "better" I mean less likely to be wrong.[19]

In 1953, Hoyle was the first to discover how nearly impossible it is to get stable carbon and oxygen from stellar nucleosynthesis. In 1980, he reflected on the odds against getting carbon, with properties so essential to life as we know it, by pure chance. He concluded that "Some supercalculating intellect must have designed the properties of the carbon atom," and that "The carbon atom is a fix."[20] Considering the chances of getting roughly equal quantities of carbon and nitrogen by stellar nucleosynthesis, we must again conclude that this is "another put-up, artificial job." Hoyle continued,

> A common sense interpretation of the facts suggest that a superintellect has monkeyed with the physics, as well as with chemistry and biology, and that there are no blind forces worth speaking about in nature. The numbers one calculates from the facts seem to me so overwhelming as to put this conclusion almost beyond question.[21]

Theism is more probable than Atheism because it provides the most reasonable and thus the best justified explanation for the existence of our kind of universe. Saying that one philosophical position is "more probable" than another is just a shorthand way of saying that the philosophical case for the one is stronger than for the other. In that sense the Teleological Argument in modern biopic dress affirms that the existence of God is more probable than God's non-existence.

The Teleological Argument tells us something important that the Cosmological Argument does not disclose, namely, that the ultimate cause of the universe is intelligent and well-intentioned or benevolent. The beneficial order of the world, especially its capacity for producing an immense variety of intrinsically valuable forms of life, including human life, reflects the stupendous knowledge, skill, and generosity of a benevolent, transcendent, personal, causal agent–God.

The central question raised by the Biopic Teleological Argument is simple. Did the ultimate cause of the universe know and care about what it was doing, or not? Did our life-supporting universe originate by intelligent choice or by blind chance? Naturalistic, atheistic, infinite worlds metaphysics says, "By dumb chance." Theism says, "By brilliant choice."

Conclusion: God Ordered Our World

Our Biopic Teleological Argument concludes that an existing God, a benevolent Supercosmic Intellect, intentionally and knowingly selected and brought about the life-supporting order of our universe; but this conclusion does not follow simply from the authority of preeminent cosmologists like Penrose and Hoyle. The *evidence* to which they call attention is crucial, and it strongly supports all the premises of our Biopic Teleological Argument. Thus, most likely, a universe ordered for the production of an immense variety of complex and intrinsically valuable forms of life was caused or created intentionally by a transcendent benevolent Supercosmic Intellect.

Alternative accounts fail to explain the stupendous cosmic coincidences that conspire to produce life in our world. The only reasonable hypothesis is that an Ultimate Cause, namely God, who knew and cared about what it was doing, is responsible.

Eliminating the competition is a perfectly respectable way to argue in philosophy. Often, the best available proof is a disproof of the alternatives; the solution that best withstands the process of critical examination is the winner. At this point in the battle, only one plausible explanation remains: God did it. God knowingly and intentionally designed our universe for life.

Many objections may be raised to the teleological argument. Atheists do not give up without a good fight, so let us turn now to some commonplace misgivings.

2. Critique and Defense of the Biopic Teleological Argument

The Teleological Argument, though very popular with religious believers, has been under assault almost from its inception in ancient Greece. Most professional philosophers today mistakenly assume that critiques by Hume and Kant made it utterly indefensible. The most serious contemporary objection to the Teleological Argument is that an infinite worlds metaphysics makes God unnecessary; but we have already disposed of that alternative, along with the difficulty that probabilities cannot be calculated for life-supporting universes. Other objections remain, and to these we must now turn.

A. Natural Creation of Order

If the universe naturally creates order, no external Divine Designer is required, some critics insist. Victor J.Stenger maintains that the life-supporting order of the universe occurred by chance, *Not by Design*, as the title of one of his books indicates.[22] In both *Not by Design* and *The Unconscious Quantum*, Stenger stresses matter's capacity to organize itself into meaningful patterns and contends that this obviates the need to resort to a divine designer to explain the order of the universe and the origin of life and mind. Eric Lerner also claims that, not divine guidance, but a "natural tendency of all matter, both animate and inanimate, to evolve toward higher rates of energy flow, toward the capture of greater currents of energy," adequately explains humanity's origin and development.[23] Both Stenger and Lerner take the creatively self-organizing capacity of the physical world as a given that needs no explanation; but it does. Why is our physical world capable of self-organization when infinitely many alternative worlds would have no such capacity? Explaining that is imperative!

According to Stenger's *Not by Design*, the natural order of things permits chance occurrences; and chance occurrences eventually add up to an ordered world in an infinite amount of time. However, an infinite amount of time is not available if the Big Bang is right; and even if time were infinite, we could not know it scientifically. Also, random or chance events presuppose order and can be recognized as such only against a background of order. Given all the laws of physics, a deck of cards can be randomly shuffled; but with no laws of physics at all, there can be no random shuffle. Randomness presupposes orderliness

Stenger has the effect before the cause, the cart before the horse. The basic natural order of things permits chance occurrences that further increase order; but without a very special basic initial order, no increase would occur. Certain types of minimal order produce more order, but absolute chance or chaos produces nothing. That a very minimally ordered universe would *not* be creatively self-organizing is infinitely probable.

Even if absolute chance could produce well ordered universes, which it doesn't, the probability is overwhelming that they would be lifeless. Why do we live in a quantum universe where chance is real instead of an old-fashioned Newtonian universe devoid of unpredictable spontaneity? Why do we live in a universe in which law and chance together contrive to produce intrinsically valuable forms of life instead of desolation? Given a minimal degree of order, why does chance (or creativity) produce more order, life-supporting order in particular, when it could produce infinite chaos or an infinite number of complexly ordered hostile-to-life environments? These questions require answers. Skeptics and naturalists like Stenger and Lerner have no good answers.

According to Lerner, there exists a "natural tendency" toward order. Yet, this natural tendency depends upon a deep structure of order in nature, which is exactly what the Biopic Teleological Argument is all about. Why is the basic order of the universe so constructed that it produces more order rather than less? In particular, why does it regularly produce beings who capture higher and higher energy flows, as Lerner puts it, when there are infinitely many other ways to order, or to disorder, a universe, some of them very complex? Why does our universe exhibit what Holmes Rolston, III calls an "upslope" to the long-range curve of evolutionary development, when neither survival nor adaptation to environment require it?[24] Why does the most basic physical order of the universe permit life to occur at all when there are infinitely many ways to fail?

Natural tendencies are existing patterns of order, but any primal creative order that produces more order requires its own explanation; it cannot just be taken for granted. Why does our universe have a life-supporting order when life-defeating orders are overwhelmingly more probable? The best explanation, according to the Biopic Teleological Argument, is that our world was designed by Divine premeditation to have a deep level of life-promoting, partly self-creative, partly self-organizing order. Atheists do not and cannot explain adequately why fundamental self-organizing propensities exist within our universe. They just take them for granted.

B. The Insignificance of Life in a Vast Universe

In a preceding chapter, we confronted the unsubstantiated view that life as we know it is insignificant because it is so small and short in relation to the vastness of spacetime. This fallacy–smallness equals insignificance–frequently resurfaces as an explicit objection to the Teleological Argument. The problem is, how can God's purpose for the universe be the creation of intrinsically valuable life when complex life exists only on earth, and merely for a few hundred thousand or million years at most? Intelligent life-forms belonging to the species *Homo sapiens* have existed for less than five hundred thousand years or so in a fifteen to twenty billion year old cosmos. Putting time scales for cosmic and biological evolution into perspective, Philip Hefner wrote,

> If we were to plot this sequence of events on a calendar with one day equaling 14 million years and one hour equaling a half million years, our natural history would look like this: on January 1 the earth's crust congealed; dinosaurs appeared on December 21; Neanderthal man arrives only at 11:50 p.m. on New Year's Eve. Relative to the overall history of the natural cosmos, the role of the human species is staggering in its minuteness.[25]

By now we know that and why the intrinsic value of our lives does not depend on vastness of size or duration, or on our utility in contributing toward some Enduring Grand Objective. God does not treasure us for these reasons, and we should not disvalue ourselves and one another for such spurious deficiencies. Our lives *as they are* exist for themselves, for other creatures, and for God. The value of life is located in the process of living and enriching it. All living things are beloved by God, and God's interests in creation are biopic, not merely anthropic, if Process Theologians are right.

Though presently unverified, that living creatures exist on numerous planets in innumerable galaxies scattered throughout the cosmos is highly likely.[26] Life in some of these places is probably far older and much more advanced than on earth. As Robert Jastrow speculated in 1980,

> If life is common in the cosmos, which is possible, then most of that life has advanced billions of years beyond us in evolution. And what does a billion years mean in evolution? A hundred years means nothing; that is only a few generations. A thousand years is not much more. A million years is the time it takes a new species to develop. What does a billion years mean? A billion years ago the fossil records show that the highest form of life on earth was the worm. So, if there are intelligent entities in space, out there, they are as far beyond us as we are beyond the worm. They may know the answer to the cosmic mysteries. They may know the meaning of the big bang.[27]

In 1997, Jastrow warned that if we ever make contact with extraterrestrial civilizations far older and more advanced scientifically than our own, our civilization might be destroyed. This seems to be the fate of all "primitive" societies that come into contact with technically advanced cultures separated by only a few thousand years of cultural evolution. Jastrow observes that "On this planet, contact between scientifically advanced civilizations and a primitive society...typically results in the destruction of the less-developed culture"; and he asks, "What may be expected of a meeting between civilizations separated by a billion years? Will we survive the encounter? I see no grounds for optimism."[28]

Perhaps we should hope that long-distance space-travel is *not* possible, that *Star Trek* and its sequels are only pipedreams (which they are), that the basic physics of the universe forbids long distance space travel even for scientifically advanced societies. Perhaps we should be thankful that stars with inhabitable planets are separated by such vast distances, and that no living beings can get from there to here. Perhaps God in his wisdom made it so!

Or perhaps we should hope that, unlike here on earth, ethical development has matched pace with technological progress in advanced civilizations beyond our corner of the Milky Way. Usually, the most ethical thing to do in relation to lesser forms of life is just to leave them alone. Far, far away, some alien civilization with a "prime directive" of noninterference may be doing just that for us. Maybe so; maybe not.

At any rate, cosmic life is probably of much greater duration and extent than earth life. If so, God knew, loved, and interacted with intrinsically valuable living things for billions of years before he had the dinosaurs or us to love. Extraterrestrial conscious beings would also be ends in themselves. So were and are members of more "primitive" human cultures and humanoid species. So were the dinosaurs. So were and are all sentient living things, including ourselves. If and when our species is sufficiently advanced, our ethical ideals and practices will reflect this ecological theology.

Life is scarce in our universe in the sense that vast regions of spacetime are unoccupied by any forms of life. No other planets in our solar system support life as far as we can tell, though Mars may have done so in the distant past. In 1996, two meteorites from Mars discovered in Antarctica were determined to contain organic compounds, best explained, some said, by bacteria-size organisms that lived on the red planet billions of years ago. In December 1996, NASA scientists announced that a large lake of frozen water has been located on the dark side of our moon. Since then, water has been found on Jupiter's moon, Europa, and in vast regions of "empty" space. Do traces of primitive aquatic life exist in some of these places? Continued space exploration may soon give us the answer.

All stars are directly inhospitable to life and many have no planets at all, much less inhabitable planets. In 1994, astronomers discovered the first solid evidence–orbital deviations of stars–for the existence of a planet in another solar system, and many others have since been identified.[29] Skeptics originally suggested that in some instances astronomers were seeing only earthquake-like vibrations on the surface of these stars, not wobbles caused by orbiting planets, but defenders justifiably did not concede defeat. In 1999, a large extra-solar planet was actually photographed as it crossed its star. *Sky and Telescope* exclaimed, "For the first time a planet of another star has been seen crossing the star's face, allowing astronomers to measure directly the planet's size, mass, and density."[30] Despite their relative scarcity, billions of sun/planet systems probably exist, and many of these may teem with life. If, on average, only one life-supporting planet is located in every galaxy in the universe, at least one hundred twenty-five billion life-supporting planets would exist!

If most of the universe is uninhabited, this does not mean that it is a wasteland. God could, and probably does, take keen aesthetic interest and delight in non-living nature, just as we do. Astronomers have discovered that the universe is much more active and violent than once thought. Perhaps God enjoys celestial fireworks as well as sunsets, grand canyons, frozen wastelands, and scorched deserts. Just because God created the universe primarily for life, it does not follow that God can have no other interests. Since God's time is limitless, God never has to rush anything. God has plenty of time for everything, including appreciating the sublime beauty of vast expanses of lifeless nature. We find great beauty and sublimity in the non-living parts of nature, including the starry heavens above, so why can't God? God's reality includes all reality, both living and non-living; but the starry heavens *above* would be no less magnificent if they were starry heavens *within*, as they are for God.

C. The Big Mess: Evil and The Religious Ambiguity of Order in Nature

Must every detail of ordered reality result from a Divine plan or design if God is as intelligent, powerful, and morally good or benevolent as the monotheistic religions profess? David Hume suggested that our evil-infested universe could have been ordered only by a stupid, weak, or demonic divinity. Perhaps it was designed and created by a Malicious Demon who formed it for the purpose of torturing its inhabitants. From the observable order of nature, can we really tell that its transcendent Divine cause is intelligent enough to know what it is doing, and that God's enduring character and intentions are morally good or benevolent? Skeptics have serious and legitimate doubts. The world is such a big mess that it really might not have been designed by an intelligent, powerful, and benevolent God. Can we conceive of a better way to introduce and organize the initial conditions of and conditions within the universe for producing life and for avoiding evil?

The reality of evil is particularly perilous to the Teleological Argument. The God of Classical Theism foreknows, causes, and plans every minuscule detail of creation. Both the general design of and every particular feature of the universe expresses God's Grand Plan. This is implausible for many reasons in addition to the problem of evil. Defenders of absolute grand design have actually claimed that God deliberately created fleas and bedbugs to be black so that people could detect them more easily on white sheets, that God intentionally created dogs multicolored so that we can distinguish them more easily from the furniture,[31] that God purposefully created more human males than females so that surplus males can be expended in war, and that absolutely every horrible thing that happens expresses God's will and God's deep but mysterious plans and purposes for creation. In the final analysis, God created the best of all possible worlds. Leibniz said so explicitly, but Voltaire vanquished the idea in his *Candide*.

Why wouldn't an omnipotent, omniscient, and omnibenevolent deity create only the best of all possible worlds–one in which every particular evil contributes necessarily, directly, and decisively to some great good? Why would God ever settle for a merely good world–one in which, on the whole good prevails, or at least has an opportunity to prevail, over evil?

The "best of all possible worlds," traditionally understood, was a universe for which God foreknew, preordained, and selected every detail of existence to achieve an absolute concord in which every distinct evil, whether apparent or real, contributes in perfect measure to the achievement of some greater good, some ultimate harmony of all things. In this best of all possible worlds, a flawless, eternal, and changeless divine plan is played out in every spatiotemporal event; and God is the ultimate decision-maker who decides and determines everything, including what we will decide. In it, absolutely everything that happens directly expresses "the will of God" and God's Grand Plan for all of creation. This feature of Classical Theology often infects popular religion.

The trouble with the best of all possible worlds, thus conceived, is that it is not the best of all possible worlds! It contains no creaturely initiative, freedom, creativity, or originative decision-making. God makes all the decisions and is responsible for absolutely everything. However, an even better world is conceivable–one in which every complex individual knowingly, freely, creatively, and responsibly chooses what is good or best and acts rightly. But this is precisely the kind of world that God alone could not create because, in this really best of all possible worlds, God is not the sole originative decision maker. In it, creativity is distributed to many if not to all creatures; and they are co-creators with God. God always has to settle for a merely good, not an absolutely perfect, world because in the truly best of all possible worlds, God is not and could not be absolutely in control of everything. Absolutely controlling free creatures is logically impossible and morally repugnant.

Creating a universe containing co-creative creatures is always risky business, for free creatures must, by definition, be able to choose without encumbrance both for and against the right and the good. The overwhelming probability is that at times they will choose against, and then there will be no perfect, calculated, and preordained harmony in which every evil achieves some particular good, in addition to the goodness of creativity and freedom themselves.

Process Theism denies that God absolutely controls and plans everything from eternity, especially particular evils. God's eternal Primordial Nature has general aims and objectives–like creating richly populated worlds in which desirable feelings and other goods are actualized in many species of living creatures; but God did and could not plan every detail of existence from eternity because the creatures are partly self-creative, are co-creative with God, and thus have some say-so themselves about what will come to be.

God did not timelessly calibrate and predestine that any individual person will exist, that he or she will have specific parents, marry a specific soul-mate, have a specific character, suffer particular ills, make very specific right or wrong moral and religious choices, and end up in Heaven or Hell. In his Consequent Nature, God makes and remakes plans to fit contingencies and choices *as they arise in the created world.* God responds appropriately to decisions made freely by spatiotemporal creatures and wills the best possible outcome for every undesirable spatiotemporal impasse, both within his own experience and within the world itself; but even this Divine will can be thwarted by co-creative decision-making creatures. God deals with unpredictable free choices and contingencies within the universe as they arise, not from eternity. God envisions alternatives for future realization while not knowing in advance which path will be chosen by co-creative creatures. No matter what the creatures decide, God wills to bring the best out of every bad situation, to assimilate evil into his own experience in the best way possible, and to bring as much good as possible into the world out of the evil within it. Even here, God's will can be frustrated and impeded by creaturely freedom.

The existence of evil is an insuperable obstacle to believing, not that a good and powerful God exists, but that God has an absolute, detailed, and eternal Grand Plan for everyone and everything, that God is a doting parent who will not let his children go and grow on their own.

When innocent persons, especially children, suffer pointlessly and die prematurely, as they so often do, could these particular events express the foreknowledge, causation, and inscrutable but still benevolent purposes of God? Classical Theism answers affirmatively, but always with the qualification that we cannot always understand how. When horrible natural catastrophes and moral atrocities occur, popular religion, perverted by Classical Theism, often assigns them to "the will of God." If this is right, however, we should conclude that God is really our enemy, not our friend, that God and the Devil are identical, that God really is a Malicious Demon after all. With an ultimate friend like such a vicious God, who needs a supreme enemy like the Devil?

If God, the transcendent cause of all creation, is timelessly omniscient as Classical Theists believe, he surely knows from eternity that horrors happen; if he is the omnipotent and sole originative or first cause of all that transpires, he surely causes all particular horrors; and if everything that occurs expresses God's purposes, he surely purposes or intends such horrors in every instance. From a world deliberately ordered to produce innumerable unthinkable harms to innocent creatures, how can we infer the existence of a divine Benevolent Superintellect? The task of *theodicy* is to reconcile the obvious presence of enormous evil in the world with belief in an all-powerful, all-knowing, all-benevolent, and all-compassionate God. Can this be done?

The eternally predetermined best of all possible worlds of Classical Theology is not acceptable. A plausible theodicy requires Process Theology's world in which good predominates over evil, or at least has a decent chance to do so, and where freedom and creativity are among the greatest goods of all. A credible theodicy must show that more good than evil exists on the whole within the universe, or at least that the universe is ordered to allow for this possibility. Some attempts to show this are clearly implausible.

i. Solutions that Don't Work

Classical Theists often argue that all evil is really good in disguise; but this makes us wonder if all apparent good isn't really evil in disguise. If we cannot identify and distinguish between good and evil when we see them, all our value judgments are suspect. If seemingly bad things can be good on the whole, then seemingly good things can be bad on the whole; and we cannot tell which is which. Theodicy tries to show that God is really good, despite all the evil in the world; but this conclusion is completely unwarranted if what we construe as goodness can be evil in disguise, and vice versa. Consider the fate of the rooster who concludes that its human feeder is a benevolent provider; but then the fatal day comes when it loses its head!

Classical Theologians often contend that evil in parts of creation is necessary for the perfection of the whole, but our finitude and ignorance prevent us from seeing how. As the dark areas of a painting contribute to the beauty of the whole painting, or as the bass notes of a musical composition contribute to its overall harmony, so evil in the world contributes to its comprehensive perfection, even if only God knows how. Despite appearances, they say, we live in the best of all possible worlds.

Unfortunately, neither dark colors nor bass notes are evil, so the aesthetic analogy is inappropriate from the start. We can understand how they contribute to the goodness of aesthetic wholes, but this cannot be said for many familiar situations containing great evil. The aesthetic analogy supposedly shows that the whole of creation is perfect in every detail, even though we cannot comprehend how. Only our ignorance can make it work. "Inscrutable" is an attribute of God in Classical Theology, including the "Westminster Confession."

Jonathan Edwards argued that sending the great bulk of mankind to eternal perdition is necessary for the perfection of creation as a whole; but, he admitted, we may not see exactly how, especially if we happen to be one of the damned! His God clearly violates Kant's Categorical Imperative and treats damned persons and all victims of tragedy merely as means to the perfection of the whole of creation, not as ends in themselves. The aesthetic analogy fails to tell us whether the transcendent Creator is benevolent or malicious in intent. If the totality is beautiful to God but inscrutable or horrible to us, this does not solve the problem of evil *for us*! Saying that God knows the answer does not provide us with an answer! We are the ones who need a solution to the problem of theodicy, especially if the Teleological Argument works.

So how can a powerful transcendent Creator be a Benevolent Superintellect, despite the presence of evil in the world? No single, simple magic bullet neatly solves the problem of theodicy. Its solution, if one is available at all, results from the cumulative weight of many considerations. The success or failure of theodicy is a matter of fallible and variable judgment. Thoughtful persons can honestly disagree about it. Massive evil in the world really is the greatest obstacle of all to belief that a good God designed the universe for benevolent purposes. Without a theodicy that is intelligible and plausible to us, God deserves our contempt, not our devotion; and the Argument from Design fails to show that a good and worshipful God designed our universe.

ii. A Process Theodicy that Works

Its handling of the problem of evil is one of the greatest strengths of Process Theism; and we will next explore the key elements in its highly credible theodicy. No one consideration can solve the problem of theodicy all by itself, so the cumulative weight of all of the following pieces of the puzzle must be considered. A credible theodicy must incorporate a. The Free Will Defense, b. The Soul-Making Defense, c. The Utility of Law and Order, d. The Conflict of Good with Good, e. Consolation, and perhaps f. Compensation.

a. The Free Will Defense

If evil decisions originate with free creatures within the world rather than with God, then God is not responsible for them. Being responsible for a choice and its consequences means originating that choice knowingly. A free and responsible choice originates with the intelligent moral agent who makes it. If, as Classical Theism affirms, God, the sole originative cause of all things, predetermined all choices ever made by created moral agents, then God is responsible for them. Human moral agents who deliberately inflict unspeakable harms on others do not originate their own malicious decisions if God programs and predestines every human choice from eternity.

Jonathan Edwards argued that all human choices are determined by our strongest desires or sets of cooperating desires, which, in turn, are ultimately caused by God. Being responsible and blameworthy, he said, means merely that a choice is evil and that the desire to do evil predominates; but *the origin* of the choice is irrelevant, he contended. Other Classical Theists appealed, inconsistently, to the free will defense, while clinging to the belief that God plans, foreknows, and foreordains everything.

Believers in free will, including Process Theologians, think that the question of the origin of our choices is highly relevant. Moral agents are responsible only for choices that they originate. Because our choices are originative or creative, the free will defense partly solves the problem of theodicy. Our choices would not originate with us if they originate with a God who programs them into us from eternity. What greater evil could a Malevolent Demon inflict upon us than to make us incapable of choosing to be virtuous or to do what is right, and then to punish us with Hell's infinite agonies for being and doing exactly as he created us? How better could we conceive of a Malicious Demon?

Process Theology says that finite agents are responsible co-creators with God. We originate our own free decisions. God is not the only existing creative or originative agent. When we freely choose to inflict evil on others, we are responsible, not God. Events in the world, including the human psyche, are influenced but not completely determined by the past. As free agents, we are partly created, partly self-creative. God makes relevant possibilities for choice available to us; but we freely select among them. Efficient causation consists of necessary but not sufficient conditions that partly structure present moments of partial self-creativity at the quantum level and in human and animal consciousness. All relatively complex created individuals are co-creative with God.

But why did God not predestine all creatures always to choose and do what is right, never what is wrong? Why did God not create the Kingdom of God in all its glory from day one? The free will defense answers that freedom is worth the price of its potential and actual abuse, and that without the potential to choose either good or evil we would not be free. Free creatures, by definition, may choose to do wrong as well as right; creating or originating our own choices between right and wrong is the paradigm instance of freedom.

Well, if freedom is so valuable that its availability outweighs the potential for its misuse, why did God not create us so that we always freely choose the right and the good? The answer is that this very notion is incoherent, like the idea of a round square. A Divine Reality who originates *all* choices must originate and pre-determine all creaturely choices to do what is right (or wrong). Creatures would not have a free and unconstrained choice between right or wrong if they are utterly constrained from eternity always to choose rightly and could not choose otherwise. The claim that God could cause free and originative creatures always to choose what is right just makes no sense. All free choices could have been otherwise.

Mainstream Process Theologians, we saw earlier, claim that God could not create unfree creatures who necessarily do what is right because being any kind of a concrete individual at all involves creative freedom. Creativity is a universal metaphysical category. Just as quantum events cannot exist at all at less than Planck dimensions, so it is impossible for any concrete events to exist at all unless their becoming is partly indeterminate and self-creative, says traditional Process Theism. The Divine self-limitation view, by contrast, affirms the widespread prevalence of creativity, even if it is not absolutely universal, because of the value of creativity itself, not because of its metaphysical necessity. Indeterminations, if not creativity, at lower levels of physical and biophysical organization is a necessary condition for freedom at higher levels. At the very lowest level of quantum events, creativity is minimal if not non-existent much of the time; but from the lowest to the highest level of physical reality, quantum physics is radically incompatible with rigid determinism. Quantum indeterminacy makes room for creative freedom at higher or more valuable and complex levels of biological and psychological organization. Even at the level of the most primitive individuals, the universe is always slightly out of control, including Divine control. Quantum level realities behave spontaneously and unpredictably occasionally. At higher levels of integration, absolute Divine control is completely out of the question. Creation is risky business, even for God.

Even the course of evolution always was and will always be slightly out of Divine control. Free decisions made by God *and* by living creatures, and indeterminateness in subatomic and atomic level quantum flukes, influenced the flourishing, survival, and evolution of all living things, including human beings. God was not the only decisionmaker directing and contributing to human evolution. We were created in part by God and in part by the free decisions made by all our pre-human ancestors throughout billions of years. If some of our progenitors had made different decisions, some species resembling us or our hominid ancestors, but not necessarily *homo sapiens*, might have resulted and prevailed. Given the power that modern medicine has given us to control and reorder genetic blueprints transmitted to future generations, we can now significantly influence the future course of evolution. Will we use this knowledge and power wisely? Not even God knows the answer!

The free will defense is possible only in a slightly wild and unpredictable universe. If the world is slightly untamed, out of control, and unpredictable at every level of actuality, the free will defense has some bearing upon the problem of natural evil as well as moral evil. The free will defense is usually applied only to evils caused by moral agents, not to evils caused by natural processes. Process Theologians repudiate this limitation. If creativity or at least indeterminacy belongs to every actual entity, then God cannot and does not absolutely control everything, including those natural processes that are hostile to living beings and their projects. Natural disasters like earthquakes, hurricanes, tornadoes, droughts, and diseases result partly from the operation of natural laws and partly from the inherently unpredictable and uncontrollable-by-others self-creativity and/or indeterminacy of all physical and biological events, including those at quantum levels of physical reality.

Quantum physics definitely rules out Classical Theology's concepts of both Divine omnipotence as absolute control, and Divine omniscience as knowledge of every minute detail of the entire past, present, and future of the universe all at once, for such definiteness about the self-creating present and the uncreated future is just not there to be known. God does not have the knowledge and will not use his power to prevent all evils, natural or otherwise, because God, by voluntary self-limitation, is not the sole originative agent functioning in the universe.

b. The Soul-Making Defense

Many desirable human traits of character would never develop in a world without evil. John Hick stresses the soul-making defense, according to which evil is necessary as a means to moral and spiritual development.[32] Process Theology can easily incorporate this as an important element in a workable theodicy, but not the sole element.

All human beings start very low (a single fertilized ovum) on the scale of intellectual, moral, spiritual, and personal development. Newborn infants begin their lives in relative ignorance, innocence, and impotence, as Adam did in the *Genesis* creation myth. Because evolution is true, the biblical Adam and Eve never really existed; but we all begin our lives as little Adams or Eves. Initially, as infants, we do not know the difference between good and evil, right and wrong, even if a predisposition is there. Our intellectual, moral, spiritual, and personal potentials develop and mature only within supportive social and physical environments. Their development also requires individual creative effort in response to real and dangerous challenges.

Adversity really is a necessary condition for the development of many highly desirable traits of character and moral and spiritual virtues. Without real evils, real dangers, we could never develop courage. Without scarcity, hardship, and suffering, we could never become generous, self-sacrificing, patient, kind, and compassionate. Without real and constant threats to happiness, security, and those we love, including ourselves, we would never develop moral conscientiousness and responsibility. Without frailty and death, we would never be able to appreciate strength, health, and life itself. The existence of evil is a necessary condition for the realization of many soul-making goods.

The soul-making defense may be carried too far; its scope is definitely limited. It solves part of the problem of evil but not all of it by a long shot. John Hick, its principal patron in our time, is well aware of this. Some attempts to make it work are very implausible.

Some theologians argue that without evil, there could be no good at all; correspondingly, without goodness, there would be no evil at all. If this were true, Heaven, lacking all evil, could not be good; and Hell, lacking all goodness, could not be evil. Even Heaven might be a challenging place to live!

Unless good and evil co-exist in a complementary relationship, it is often said, we would not be able to recognize either one, for contrast is essential for recognition. Without evil, the argument goes, we could not recognize goodness; without pain, we could not recognize pleasure, a significant ingredient in human happiness. But this is not true. There are degrees of pleasure; and contrasts between high and low degrees of it would be quite sufficient for recognition purposes; so would the contrast between any given degree of pleasure and a neutral state of consciousness that is neither pleasurable nor painful. Pleasure definitely could be recognized and appreciated in a world without pain, so the reality of pain cannot be justified merely because it is essential for the recognition of pleasure or happiness. However, pain has other constructive uses.

Pain has tremendous survival value. By alerting organisms to danger, it often prevents much greater injury or harm. Its presence in the world has considerable justification. Thomas H. Huxley, following Alfred Lord Tennyson, characterized the world of nature as "red in tooth and claw;" but many of the horrors of the "struggle for existence" may be greatly exaggerated.

In the wild world of non-human animals, much less pain and struggle, and much more joy, empathy, and cooperation, exist than we often suppose. Many of the cooperative and self-sacrificial features of human morality may be found also in the non-human animal world;[33] but non-human animals can foresee relatively little of what is to come and do not suffer greatly from the anxieties about the future that trouble members of our species with more foresight.

Most living species and individuals are plants, and so are most things that are killed and eaten for food. As Mary Midgley indicates, "In fact, nature is green long before she is red, and must be green on a very large scale indeed to provide a context for redness."[34] Most animals, both individuals and species, including the dinosaurs, were and are plant-eating herbivores, not carnivores or omnivores who consume the flesh of other living creatures. Carnivores, dinosaurs included, have a vital role in maintaining ecological balance and the vigor and zest of individuals and species, as does death itself. Much of the animal pain that we imagine to be involved in being killed and eaten by predacious carnivores is actually suppressed by endorphins and other natural analgesics secreted during the chase and attack, especially when predators kill quickly. Unfortunately, they do not always kill quickly; and we do not know how long a dying animal can benefit from "stress induced analgesia."[35]

Pain is genuinely troublesome for theodicy because all too often its intensity and duration are way out of proportion to its usefulness for soul-making or any other rational purposes. Agonizing bodily pain and mental anguish suffered while dying from cancer and many other diseases and injuries often do not correlate with valuable practical or spiritual lessons learned from suffering or with virtues developed by enduring anguish. All too often, both human and subhuman animals are simply and speedily crushed by overwhelming adversity, and they neither learn from it nor grow in moral and spiritual character as a consequence of it. God cannot and does not always bring from evil the sort of good that we would like to see. Tragedy and loss are very real in our universe.

Clearly, despite its importance, the soul-making defense alone does not resolve all the problems of theodicy. Too many individuals suffer and perish with little or no opportunity for soul-making, even if on the whole the world-system tends toward the best. In many instances, soul-making is a good brought forth from evil; but all too often it does not happen. If the God who designed and made our world-system is to be regarded as worthy of worship, the soul-making defense must be bolstered by further considerations.

c. The Utility of Law and Order

Many very real evils cannot be explained and justified by the free will and soul making defenses. Premature deaths, grave incapacitations, and grievous sufferings that serve no legitimate purposes of the sufferer appear to be pointless evils; but are they entirely pointless in the great scheme of things? Some great good other than soul-making might justify the existence of many seemingly pointless evils.

The immense usefulness and practical value of natural law and order resolves many of the remaining difficulties. Laws of physics, chemistry, biology, sociology, history, and psychology are neutral to good and evil in the sense that they are necessary conditions for both. Beneficial sunshine and rain fall upon the just and the unjust alike; so do harmful diseases and other natural adversities. The same natural laws are in effect no matter what.

A category mistake is made when moral categories like "unjust" or "unfair" are applied to the undesirable effects of the workings of natural laws and processes. When bad things happen naturally to good people, only non-moral categories like "tragedy" or "misfortune" are appropriate. Tragedies do not call the moral goodness of God into question. Ours would be a strange world, indeed an impossible world, if one set of physical or natural laws were operative for righteous people and another set for the wicked.

On the whole, natural laws and all the initial components of our life-supporting universe are more beneficial than harmful. Without them, we would not be here at all. Individuals who master the laws of nature and learn how to use them can, within broad limits, control their own destinies and greatly improve their chances for having a good life relatively free from suffering, incapacitation, deformity, and premature death. Informed individuals may use the laws of nature to enrich their lives by cultivating happiness, adventure, beauty, virtue, and other consciousness-enriching goods in effective ways.

Many life, health, and happiness enhancing behaviors are wired into non-human animals as instinctive responses to environmental conditions; and, to our great benefit, we human beings can learn the laws of nature and adjust our activities accordingly. Laws of nature can and do work for us rather than against us most of the time, but not always.

Steven Weinberg complains that we will find no evidence of a God who cares for life, intelligence, morality and beauty in the laws of nature because "The God of the birds and trees would have to be also the God of birth defects and cancer."[36] But what else would one expect from a wise and benevolent Deity? How otherwise should God have created the world? How could and should the order of nature be different? Be specific! If the laws of nature were changed very much, no life would be possible. It is very doubtful that either we or God could create a significantly better world if we tried. Birds, trees, birth defects, and cancers all exemplify the same basic laws of nature.

Should an intelligent, powerful, and benevolent Deity suspend the laws of nature to prevent harm to beloved creatures when natural laws work against them? Some Theists believe that God occasionally works miracles for such purposes. Miracles result, by definition, from an influx of transcendent energy or efficient causation that temporarily interrupts, suspends, or redirects existing laws of physics and chemistry. For the sake of the argument, let us assume that miracles can happen, sometimes, perhaps, in response to prayer. Why then does God not work many, many, more miracles to prevent much more pointless suffering, incapacitation, deformity, and premature death? Loving parents would protect their children from these ills if they had the knowledge and power; so why does a God who is supposed to be much more loving, powerful, and knowledgeable not do as much?

The answer, in part, is that a lawful and orderly environment, one that loving human parents are in no position to change, is itself a very great instrumental good; and the dependability of nature is worth the price of most if not all evils that result from the orderly workings and habituated activities of concrete entities. We abstract and generalize these regularities into natural laws. The advantages of natural laws or regularities clearly outweigh their disadvantages most of the time, but not always. So why aren't they occasionally suspended?

If we knew that we could expect God to solve all our problems for us and save us miraculously every time we get into a jam, we would never develop into conscientious and responsible persons. People who expect too many miracles are usually not very responsible individuals! Soul making reenters the picture unexpectedly at this point. The laws of nature enable and promote it. Even nonhuman animals who learn and generalize from experience, as most do, would not learn and grow in their own more limited yet significant ways if miracles were commonplace,

To announce God's presence and concern, miracles may happen occasionally. Another process heresy! If they do they are exceedingly rare; and we cannot count on them. If miracles were so frequent that we could predict and rely upon them to deliver us from all evils, then they would be the laws of nature, and we would assume that they had natural causes! Natural laws just are the regularities that we can count on and predict. But little or no soul-making would exist in a world where miracles were laws of nature.

d. The Conflict of Good with Good

Process Theologians point out that much seemingly pointless evil actually results from the conflict of good with good in a pluralistic universe, one that contains a significant number and variety of living beings, where legitimate interests are bound to conflict. No moral evil is involved when legitimate interests are pursued at the expense of others, but tragic non-moral evils may result.

Our world manifests many examples of the conflict of good with good. When two people apply for the same job, both cannot have it; but the winner normally does nothing immoral in taking it at the loser's expense. A violent death for a wildebeest or gazelle provides food for lions, cheetahs, wild dogs, and their young. They would otherwise die painfully of starvation. Seeming inefficiency, waste, and death for some means vibrant life for others. All individuals must die eventually to allow room and resources for others to live. Organic matter is never wasted; in the struggle for life, evil is transformed into redeeming, sacramental goodness.

Holmes Rolston, III[37] and M. A. Corey[38] forcefully developed the thesis that the apparent randomness, wastefulness, and inefficiency of the evolutionary process is actually God's deliberate rational strategy for assuring species survival, adaptability, and complexity under environmental conditions that are likely to change drastically over the course of time. A creator God who cherishes biodiversity must consider the design and well-being of vast unfolding ecosystems within which individuals and species exist and interrelate, as well as the design and well-being of the individuals and species themselves. At a systems level, earth's ecosystems, where one individual's loss is always another's gain and nothing is ever wasted, are remarkably efficient and well-ordered. They are models of rational foresight and planning.

Still, individual losers in the conflict of good with good can be greatly frustrated and disturbed; some suffer unbearably; and many perish prematurely. Grief is commonplace over losses in love, athletics, business, the struggle for life, and every legitimate competitive interest. Painful conflicts, serious frustrations, deep disappointments, unbearable sufferings, irrevocable losses, and premature deaths are inevitable in an orderly world containing a great diversity of consciously active and creative individuals. This is the price that must be paid for richness, diversity, freedom, and creativity.

Conflicts of good with good may also evoke positive virtues like wisdom, patience, resoluteness, inventiveness, heroism, and sacrifice. The immense worth and extent of both human and non-human cooperation, ingenuity, fortitude, and voluntary self-sacrifice should not be underestimated. Fortunately, shared interests do not always conflict; and innovative strategies for minimizing conflict are often available.

Could God, should God, do anything to prevent evils resulting from the conflict of good with good? Not if the price is too high. What exactly would God have to do? Eliminating all or most conflict of good with good would necessitate either abolishing or preventing the existence of great numbers of active individuals, or else significantly diminishing their creative power or freedom. If the existence of many and diverse individual centers of conscious experience, creative activity, and valuation has sufficiently great worth, lesser evils resulting from the conflict of good with good are justified.

God does not deliberately cause every bad thing that happens to good people; in this sense, evils are no part of God's grand plan. But God always aspires and tries to bring as much good out of evil as possible, so there is this much truth to the common belief that everything is a part of God's plan. God always plans to bring good out of evil! For example, mad cow disease is a horrible thing for its animal and human victims; but great good may come from it in the long run. It may be one of the best things that ever happened to further vegetarianism! God may use this horrible disease to inspire us rethink the exploitative destructiveness of our relations with other living things and our environment as a whole, when otherwise we would be oblivious.

Theodicy requires an ideal balance between divine and creaturely power. A loss of desirable creaturely power correlates with every increase in desirable divine power and control; and every increase in creaturely power must be accompanied by diminished divine power and control. To empower his creatures, God voluntarily limits his own power by divine choice. Every desirable reduction in conflicts between individuals would require undesirable decreases in the number, variety, intensity, zest, vigor, virtue, and creativity of creatures, as well as radical and inefficient modifications of the inclusive ecosystems that support them. All good things have their price. If we were God choosing a design for a good world as complex as our own, could we do any better? That we could is very doubtful.

e. Consolation

In Process Theology, God is "The great companion–the fellow-sufferer who understands," as Whitehead put it.[39] The profound conviction that an all-compassionate God genuinely understands and suffers with all creaturely suffering and loss gives great consolation in times of sorrow and woe. The God of Process Theology literally suffers with all who suffer; and Jesus on the cross is the supreme historical symbol of this. All evils inflicted upon all sentient creatures are ultimately inflicted upon God, who endlessly bears all suffering, loss, and every creature's cross so that an orderly world can be rich in intrinsically valuable, creative, responsive, and responsible individuals.

In Classical Theism, all forms of feeling are judged to be unworthy of God, even the highly desirable feelings necessarily involved in love, mercy, empathy, and compassion. God loves without feeling, is merciful without passion, and has compassion without pity, said the Classical Theologians. Kant considered any love composed even partly of feelings to be "pathological."

To the contrary, total affective insensitivity robs love, mercy, empathy, and compassion of most of their meaning and significance. Yes, cognitive elements are present in love, mercy, empathy, and compassion; but mere cognition is not enough. Divine attributes of immutability and impassivity betoken emotional insensitivity and moral impoverishment; but God truly bears our burdens.

Classical Theism actually affords no deep religious consolation for evils suffered because in himself God is really not compassionate, as St. Anselm forthrightly acknowledged. God's works make it appear that he is compassionate, but he really isn't! For Classical Theism, the discrepancy between divine appearance and divine reality is absolute. In Process Theism, by contrast, desirable kinds of feeling are perfections; stony immobility is an imperfection; and God quite literally suffers with all who suffer. God cares deeply, fully understands, empathizes with, and is all compassion, pure unbounded love. God bears all our sorrows and griefs. In that insight is deep comfort and consolation.

f. Compensation

Do the elements of theodicy presented thus far really succeed in reconciling belief in an immensely knowledgeable, powerful, caring, and worshipful God with the hard facts of evil in the world? There is still room for honest doubt. Many individuals, human and non-human, are crushed by the evil of overwhelming suffering. Others are profoundly retarded or irreversibly comatose from birth. Others are gravely incapacitated or struck down as infants, children, or in the prime of life. Many die so prematurely that they have little or no chance for a worthwhile life or for any soul-making experiences, efforts, and adventures. If a sadistic maniac entertains himself by shooting a tiny baby to death before the eyes of its mother, the poor infant has little or no opportunity for happiness, spontaneous creativity, self-development, and soul-making. How could a good, powerful, and benevolent God allow such things to happen? How could everything that happens be a direct expression of the will of God, as Classical Theists claim? Do we never fail to do God's will, not even when we perpetrate unthinkable atrocities like the Holocaust?

In this world, evils suffered pointlessly are sometimes compensated. Theologians like John Hick hold that God will eventually compensate unfortunate individuals for their tragic sufferings and losses in a better life to come. Heaven will make all things right in the sweet by and by. Hick's book, *Death and Eternal Life*,[40] defends this theological perspective in depth and effectively replies to philosophical charges that the idea of survival after death is empirically meaningless or nonsensical.

Hick shows that the idea of compensation after death for evils suffered in this world makes sense, at least to the extent that we can imagine experiences that would verify survival after death, but not in the sense that we now have experiential access to other worlds. Hick argues convincingly that compensation in a life after this life would go very far toward solving all the residual problems of theodicy. However, sensitive rebels like Dostoyevsky's Ivan Karamazov may want no part of such an "eternal harmony." By the time we reach the end of Hick's book, we discover that he provides no evidence whatsoever that other worlds exist or that we survive within them after death. Intelligibility is not equivalent to knowledge and truth, even if it is a first requisite.

Actual evidence for survival after death in another world is indecisive but not totally nonexistent. The transcendent spacetime of Heaven is vulnerable to the same objections previously raised against world-ensemble cosmologies. We have neither direct observational nor indirect inductive evidence for other worlds, including Heaven, antecedent universes, or contemporary bubble universes. If Heaven exists, we could never know it rationally in this life. Yes, a compensatory Heaven is logically possible; but this concedes only that the concept is intelligible and free from logical self-contradiction.

Compensation theodicy resolves residual problems of evil in this world by appealing to worlds about which we now know nothing; but a successful Teleological Argument requires that an adequate solution to the problem of evil be based upon the observed order of this world alone. It contends that the order of *our* world implies an intelligent and beneficent world designer. If the problem of theodicy cannot be solved by reference only to the discerned design of our universe, the Teleological Argument fails, for it aspires to infer God's existence from the observed order of our world. Uncompensated evils permitted by the observed order of this world may (or many not) be incompatible with the existence of a supercosmic, benevolent, and intelligent God. We do not know with certainty.

This author believes that elements a. through e. in the preceding discussion are sufficient to resolve the problem of theodicy; but if they fail, the Teleological Argument is not successful. Whether any theodicy succeeds or fails depends on fallible and variable human judgments that weigh existing goods against existing evils. People who sincerely want to be rational about such immense complexities and uncertainties must often just agree to disagree without questioning one another's integrity.

We have no knowledge of or control over what happens to us after death. At that point, everything is in the hands of God. We have no legitimate claim to anything more, but God's love may give us more than we deserve. There may or may not be compensatory survival after death, just as Quantum Cosmology's many worlds may or may not exist in Divine Superspacetime. We cannot know such things now, even if they are so. Theistic religions usually teach compensatory survival after death; faith often affirms it; but rational evidence for individual survival after death is presently inconclusive. If God chooses to give it, there could be ultimate compensatory justice; but about this we now have no sufficient rational knowledge.

Evil in the world menaces the Teleological Argument for the existence of God; but even without postmortem compensation, the threat is significantly abated by the collective weight of the free will defense, the soul-making defense, the usefulness of natural law and order, the inevitability of the conflict of good with good in a pluralistic universe, and the consolation of Divine compassion. Compensation for evils suffered may or may not be operative in some Other World. About that we really do not know; but we can hope.

Twelve

THEISM AND COSMIC CONTINGENCY

The Cosmological Argument for the existence of God begins with our awareness of the contingency of things within our world; from this it tries to reason philosophically, not scientifically, to the necessary existence of God. In Process Theism, God's Primordial Nature has necessary existence. Classical and Process Theologians agree that God exists necessarily. The Cosmological Argument expresses a widely shared deep cosmological intuition to the effect that the existence of any contingent thing–anything that might or might not be, ultimately implies the existence of a Necessary Being–one that could not not be. The concept of Necessary Being that applies to God in the Cosmological Argument is the same as the one that applies in the Ontological Argument (which is not here discussed in any detail). Kant maintained, correctly, that the Cosmological Argument depends on the Ontological Argument in one important respect–both employ an *a priori* concept of God, that is, one that is not derived from sense experience. Thus, if the Cosmological Argument is valid, so is the Ontological; for the Necessary God on whom all actual and possible worlds depends is an everlasting, uncreated, indestructible Reality who could not not exist and whose existence cannot be denied without self-contradiction.

As indicated in Chapter Ten, by definition, contingent beings might or might not exist; if they do exist, they have causes, can be created or destroyed, and normally come into being in or with time. A Necessary Being could not not exist and has no cause, exists self-sufficiently, and is everlasting, uncreated, and indestructible. St. Thomas Aquinas reasoned from the contingency of perceived motion and change, and from contingency itself, to the necessary existence of God. Contingency involves possible or actual nonexistence. Descartes reasoned cosmologically from the contingent existence of his idea of God to the necessity of its referent as its adequate cause; he also reasoned ontologically from the very meaning of this idea of God to the necessity of its referent. If the Cosmological Argument is any good at all, it should be possible to start with any experience of anything that exists contingently, including the whole contingent universe, and infer the reality of God. Formulating and defending an argument that gives adequate expression to this cosmological intuition is difficult. Since Plato, philosophers and theologians have tried and generally failed for a variety of reasons.

This chapter will proceed directly to a contemporary formulation of the Cosmological Argument that has a decent chance to succeed, partly because it avoids suspicious premises of traditional formulations like "There is a Great Chain of Being that leads from the earth at the center of the universe to God above," or "An infinite regress of causes either in time or in the Great Chain of Being is impossible."[1] Saint Bonaventure argued for the impossibility of an actualized temporal infinity. A more influential St. Thomas Aquinas thought that the finitude of the past is disclosed only by revelation, not proved by reason. He insisted that the impossibility of an actualized infinity applies only to causes that simultaneously coexist and act within the Great Chain of Being, not to causes that could exist and act successively in an infinite cosmic past.[2]

The following Cosmological Argument from Contingency makes no appeal to these antiquated and highly suspect elements of Thomistic metaphysics. It is perfectly compatible with and builds upon the empirical discoveries and well tested theories of contemporary scientific cosmology; but it goes beyond them. It is philosophy, not natural science.

1. A Cosmological Argument from Contingency

The basic premises of a plausible contemporary Cosmological Argument from the Contingency of the world to the existence of God are:

Premise 1: If each of the parts of any whole has contingent existence, then the whole itself has contingent existence.

Premise 2: Each of the parts of nature or the universe as a whole has contingent existence.

Deduction and Premise 3: Therefore, nature or the universe as a whole has contingent existence.

Premise 4: Definition: If something exists contingently, it is causally derived from or dependent on something other than itself.

Deduction and Premise 5: Therefore, nature or the universe as a whole is causally derived from or dependent on something other than itself.

Premise 6: The something on which nature or the universe depends is either the Principle of Plenitude, which requires infinitely many worlds in time and/or space; or it is God.

Premise 7: The something on which nature or the universe depends is not the Principle of Plenitude or infinitely many worlds.

Final Conclusion: The something on which nature or the universe depends is God.

A sound deductive argument actually proves its conclusion if it has both a valid form and all true premises. The above argument is deductively valid. Following discussions will argue that its premises are all true. Thus, the conclusion is true: The something on which nature or the universe depends is God.

A. Naturalistic Metaphysical Options

If Naturalists wish to attack this argument and establish their own metaphysics, they must prove the truth of at least one of the following propositions that contradict one or more of the premises of this Cosmological Argument from

Contingency. Not knowing whether one or more of these premises is true or false makes us agnostics, not Naturalists. Lack of knowledge does not establish naturalistic metaphysics or anything else. Naturalists themselves must prove something if their own metaphysics is rationally justified. The burden of proof in philosophy is on anyone who has anything to say, so Naturalists must establish the truth of their own alternatives to Theism. Naturalists could disprove the premises of our Cosmological Argument from Contingency only by showing that any one, perhaps all, of the following propositions are true; but they cannot do so. All of these Naturalistic metaphysical claims are false, as demonstrated in what precedes and in what follows.

1. Each of the parts of some whole has contingent existence, but the whole itself has necessary existence.

2.Some part (or parts) of nature or the universe as a whole has (or have) necessary existence.

3. Nature, the universe as a whole, has necessary existence.

4. Something may exist contingently without being causally derivative.

5. Nature, the universe, is not causally derived from or dependent on anything other than itself; it was caused by nothing.

6. Our system of nature, our universe, ultimately depends on either infinitely many diverse worlds in Supertime and/or Superspace, as required by the Principle of Plenitude, or on God.

7. Nature, the universe, ultimately depends on the Principle of Plenitude's infinitely many worlds.

Therefore, the universe does not depend on God.

Which cosmological claims are best justified, those of Theism, or those of Naturalism? What is the evidence?

The case against naturalistic alternatives to theistic cosmological premises 4 through 7 above has already been made and does not need to be repeated. With respect to 4 and 5, we already know from earlier discussions that if something exists contingently, specifically our system of nature, it is causally dependent on something other than itself, and that no plausible case can be made for the Big Accident contention that contingent beings can come to exist without a cause, specifically, that our universe just popped into being from absolute nothingness without a cause. With respect to 6 and 7, we know from Chapters Four through Eight and Chapter Eleven that variations on many worlds metaphysics cannot be defended, and neither can the Principle of Plenitude, so the best explanation for the existence of our world is that God created it. If the Big Accident, the Principle of Plenitude, infinitely many worlds in time and/or space, and so on, are indefensible, as previously established, the God hypothesis remains as the most plausible account of the supreme transcendent, ultimate reality on which our universe depends.

Only the first two premises of our Cosmological Argument from Contingency need further defense. The third premise is but an intermediate conclusion from these that functions as an additional premise in the wider argument. We must now dispose of the naturalistic options that some necessary wholes are composed entirely of contingent parts, and that some parts of our universe exist necessarily. The remainder of this chapter first develops and defends this part of the Argument, then explores some common objections.

B. Contingent Parts and Wholes

That all wholes exist contingently if they are composed entirely of contingent parts is universally verified in experience. This is the best reason we could possibly have for thinking that the first premise of our Cosmological Argument from Contingency is true. In absolutely no circumstances do we find wholes having necessary existence when they are composed entirely of contingently existing parts. In fact, we experience no necessary wholes at all.

The notion of a "whole" in this argument is not necessarily limited to well integrated organic wholes; any collection or totality, integrated or not, would probably do; but we will assume hereafter that relevant wholes are integrated in the sense that all their parts have a common ultimate origin and all these parts have causal relations with some other parts of the whole. All the parts of a unitary universe have a common ultimate cause and somehow hang together in linked causal cones. Everything in our universe has a common origin in the Big Bang and, despite relativity and the mutual independence of contemporaries, all existing entities have linked spatiotemporal and cause/effect relations with at least some other members of our universe.

The empirical first premise of our Cosmological Argument is an inductive generalization from observed wholes to all wholes. Like all inductive empirical claims, it might prove false in some as-yet-undetected situation. Admittedly, we have not empirically tested absolutely all existing wholes. Cosmological arguments that contain empirical premises are infected by the same sort of uncertainty that plagues all inductive inferences. Nevertheless, our first premise is just as well established as any and all other scientific generalizations like Einstein's "$E = mc^2$," or Hubble's "The whole universe is expanding," or thermodynamics' "Disorder always increases in closed systems." After all, we have not empirically tested all the mass/energy in the universe, or its universal expansion, or the entropy of all closed systems; but as far as experience takes us, these generalizations are universally confirmed, without exception. Hubble's law of uniform expansion is found to be true one hundred percent of the time because, excluding galaxies that are gravitationally bound to our Milky Way or to other galactic systems, all galaxies are moving away from us and from one another at speeds proportional to their distance.

Surface appearances may suggest that the contingency of all wholes composed entirely of contingent parts is even better verified than Einstein's energy/mass equation. Some astrophysicists affirm that photons, which carry the energy of light, have no mass; and this may be true of some other strange particles as well.[3] If photons have energy but no mass (even if only in an idealized rest state), then it is not universally the case that $E = mc^2$, for E cannot equal mc^2 where there is no m. Actually, photons have zero mass only in a purely hypothetical state of rest; and in that ideal state of masslessness, they are energyless as well. This purely hypothetical state must be expressed by a zero on both sides of Einstein's equation, but *real* photons are never purely at rest. All real photons have both mass and energy; so no anomaly here challenges $E = mc^2$.

Similarly, one hundred percent of the time, as far as experience takes us, if each of the parts of any integrated whole has contingent existence, then the whole itself has contingent existence. This empirical truth could actually count as a scientific truth except that scientists traditionally have not been interested in it. Wolfhart Pannenberg points out that scientists ignore contingency and concentrate instead upon formulating natural laws or uniformities.[4] As Aristotle indicated, the sciences carve out limited domains of being as their subject matter, while philosophy is concerned with truths that apply universally to everything. The contingency of all wholes with completely contingent parts is true of all empirical subject matter; it applies to all the parts of and to the whole of contingent existence. In my "Philosophy of Religion" courses, I regularly offered to give students an "A" on the spot if they could identify a directly or inductively verified instance of an integrated whole that exists necessarily even though all of its parts exist contingently. I never gave the "A."

Critics may object that our first premise applies only to finite wholes but not to the universe as an infinite whole. All finite wholes composed entirely of contingent parts are themselves contingent, but this may not be true of infinite wholes. We cannot reason inductively from the finite to the infinite. Thus, if the universe is an integrated infinite whole, it could have necessary existence even though composed entirely of contingent parts.

That the universe is an infinite whole is logically possible, but this could never be verified either directly or inductively, and the evidence to the contrary is overwhelming. Remember the Big Bang! Verification and falsification move far beyond toying around with possibilities. Possibilities are not actualities, and they should not be confused with probabilities. We definitely do not know that the universe is an infinite whole. Even if it is, we could never know or verify the claim, and Big Bang Cosmology provides convincing evidence that the universe is not an infinite whole. Temporally, the universe is only fifteen billion or so years old; spatially, though astronomically immense, it is finite but expanding–finite in actuality, even if infinite in potentiality.

Mathematicians tell us that any line has an infinite number of parts (Euclidean points); and each sub-section of a line contains just as many parts as the whole. Since lines depend for their existence upon the causes that draw them, or on the mathematicians that imagine them, we know that theoretically infinite wholes, those composed of an infinite number of parts, exist contingently.

Yet, according to quantum discreteness, mass/energy cannot actually be sub-divided below the spatiotemporal minimals of Planck dimensions. Particles exists and energy is transmitted in minimal quanta or not at all; even space has an irreducibly granular quality. Thus, no actual line drawn in the objective world is an infinite physical continuum. Observed quantum-atomized lines are not infinite as wholes actually having an infinite number of parts; but this is not incompatible with our first premise.

Conceptually constructed imaginary lines are only potentially infinite wholes with potentially infinite parts; but they depend for their existence upon conscious thinkers or imaginers. Since infinite wholes with potentially infinite parts exist contingently, and we know of no instances to the contrary, the universe as a whole should be contingent if it has an infinite number of parts only potentially. We cannot generalize from potentiality to actuality, and the *thought* of infinity, dependent on the mind of the thinker, is not an actual infinity. The actual thought of "infinity" is not an actual infinity of thoughts; it is only one thought, just as the actual thought of "blue" is not a blue thought.

Note carefully that our Cosmological Argument from Contingency makes no appeal to the contingency of actually infinite wholes, for our universe is definitely finite. It is not actually infinite in any respect. According to Big Bang Cosmology, our universe is vast; but it is finite in expanse, past duration, mass, and in all other denumerable ways. The Planck-dimension minimals of quantum physics prohibit this finite mass from being actually divisible into an infinite number of real parts. Naturalists can identify no necessary infinite wholes composed entirely of contingent parts. Our universe is certainly not such an infinite whole, and it does not have an infinite number of parts. All experienced finite wholes composed of contingent parts are themselves contingent. This empirical truth can be generalized inductively to include our universe. Protests will be explored later.

The Cosmological Argument from Contingency may be defective just because it reasons from parts to the whole of nature. Logic texts say that reasoning from parts to wholes commits the informal logical Fallacy of Composition, according to which inferring that a whole possesses a certain property merely because each of its parts possesses that property is erroneous. Despite what introductory logic texts say, it is not always fallacious to conclude that a whole possesses a property because all of its parts have that property. Sometimes so to reason really is a mistake, but sometimes not. How can we tell the difference?

Arguments from parts to wholes can be formulated, or reformulated, as deductive arguments. Unjustified conclusions can be drawn from deductive arguments in at least two ways. First, the argument form or pattern may be invalid; second, one or more of the premises may be false or not known to be true. The first three of the following arguments are obviously erroneous; but what kind of a mistake do they make?

Argument I.
(1) If all the microscopic parts of a machine are invisible to the naked eye, the whole machine is invisible to the naked eye.
(2) All the microscopic parts (atoms) of this car are invisible to the naked eye.
Therefore, this whole car is invisible to the naked eye.

Argument II.
(1) If all the microscopic parts of a body (the cells, molecules, atoms, and so on) are devoid of consciousness, the whole body is devoid of consciousness.
(2) All the microscopic parts of my body (the cells, molecules, atoms, and so on) are devoid of consciousness.
Therefore, my whole body is devoid of consciousness.[5]

Argument III.
(1) If each of the elemental parts of a chemical compound is a gas, the whole compound is a gas.
(2) Each of the elemental parts of water (hydrogen, oxygen) is a gas.
Therefore, water is a gas.

These three arguments from parts to whole obviously fail to prove their conclusions. The next two arguments also move from parts to wholes, but they are not faulty. They commit no fallacy; they involve no falsehoods or uncertainties; and they show that we can reason successfully, deductively, and correctly from parts to wholes.

Argument IV.
(1) If all the macroscopic parts of a machine are made of metal, the whole machine is made of metal.
(2) All the macroscopic parts of this water pump are made of metal.
Therefore, this whole water pump is made of metal.

Argument V.
(1) If each island in a group is in the Pacific Ocean, the whole island group is in the Pacific Ocean.
(2) Each island in the Hawaiian group is in the Pacific Ocean.
Therefore, the whole group of Hawaiian Islands is in the Pacific Ocean.

No fallacious pattern of reasoning is involved in any of the above arguments. They all involve an instantiation with respect to particulars and manifest the valid pattern of *modus ponens*: If p, then q; and p; therefore, q. What then is the error in arguments I, II, and III that is absent in IV and V? The first three do not contain any ambiguous concepts, though logic textbooks regularly but mistakenly classify the Fallacy of Composition as a Fallacy of Ambiguity. The only difference is that the first premises of I, II, and III are false, as shown by experience; but experience confirms the first premises of arguments IV and V.

Our Cosmological Argument from Contingency is a valid deductive argument that commits no informal logical fallacy. Its first premise resembles the first premises of IV and V, not those of I, II, and III.

Some reasoning from parts to wholes is illegitimate, some not, depending on the formal pattern of reasoning and/or the truth or falsity of the premises. We cannot correctly infer that a whole machine is invisible because each of its atomic parts is invisible, or that a whole body lacks consciousness because each of its cells lacks consciousness, or that a compound is a gas because each of its elemental parts is a gas, or that a whole machine weighs one pound because each of its ten parts weighs one pound, or that a whole painting is beautiful because each of its parts is beautiful. Yet, we can infer correctly that a whole macroscopic machine is made of metal because each of its macroscopic parts is made of metal, that an island group is in the Pacific because each island is in the Pacific, and that a whole chair is painted green because each of its visible parts is painted green. With a valid argument form, these inferences commit no logical fallacy. The only legitimate question is whether the premises are true. In IV and V, experience shows that the premises are true; and because the forms are also valid, the conclusion is proved.[6]

Some properties like being metal, being in a geographical area, and being a certain color can be inductively extended or extrapolated from parts to wholes; and others, like being invisible, being conscious, weighing a pound, and being beautiful, cannot. *Experience tells us which is which.*

Contingent existence is an extrapolatory property, as experience also invariably shows. If each of the parts of any whole has contingent existence, we know from experience that the whole itself has contingent existence. We know that its non-existence is logically conceivable, that its mode of being is not self-sufficient but depends on some cause or causes, that it has not existed forever, and that it is destructible and will probably cease to exist at some point in the future. This includes wholes that are combined into even larger wholes up to infinity, and perhaps including infinity (though infinity is irrelevant to our finite cosmos). Verifying experiences support the first premise of our Cosmological Argument from Contingency one hundred percent of the time; and the form of the argument is perfectly valid. This is as good as philosophical arguments ever get! Our intermediate conclusion–nature, the universe, has a contingent form of existence–will follow if the second premise is true, that is, if there are no necessary parts of the universe. Naturalistic metaphysics from the time of the Greek atomists has been based partly upon the possibility of finding some necessary part(s) of nature; but the enterprise is futile!

C. No Necessary Parts of the Universe

The second premise of the Cosmological Argument from Contingency affirms that all parts of nature are contingent. Is this false or at least not known to be true? The universe would not be a whole composed entirely of contingent parts if some of its parts exist necessarily. Naturalists often make this claim, but they must do much more than just proclaim it. They must actually show that some parts of the universe exist necessarily. The burden of philosophical proof is not always on the Theist. It is on anyone who has anything to say.

Naturalists agree with everyone else that all macroscopic bodies like stars, planets, and human bodies are contingent. They locate the necessary parts of nature in its microscopic innards. The naturalistic presumption that the necessity of the universe is situated in the subject matter of particle physics was never effectively challenged before the twentieth century. The Greek Atomists and naturalistic Materialists through the centuries believed that spacetime itself and its most primitive contents, the atoms, exist necessarily; the atoms and the infinite void are self-sufficient, everlasting, uncreated, and indestructible. Atomists once called the smallest particles of matter "atoms" and pronounced them necessary beings, but we now know that they are not. Exactly which parts of matter get classified as "smallest" changes as our knowledge of sub-atomic physics advances; but *nothing* at atomic or sub-atomic levels exists necessarily.

Particle physics gave naturalistic metaphysics an enormous shock in the twentieth century. Traditionally, naturalistic confidence in the eternity and necessity of the world presumed that the most primitive particles of matter are self-sufficient, everlasting, uncreated, indivisible, and indestructible. Physics today finds no physical particles or sub-particles that have such attributes, not those now called "atoms," and not their more primitive sub-atomic components. All physical particles exist only derivatively or contingently. Some endure only for only thousandths or millionths of a second. Nucleosynthesis and nuclear fusion bring all atoms and all their sub-atomic parts into being, so they haven't been around forever; and nuclear fission can destroy them all and convert them into energy or other particles or forms of radiant energy.

The Greek Atomists were totally wrong in thinking that the smallest physical particles have always existed unchanged and uncreated. We now know that even the smallest parts of atomic nuclei, the quarks, exist contingently, and so do the electrons that orbit atomic nuclei. Quarks always come in pairs (in mesons) or three at a time (in protons and neutrons). Pairs or triplets of quarks cannot be separated, but they can be created and destroyed. Bombarding them with larger particles in accelerators to try to separate them merely creates additional pairs or triplets of quarks out of the exchange of kinetic energy. Quarks can thus be created in the laboratory out of energy; and anything that is or can be created in the laboratory or elsewhere has a merely contingent form of existence. Quarks are not necessary beings, and neither are electrons, which can also be produced and destroyed by the same processes. One type of quark is called the "Higgs particle," or "Higgs boson." In defending Naturalism, specifically its claim that our universe is eternal, Kai Nielsen wrote,

> Even if the big bang theory is true, it does not show that the fundamental particles were brought into explosive existence by the big bang. The Higgs particle need not be thought to have so come into existence, or to have come into existence at all, and it stands as a candidate for a necessary being and a more plausible one than God.[7]

On the contrary, innumerable articles in the literature of physics describe the production of Higgs particles in cyclotrons under many different conditions.[8] No kind of entity that is or can be produced by something else is an eternal, self-sufficient, and necessary being. Also, Big Bang and Grand Unification Theories say that in the beginning *all* existing particles emerged from a more primitive state of unified pure energy, which in turn had its own cause.

The most stable and enduring sub-atomic particle is the proton. Once created, protons last for billions of years. The protons in the nuclei of most atoms existing in the material world today were created within the first three minutes after the Big Bang, but they were created. Most physicists believe that protons eventually decay; but elaborate experiments designed to detect proton decay have failed thus far. Nevertheless, protons are not eternal and necessary beings. They are composed of quarks, and quarks are not eternal and uncreated. Protons may naturally decay very slowly, but they can be destroyed in proton/antiproton collisions and in nuclear reactions. In fact, any kind of particle can be destroyed by its antiparticle; and all particles and antiparticles were created *after* the onset of the Big Bang by the primordial grandly-unified matter-less, particle-less energy that preceded them. The same is true of the four basic forces of nature that once were one, says Grand Unification Theory. No created and destructible entity or force is a necessary being.

Though all particular physical entities exist contingently, perhaps spacetime and/or pure physical energy exist necessarily. No, spacetime and physical energy also exist contingently if the Big Bang originated in a singularity of nothingness totally devoid of spacetime and energy, or if the Big Bang and all its contents were created at some point in the past by anything whatsoever. So what caused the Big Bang?

In contemporary physics, space, time, and physical energy are inseparable; but the emphasis is on time. Time has not been spatialized; rather, space has been temporalized. Neither space nor time are merely empty forms into which actual events and particles are infused. Space is not vacuous nothingness; it has its own energy density; it is an elastic physical reality that may be warped, stretched, condensed, straightened, curved, expanded, and contracted to nothingness (almost?). All physical particles and configurations of energy are constituted by both time and space. Particles, waves, and all microscopic entities are warped, knotted, or stringed concentrations of spacetime. In contemporary physics, neither atoms nor any sub-atomic particles, waves, fields, or other entities are basic necessary beings.

As Werner Heisenberg indicated, in contemporary science, energy is "The primary substance of the world."[9] Is energy the necessary being? No. According to the Standard Model of Big Bang Cosmology, space/time/mass/energy was created in the beginning out of what was empirically nothing–often called an initial singularity. No alternative cosmology actually proves otherwise. If our universe emerged from an initial singularity, neither spacetime nor physical energy exist necessarily. Yet, other possibilities remain to be examined.

Perhaps pure energy as such exists necessarily, though all particular forms of energy exist contingently (as suggested by the quotations from Corliss Lamont with endnote numbers 15 and 16 in Chapter Two). Electrical energy is derived from dynamos; atomic energy is derived from nuclear reactions; solar energy is derived from the sun. *Every empirically observable* manifestation of energy is derived from some energy source other than itself. Having a derivative existence is precisely what existing contingently means.

Does pure (non-empirical) energy as such exists necessarily? If so, it is too pure to count as something empirical or observable! All *observable* forms of energy exist contingently, being derived causally from something else, from other forms of energy. Furthermore, *all of these* emerged initially from, were caused by, the Big Bang. So we are back to our original problem, What caused the Big Bang? It was something transcendent, something outside our spacetime system, something beyond nature! Naturalists, like Theists, resort finally to explaining things visible (natural) in terms of things invisible (non-natural). Not all natural effects have natural causes. Naturalist and theists have their backs to the same wall.

One hundred percent of the time, verifying experiences confirm that all parts of the universe exist contingently. True, we have not tested every last part of the universe, just as we have not tested every last speck of energy to verify that $E = mc^2$. Neither proposition is absolutely certain, but both are as well confirmed as anything that we know.

Indisputably, if space/time/mass/energy originated at some point in the finite past either from nothing or from extra-natural antecedents, our whole universe and all its parts exists contingently. This is exactly what Big Bang Cosmology proclaims. Neither any parts of nature nor nature as a whole are necessary, self-sufficient, uncreated, everlasting, and indestructible. Even a reified infinitely transcendent Superspacetime/mass/energy would exist contingently if it were wholly comprised of contingent parts. An infinite whole composed of an infinity of contingent parts is an infinitely contingent whole, not a necessary being, but this can be disputed. Let's try to imagine how.

Assume for the sake of the argument that Oscillation Cosmology's claim that later universes are caused by antecedent universes at least makes sense, even if unverifiable. Even so, "Any whole is contingent when composed entirely of contingent parts," applies to all combinations of contingent antecedent universes linked by a common thread of causation. Let our world of nature, our universe, be U1; and let it be our first whole, W1. Let U1 be caused by an antecedent universe, U2, which is our second whole, W2. Together they will form a Third whole, W3. If both universes are themselves contingent, W3 will also be a contingent whole. This process can be repeated any number of times, perhaps even an infinite number of times, and all repetitions will be swallowed up by the premise: if any whole is composed entirely of contingent parts, the whole itself is contingent. We will return to this later in replying to objections to the Cosmological Argument.

Our universe as a whole does not have a necessary mode of existence. The first two premises of our Cosmological Argument from Contingency are true, and its logical form is valid. Our universe does not exist necessarily, and Naturalists are wrong in believing otherwise. All wholes, including our universe, composed entirely of contingent parts are contingent wholes; and our universe is composed entirely of contingent parts. Therefore, our whole universe exists only contingently. Once established, this deduction is used as an additional premise in our more extended Cosmological Argument from Contingency.

Theology is not tied inextricably to creation *ex nihilo*, even if it now seems to be a very reasonable position. A theology for rational persons must be compatible with the most firmly established conclusions of the natural sciences, but theology should proceed with caution in binding itself too tightly to particular scientific theories, concepts, and discoveries. Sometimes, a healthy agnosticism, an open mind, is the most reasonable theological response. Still, the preponderance of evidence as we see it cannot be ignored. All philosophical knowledge falls short of certainty; but we want the most reasonable belief about what caused the Big Bang that we can find; and our best informed guess is that creation *ex nihilo* by God is it.

Two basic options, (1) creation out of nothing and (2) out of spatiotemporal antecedents, are now both available to Process Theologians, who traditionally presupposed something like theistic oscillationism. Whitehead said that God does not exist *before* the world but always *with* the world–some world. Yet, creation *ex nihilo* is a viable process option, as explained in Chapter Ten. From the perspective of Superspacetime, God could create all worlds either out of nothing, out of his own Superspatiotemporality, or out of the ashes of antecedent universes. No matter which, we could expect God to choose laws and initial conditions that would make each world interesting and worthwhile.

But our own world is our most immediate problem. Do we know enough at this point to draw any conclusions about it? Is it a necessary being? A contingent being? Did it have a cause? What really caused the Big Bang?

D. Conclusion: The Dependence of the Universe on God

Our Cosmological Argument from Contingency concludes that nature, the universe, our system of spacetime, ultimately depends on God. St. Thomas Aquinas said that the ultimate cause and designer of the universe is what "all men call God"; but this is really not true according to Naturalists. They don't call it God! St. Thomas also said that the Cosmological Argument can tell us that God exists but not what God is like, nothing about God's essence. If so, it generates very little useful or valuable information about God. Actually, taken with the Teleological Argument, it does tell us some religiously significant things about God. Together these arguments inform us that our contingent world of nature requires a cause that is (1) transcendent, (2) immensely powerful, (3) creative, (4) necessary, (5) infinite, (6) intelligent, (7) purposive, and (8) benevolent. All arguments for the existence of God must work together to give us this much information. The Cosmological Argument gives us (1), (2), (3), (4), and (5); and the Teleological Argument gives us (6), (7), and (8). Only the Ontological Argument (not explored in depth in this book)[10] can give us all divine perfections, including (9) perfect virtue or optimal righteousness, (10) supreme holiness, and so on. All three of these arguments taken together bring us to an ultimate Divine reality in whom all our questions about why there is something rather than nothing finally come to rest. On rational grounds, they bring us to God as the supremely worshipful, creative, everlasting, uncreated, and indestructible Ultimate Reality who could not possibly not be.

2. Critique of the Cosmological Argument from Contingency

Replies to several important objections to the Cosmological Argument were implicit in many earlier discussions. We now know that our universe is not a Big Accident having no cause at all, that it is not a chance occurrence within infinitely many worlds, and that cosmological reasoning from the world to God does not commit the informal logical Fallacy of Composition. The following additional charges must now be confronted. A. An everlasting universe needs no God. B. Contingent wholes imply only contingent causes. C. God must have a cause if everything has a cause. D. The concept of the universe as a whole, and thus of God's creating it, is meaningless. E. Transcendent reality is unknowable. Can plausible replies be given?

A. The Universe Needs No God

If creation *ex nihilo* is true, the universe as a whole and all its parts depend on God for their very existence since God created all the primordial spacetime/mass/energy of our universe from nothing. Many atheistic oscillationists assume that the universe, some universe, has always existed, and that an everlasting set of successive universes is itself a necessary being that needs no God. As we saw in Chapter Four, Oscillation Cosmology is often adopted mainly to avoid God, but it may not be true that an everlasting universe or set of universes would not be contingent upon God in any respects, and it is definitely not true that ours is an everlasting universe or that it belongs to an everlasting set of sequential universes.

The evidence already given is overwhelming that our universe is finite but expanding spatiotemporally, and it is not divided into an infinite number of tiny but real parts, some of which exist necessarily. The contingency or derivative existence of the universe might not depend totally upon its finitude in time or space. Naturalists suppose that the universe (or some universe) has infinite temporal duration. This means, they conclude, that our universe exists necessarily; but they are mistaken. Unlike ours, an eternal universe might resemble necessary existence in being uncreated and everlasting, yet it would still be causally contingent if it depends everlastingly in some way on something transcendent.

Traditional theism says that God both created the world and continues to preserve or sustain its existence. The world depends on God originally and at every moment. Either (or both) implies that the universe has a contingent or dependent form of existence and that God is the world's ultimate necessary condition. Theism is not inescapably tied to creation out of nothing in the finite past or to Big Bang Cosmology. The universe would have a derivative or causally dependent mode of being if either of the following propositions is true:

(1) The whole universe came into being at some point in the finite past; and/or

(2) The universe, or other universe(s) to which ours is connected, depend somehow on God at some or all moments throughout an infinite past and/or an endless spatial expanse.

If all contingent wholes in a finite or an infinite series depend on God either originally, continuously, or historically in any way at all, the claim that "All wholes within the universe are caused by contingent beings" is only part of the truth. The whole truth adds, "and by a necessarily existing God."

St. Thomas Aquinas believed on the basis of divine revelation (as he interpreted it) that (1) is true; but, he contended, the Cosmological Argument for God hangs on the second type of contingency, not the first. Aquinas thought that Aristotle, for whom the world co-existed eternally with God, set the definitive standard for what reason could or could not accomplish. In deference to Aristotle, Aquinas held that reason cannot prove that the world has a beginning or disprove its everlasting co-existence with God. Revelation may say otherwise, but not reason. A world co-existing with God throughout an infinite past would nevertheless depend infinitely on God for its order and being, just as rays from the sun would depend infinitely on the sun if the two co-existed forever. Though Aquinas did not say so, a spatially infinite universe would also be contingent if it somehow depends on God at any or at every real unit of spatial extension. An infinitely prolonged or extended set of contingencies does not add up to necessity, especially where causal dependence exists at spatial and/or temporal units along the way. An infinite spatially and/or temporally contingent world is not a necessary being; it is an infinitely contingent being. For many good reasons, now we know that we do not live in an infinite world.

St. Thomas Aquinas anticipated neither contemporary Biblical criticism nor Big Bang Cosmology. Perhaps he was wrong in thinking that revelation teaches creation *ex nihilo* in the finite past, and that reason cannot show that the universe originated at some point in the finite past. Modern Biblical scholarship affirms that creation *ex nihilo* is not clearly taught in *Genesis* or elsewhere in the Protestant *Bible*; and Big Bang Cosmology shows that the universe, the whole shebang, was created in the finite past–about fifteen billion years ago.

St. Thomas was convinced that creation *ex nihilo* is a clear deliverance of Christian revelation,[11] but this is far from certain. Modern Biblical scholars generally agree that the first chapter of *Genesis* does not clearly affirm creation out of nothing. In the King James translation of the *Bible*, the first verses of *Genesis* read, "In the beginning, God created the Heavens and the earth; and the earth was without form and void." Biblical scholarship in the twentieth century concludes that these verses may, with perfect faithfulness, be translated to read, "When God began to create the Heavens and the earth in the beginning, the world was without form and void." In margins or notes, current translations of *Genesis* give this as a perfectly legitimate rendition of *Genesis* 1:1. *Genesis* declares that God created the universe, not *ex nihilo*, but out of the formless chaos that already existed "when God began to create." Creation itself was initiated only when God said "Let there be light." In *Genesis*, at the beginning, God brings order to the world but does not bring it into being absolutely. God is a world-designer but not a world-creator.

Although not clearly taught anywhere in what Protestants recognize as the *Bible*, creation *ex nihilo* is taught perhaps in the intertestamental literature that Protestants do not regard as scriptural, though Catholics do. Specifically, the idea seems to occur for the first time in *II. Maccabees*, Ch. 7, verse 28, which reads, "I beg you, child, look at the sky and the earth; see all that is in them and realize that God made them out of nothing, and that man comes into being in the same way."[12] This was written perhaps as early as 125 B.C., but scholars debate its date. St. Thomas Aquinas regarded *II. Maccabees* as scriptural, but the authors of the "Westminster Confession," along with most Protestants, did not and do not so regard it. In any event, Eric Lerner's assertion that creation *ex nihilo* originated with Tertullian in the third century A.D. is incorrect.[13] As far as biblical religion is concerned, God ordered the universe from pre-existing chaos, which may have co-existed everlastingly with God. The God of *Genesis*, like Plato's Deimurge in the *Timaeus*, ordered the world out of pre-existent materials. Neither created the universe out of nothing. Still, creation *ex nihilo* might be true in spite of *Genesis*.

i. Creating and Sustaining

To return to philosophy again, how might a created or a continuously existing universe or everlasting set of universes depend on God, and how much of it makes sense within the framework of Process Theology?

(1) The first and most obvious way in which the universe depends on God is with respect to *its creation, its existence*. If God created our universe out of nothing, Process Theologians must join with Classical Christian Theologians in recognizing the most obvious way in which our universe depends upon God. God produced from nothing the very stuff of physical existence, all the space/time/mass/energy of creation. Except for God's creative activity, there would be no space/time/ mass/energy, no chain of secondary causes within the world, no world to sustain. Creation out of nothing does not explain how God sustains all creation, but it is the most conspicuous way in which the world depends on God.

Traditionally, Process Theologians repudiated the claim that God creates our world out of nothing; but from the revised process perspective developed in Chapter Ten, Process Theology can incorporate creation *ex nihilo*. Thereby it can further enrich its understanding of how the world depends on God. Creation out of nothing and ongoing providential directing were emphasized traditionally by Classical Theologians, but Process Theologians can also make a prominent place for them.

(2) Closely related to depending on God for its existence is dependence on God for *its form, its life-supporting law and order or structure*. Some important features of original creation are ongoing. Our teleological inquiries show that our universe depends on the intelligent and purposive foresight of a being of divine proportions for its life-producing and preserving law and order, its life-favoring natural laws or habits. If natural laws are only statistical abstractions from the collective habits of massive numbers of kinds of things like electrons, protons, photons, paramecia, and people, then claiming, as Charles Hartshorne did,[14] that God chooses natural laws for our universe or cosmic epoch is just a roundabout way affirming that God created a plurality of creatures having certain natures which, on average, behave in certain calculable ways. Hartshorne was well aware that natural laws are formal, statistical, and changeable; but he might not have appreciated fully the very concrete causal efficaciousness involved in this claim. This means that God selects and produces not the lawful forms as such, but the basic kinds of entities that exist. God gives them the natures that their habits express statistically. The basic structures and habits of existing things can evolve, along with the laws that express these changes. Today's "scientific" cosmologists agree that no laws for physical particles

existed before particles separated from the grand initial unification; and no laws of human psychology existed before human beings evolved. Still, the "hand"–the anticipations, intentions, and influences–of God is in all of this. In formatively influencing the natures of things that act lawfully in nature, God acts as an efficient cause of particular kinds of entities, not just as the final cause of the abstract lawlike patterns that describe their activities.

Our universe depends on God for its basic life-supporting structural features, its original life-assuring initial conditions, including its initial low entropy, the kinds and strengths of its basic physical forces, its asymmetry of matter over antimatter, and many other initial conditions previously discussed.

If ours belongs to an extended set of oscillating universes (as seems unlikely), there might be a place after for a God of the Gaps, especially if the gaps fall between or at the beginning of universes. Science can know only our own cosmic epoch. Existing originally only in the Primordial mind of God, beneficent life-favoring natural laws (general patterns) and initial conditions could vary in and need to be chosen anew for each cosmic epoch. As we know and discover them, the formal patterns of the enduring habits of primitive physical entities in our universe are expressed in the laws of physics, chemistry, biology, psychology, and the other natural and social sciences. The earliest emergent laws of nature and the original conditions of the universe belong more to initial creation than to continuous sustenance, but they or their effects persist even today. Here the Cosmological Argument (dependence) merges with the Teleological Argument (purposefulness). The beneficent order of nature is both ongoing and evolving.

(3) Theologians have held that God *sustains the world in being* whether created or everlasting. If the universe in some form existed infinitely into the past, might it still depend on God? Many cosmologist seem to believe that this would be a way avoiding God. Not so! St. Thomas Aquinas, the most influential classical theologian, strongly believed that if our world existed throughout an infinite past and depended on God for something through it all, it would definitely be contingent upon God, just as rays coming from the sun would depend infinitely upon the sun if they co-existed throughout an infinite past. This analogy would also hold for a modern process theistic oscillation cosmology in which each member of an infinite set of successively existing worlds somehow depends on God. But how could an infinite past actuality or chain of past actualities depend on God in the absence of original creation from nothing?

The classical answer that God constantly *sustains or preserves* the world in existence might do the job if we really understood what "sustains" means and were sure that an infinite series of contingencies lacks the ability to sustain itself, or that physical mass/energy, once created, lacks the intrinsic self-existence attributed to it by the First Law of Thermodynamics. The meaning of "God preserves" has been difficult to specify. We definitely cannot argue without circularity that the world cannot sustain itself because it depends on God, for that is precisely what is at issue. That is what must be clarified.

Most attempts to explain the meaning of "God sustains the world" get no further than identifying sustaining with continuous creation, preservation, or dependence, all of which are then defined as sustaining. Getting beyond this conceptual circularity is extremely difficult.

Consider briefly some implausible analyses of what it means to say that "God sustains the existence of all creation." Since sustaining applies to everything in the universe, an account of how it works must have universal significance and application. Because they have only local significance and application, particular miracles (if any) and acts of providence (which are not necessarily miraculous) do not count as sustaining. Sustenance belongs to general providence, not to particular providence.

Some explications of how God sustains the universe are clearly implausible. (A) Perhaps new matter/energy is constantly being created everywhere, as suggested by Fred Hoyle, except God rather than matter is its cause. Yet, little or no evidence supports such ongoing violations of the principle of the conservation of energy. This may or may not happen occasionally at that fuzzy borderline between potentiality and actuality that the physicists call the space/time "vacuum" where the world merges with God. If it happens constantly, the pace of it is too slow to account for the entire mass of our universe within finite Big Bang time.

(B) Perhaps, as Bishop Berkeley and Quantum Observer Theorists maintain, an actual world exists only as something being perceived either by ourselves, by other finite minds, or by God, who is always around to keep things going when no one else is looking. The universe would cease to exist if God quit thinking or perceiving it if Idealists are right. For Idealists, God's sustaining the world just consists in his continuing to think it. Yet, this book opts for and defends a critically realistic theory of knowledge and a corresponding realistic metaphysics.

(C) Perhaps sustaining is just constant recreation every instant, as Descartes and a few others maintained. Jonathan Edwards, who held this view, argued that no real causal relations obtain at all between events within the world; the world continues to exist because God recreates the whole of it from nothing at every moment.[15] Edwards anticipated David Hume's empirical reduction of causation to mere temporal succession and spatial contiguity devoid of "secondary" causal efficacy. Yet, despite Edwards and Hume, things within the universe do seem to have real causal relations and connections. Existing entities really do transmit their energy, structured patterns, and purposes to other things; and constant recreation just seems like a lot of unnecessary repetitive work for God. So does creating a world that is not self-sustaining once created.

(D) Traditional theology simply decreed that created things in themselves lack the power to continue to exist and that God alone supplies that power.

Naturalists regard this as question-begging; and Theists who say this must explain how God supplies things with the power to exist, which is not accomplished simply by declaring that he does. Theists might hold that God made the universe to be self-sustaining, once created, that God himself is behind the first law of thermodynamics. Once created, the world's mass/energy cannot be destroyed. Perhaps making it so from the outset is one of God's ways of sustaining creation.

"Sustaining" is a difficult concept to make intelligible. Process Theology's distinctive accounts of how the world depends on God can be construed as ways in which God constantly sustains creation. According to process thought, God does not act on the world merely at the beginning. Rather, God continuously interacts with it, and it constantly depends on God in a variety of ways. God's influence on the world is ongoing. How so?

ii. Influencing and Saving

How else might a universe or many universes depend on God? Are other accounts of "sustaining" more plausible? Some forms of dependence belong to initial creation, some to ongoing preservation. Perhaps ongoing preservation is ongoing creativity. The several ways in which the world depends upon God according to Process Theism may help to make sense of the vague notion of "sustaining the world." As mainstream Process Theologians contend, God continues to influence creatively and include within himself the course of events within the world, so sustaining is ongoing creation and preservation. But how does this happen?

(1) Temporal entities within the world rely on God to supply them with an "initial aim," consisting of an awareness of *novel possibilities* for creative activity and self-development. Efficient causation within the world is mixed with final causation, purposiveness, or teleology derived from God.

(2) Temporal entities depend on God to *lure them gently toward goodness*, without overwhelming their ability to choose otherwise. Alluring visions or intuitions of the true, the beautiful, the right, and the good within the world ultimately come from God. Higher or more complex organisms like us clearly have them, but their pervasiveness is open to debate. In our awareness of and sensitivity to values that transcend time and place, God's presence is ongoing.

(3) All individual events depend on God's memory for the *preservation of their concrete being and value* once they perish in time to themselves. Every universe, if others exist, depends on God to preserve all intrinsic, extrinsic, and systemic worth achieved within it. Without God, contingent goodness perishes altogether with the passage of time. Divine value-conserving activities are ongoing, universal, and count as "sustaining." By remembering them flawlessly, God sustains and conserves within himself all values actually achieved in

ongoing creation. Plausibly, sustaining is God's remembering. This way of sustaining the world is its eternal salvation.

(4) Finally, once the mass/energy of existence is created, God *voluntarily shares creativity itself*. Newtonian universes devoid of creativity are logically possible, but God grants freedom and co-creativity to his creatures. The transition, transmission, and transformation of mass/energy belongs to the subject matter of physics; the internal self-creation of actual entities belongs to psychology, axiology, metaphysics, and theology. Self-creation and final causation or persuasion may not apply everywhere, but they are widely distributed. In originating universes, in creating the intensely concentrated stuff out of which partly self-creative actual occasions emerge, God acts only as an efficient cause without imparting final causation to existing subjects, without giving initial aims to pre-existing individuated occasions, for none exist at the initial moment of grand unification and perhaps for some time thereafter. Non-existence and non-individuation cannot be co-creators with God; but very near the outset of a Big Bang, individuated entities in a newly created *ex nihilo* universe could be. God caused and designed the Big Bang, and the Big Bang produced both freedom and order.

(5) Plausible versions of both Classical or Process Theology may make a place for *special acts of divine providence and self-disclosure* to particular individuals in the course of history. Special acts of God and Divine self-disclosures have huge moral and religious significance and make momentous differences in the course of human events; but this type of dependence is particular, not universal.

Acts of God and Divine self-disclosures involve interactions between God and created individuals in specific situations and lack the universality required to count as sustaining the universe; since they occur within an established universe, they do not count as world-creation; but they are instances of ongoing Divine creativity and the world's contingency upon God. Some critics suggest that Process Theology cannot allow for personal historical interactions between God and human individuals,[16] but this is not true. The world may depend in special ways on God for what it knows about how God has related himself to particular people in particular times and places. As Charles Hartshorne explained:

> With Crisis Theology...our theory can agree that God is personal and self-related to the creatures, and that his acts of self-relationship are not rationally deducible, but require to be "encountered." However, as Barth and Brunner seem not to see, this is compatible with there being an essence of God which is philosophically explicable and knowable. The concrete volitions of God may be contingent...For each man–religion is a matter of the actions of God as self-related to him, that is, to a wholly contingent being, or to humanity, likewise contingent. Relations whose terms are

> contingent can only be contingent. Philosophy seeks that general principle or essence of the divine being of which such concrete actions of God are mere contingent illustrations. But from a religious point of view, it is the illustrations that count. Thus the religious and the philosophical attitudes are complementary, not conflicting. Our doctrine appears, then to effect a peculiarly comprehensive synthesis of past and present thought concerning theism.[17]

The success or failure of any act of divine self-disclosure depends upon the openness of human receivers to God. Some Christian Process Theologians hold that Jesus was both fully man and fully God in the sense that he was the one man in all of human history who was most fully open and sensitive to God's nature, thoughts, values, sensitivities, emotions, desires, and decisions. The historically limited and conditioned intellectual, aesthetic, moral, and religious capacities, predispositions, and perceptivity of human receivers and interpreters always color and may even distort divine disclosures. Sinners who may distort the message are always on the receiving end of divine communications.

To return to the objection that an everlasting universe would be a necessary being, recall that if a thing exists contingently, this means that (1) its non-existence is possible or conceivable without contradiction; (2) it is caused to exist or created in some or all respects by something other than itself; (3) it is corruptible and destructible; and, normally, it (4) comes into being in time and (5) perishes or is capable of perishing in time. A necessary being has just the opposite properties. A temporally infinite set of contingent things, for example, contingent but successive universes, would not be a necessary being because (1) its non-existence is possible or logically conceivable; and thus (2) its existence is not fully self-explanatory. It is (3) destructible in principle unless some Necessary Being sustains it, or causes it to be self-sustaining once created. If a temporal set of universes is infinite, (4) (a) *each part* (each epoch) would come into being in time, albeit infinite time; (b) no part of it would be absolutely uncreated and persist through all time. (5) Still, the whole of it would embrace an everlasting past; and it would in this respect resemble the everlastingness of a single necessary being. If (6) it continues infinitely into the future, the whole would not perish in time, even though each of its parts does, so in that respect also it would also be everlasting. Still, it fails in many important respects–(1) through (4)–to be a fully necessary being.

Most importantly, ours is not a temporally infinite universe. Remember the Big Bang! The most crucial cosmological fact about our universe is (2) above. The one and only universe that we know to exist was caused to exist by something other than itself. The first premise of our cosmological argument applies to it. If any whole is composed entirely of contingent parts, that whole is itself contingent. Many ways in which our universe depends on God have now been identified.

Would an infinite string of universes depend on God the way ours does? No available evidence supports the claim there is or was an infinite string of universes in either space or time. Infinity is not a real problem for our spatio-temporally finite universe, the only one we know to exist. Epistemologically, infinite universes are just postulates proffered to avoid God by atheists, or to imagine how God could be infinitely creative in space as well as time by Process Theists. Our problem is, What caused *our* Big Bang?

B. Contingent Wholes Do Not Imply Necessary Causes

At least one difficulty with applying "If the parts are contingent, the whole is contingent" premise to our universe is serious. Applying the Principle of Contingent Wholes to the universe as a whole enables us to conclude that it, too, is a contingent whole that has a cause. Yet, one way of applying it–as an inductive generalization–seems highly problematic

Recall that inductive logic allows us to generalize to more of the same, but it does not allow us to infer something different. The Principle of Contingent Wholes is an empirical truth derived inductively from what is "evident to our senses" about relations between parts and wholes *within our own world* of spacetime. When we extend it to apply to the universe as a whole, we are still reasoning from parts to wholes, so what is the problem?

The difficulty is with respect to the nature of that cause. Here, even if so-far-so-good with respect to reasoning from parts to wholes, *other inductive generalizations* seem to count against a Necessary Cause. Empirically, *all experienced wholes derive their existence from other contingent beings*, not from a necessary being. Thus, it seems, we are warranted inductively to generalize only that all contingent empirical wholes are caused by other contingent entities. Applied to the universe itself, this means that even if our contingent universe had a cause, we are warranted inductively to infer only that a contingent being (or beings) brought our world into existence. If this is all we are entitled to infer, the first premise of our Cosmological Argument from Contingency cannot generate the conclusion that a set of contingent beings or wholes, whether finite or infinite, depends ultimately upon a necessary being that transcends that set.

So, how does omnipresent contingency entail the reality of a transcendent necessary being? Inductive logic will take us from wholes to their causes, but will it take us from contingency to necessity? If not, Naturalism is in the same boat; it can't get the necessity of Nature by inductive reasoning either. Recall Naturalism's claim that "Nature as a whole exists necessarily."

In response, the existence of a necessary Divine Being is not an inductive inference. Even though it cannot be inferred inductively, it is still the most plausible *explanatory hypothesis* available to us. From contingency and finitude alone, neither Theists nor Naturalists can derive the infinity and necessity of

God or Nature by direct observation or inductive inference. Still, they might get them as the most plausible explanatory hypothesis, that is, by abduction. Who has the best case?

Some analogies between nature and its parts seem to hold, for example, *if any whole is composed entirely of contingent parts, the whole itself is contingent*. Many other analogies cannot hold between intra-universe and extra-universe causation. Not counting miracles, if there are any to count, experience shows that all empirical contingent wholes are caused by other empirical contingencies; but we have good reasons for not applying *this very different generalization* to the ultimate cause of the universe as a whole.

Many valid inductive generalizations cannot be applied to the ultimate cause of our total universe. Experience shows that *all wholes within our universe are caused by other beings inside the universe*. Yet, the universe as a whole cannot be caused by something within itself. The universe as a whole could not be caused to exist by another being (or beings) inside the universe because they are effects of the universe's existence, not its cause. Even cosmological Naturalists do not resort to that. Before Big Bang theory, they said it has always been around in some form, which isn't true. Now, they say, something transcendent like Superspacetime or an Antecedent Universe caused our Big Bang about fifteen billion years ago, but they do not claim that something within it caused it to be. Even Big Accident Theorist do not claim that something within our universe caused it to be; rather, it had no cause at all.

Another inductive generalization that applies to all wholes within our universe could not hold true of the ultimate cause of the universe itself. *All wholes within our universe are caused by a finite set of natural events that go back no further than the Big Bang*, but naturalistic oscillationists and all other infinitely many worlds metaphysicians postulate *an infinite set* of antecedent conditions as the ultimate cause of our universe. Naturalists deny both that nature as a whole is caused either by a finite set of causal conditions or by conditions that are parts of our own natural system of spacetime. They think that as whole it just isn't caused at all; it is itself the Necessary Being; like God, nature is "self-caused." But this is where the first premise of our Cosmological Argument from Contingency shows them to be wrong: all contingent wholes, nature included, have causes.

Many other inductive analogies of causation do not hold when dealing with the origin of the universe as a whole. *All causes within nature are spatiotemporal*; but singularities are not spatiotemporal. *All causes within nature obey the laws of quantum physics*; but antecedent and contemporary "many worlds" may have their own very different laws. The truth is, characterizations of the very nature of the ultimate cause (an infinite set of crunching antecedent universes, perhaps correlated with an infinite set of initial singularities, infinitely many co-existing worlds, an infinite and self-sufficient Nature, or what have you) are explanatory hypotheses, not inductive generalizations.

The same is true of God as the ultimate Necessary Being. So the question is, which explanation is best? Preceding chapters demonstrated that non-theistic explanations just do not work.

That *our universe as a whole had a cause* is denied by Big Accident Cosmologists, who say that it just popped into being without a cause, and in a peculiar way, by Naturalists, who say that it is itself the self-sufficient Necessary Being. Both have been refuted in earlier discussions. Our reasoning about nature as a whole cannot be entirely nature bound. The cause of our universe was not the universe or some part of itself, not some finite set of causal conditions, and not a contingent set of conditions because no such entities are ultimately self-explanatory.

In one respect, Process Theology can easily accept contingency as an element of that which created our universe. As indicated earlier, most Process Theists affirm the reality of infinitely many temporally successive worlds or cosmic epochs on theological (but not empirical) grounds. It is reasonable to expect that an infinitely loving, social, and creative God would be infinitely creative of creatures to love and with whom to socialize. In a very important sense, "All contingent wholes have contingent causes" applies even to God; but this is not embarrassing to Process Theology as long as it is not the whole truth–as next explained.

An element of contingency pertains to God's creating all particulars. God's decisions to create particular actualities belong to God's contingent Consequent Nature, not to God's necessary Primordial Nature. God's decisions to create particular worlds are freely and contingently made. Particular manifestations of God's causal efficacy are contingent, not necessary. That *the Divine cause of our universe is contingent, in part* is compatible with the dominant process view that God creates new contingent worlds out of old contingent worlds, as well as with the oscillationist position that contingent old worlds enter into the creation of new worlds. In fact, most Process Theists have been implicit if not explicit oscillationists who think that God had important contingent roles to play in an infinite string of previously existing contingent universes or cosmic epochs.

The element of contingency in the God who created our finite and contingent world cannot be pushed too far. Not everything about God is contingent, or God himself would fall prey to the first premise of our Cosmological Argument from Contingency: God would also be a totally contingent being if everything about God is contingent; but there is an important disanalogy between Divine and mundane causation. If a totally contingently God caused our universe, then the world plus God would constitute another contingent whole requiring a higher order God as its cause. This higher order God is either contingent or necessary. If contingent, the Principle of Contingent Wholes applies again. No matter how many totally contingent Gods exist, the principle continues to apply until we come at last to a necessarily existent God. All contingencies, no matter how many, are swallowed up by the Principle of Contingent Wholes; and the process comes to rest only in a Necessary Being. No contingent wholes are self-existent or self-explanatory.

If our finite universe as a whole depends on God either originally, continuously, or historically in any way, the claim that "All contingent wholes are caused by contingent beings" is only part of the truth. The whole truth requires us to add, "And by a necessarily existing God." The most plausible explanation is that our universe depends for its existence and order on the reality of a transcendent Divinity who could not not exist, who knew what it was doing, did it intentionally, and did it well.

C. "Cause" Cannot Apply to the Universe as a Whole

With no necessary natural wholes or parts, our world is doubtless derived causally from something other than itself; but another small problem remains. Perhaps the very notion of causation makes sense only when applied within the universe and cannot be applied to the universe as a whole. Experienced causes are always associated with space, time, and natural laws; but all of these disappear at $T = 0$ if the universe as a whole is created by a Divine transcendent reality out of absolutely nothing or the empirical nothingness of a singularity.

An analogy between God and an initial singularity as the cause of the universe may help us to understand how God can be its cause. Recall that some versions of "scientific" Oscillation Cosmology affirm that an antecedent universe collapsed into a singularity and then caused our universe on the rebound. The existence of neither a single antecedent universe nor a self-sufficient infinite totality of prior universes can be established scientifically. Everything to which science appeals in tracing causal connections just plays out at an initial singularity. A singularity is spaceless, timeless, and lawless; and without space and time, the concepts of physical energy, causation, and natural laws are meaningless. Thus, even in this "scientific" account, our system of spacetime erupts from something nonspatiotemporal, nonphysical, and nonnatural.

Given an initial singularity, our space and time begin just this side of $T = 0$; and it makes no sense to say that the initial singularity was earlier than the onset of the primordial fireball itself. Nothing can be temporally earlier than the first moment of time, neither a singularity, nor an antecedent universe, nor an atemporal God. According to George Mavrodes, an initial singularity must have the same relation to creation that God has in Augustinian/Thomistic Classical Theism.[18] It is a logical, theoretical, hypothetical, metaphysical–but not a temporal–prerequisite for all space, time, and physical energy and causation.

Natural or physical causation also plays out at an initial singularity. Being the effect of something natural, of spatiotemporal energy, can be traced back no further than the very first "products" of an initial singularity. Scientific or empirical knowledge of natural, that is, spatiotemporal causes terminates abruptly at that point; no spatiotemporal or natural causes exist where no spacetime exists at all. Looking backward, all scientific knowledge based upon natural causal relations ends where natural causes begin. A singularity is a non-physical, non-spatially extended cause or ground of everything physical.

If, as many scientific cosmologists hold, the habits and laws of nature are themselves products of a colossal cataclysmic explosion that emerged from we know not what, there can be no natural law for–or underlying–creation itself. Natural laws themselves are (very abstract) creatures, no matter whether God, an initial singularity, or an antecedent universe produced them. As John Wheeler cautioned, "There never has been a law of physics that did not demand 'space' and 'time' for its statement....With the collapse of space and time the framework falls down for everything one ever called a law of physics."[19]

If we extrapolate mathematically from observationally confirmed laws of Hubble expansion, the redshift of the galaxies, and cosmic entropy, then make proper allowances for an increasing (once thought to be declining) pace of expansion as the explosion winds down (or up!), and finally calculate retroactively the past natural history of the cosmos, we arrive at zero space and time somewhere close to fifteen billion years ago. Natural laws take us back that far, but they can carry us back no further than the point at which they themselves originate. Thus, we cannot extrapolate scientifically back to even one antecedent universe, much less an infinite number of them. Scientific knowledge ends at $T = 0$, or just this side of it. From our perspective within this world, an initial singularity, antecedent universes, and a transcendent Superspacetime are supernatural, atemporal, non-natural-causal entities, just as God is in Classical Theology. They may be just two ways of thinking about the same thing–an initial non-physical cause or pre-condition of the universe.

Nothing remains of the notion of physical or natural causation once space, time, physical energy, physical causation, and natural laws are eliminated; but this is no embarrassment to Theologians, for whom the creation of the universe *ex nihilo* had a transcendent Divine cause, not a natural cause. If cosmological reflection comes to this, the theologians can say: "I told you so."

Natural laws can not "govern" or enable us to predict something that happens only once. The unique creation of a unique universe is more like the expression of choice or will than an effect of lawful physical regularities; but a puzzle lingers. Does any meaning remain for "cause" after its natural associates–space, time, physical energy, and natural laws–are altogether expunged? Can "cause" still have metaphysical meaning when stripped of all natural or physical meaning? Some critics think not. A Classical Theist might reply: "Well, I hope so; but if it doesn't, the singularity of Antecedent Universe Cosmology is in the same boat. God, who caused the Big Bang, is somehow a necessary condition for the existence of our universe." Yet, just what this "somehow" means is less than clear.

Process Theism, or some form of temporalistic theism, has a much easier time with such puzzles of causation and creation. George Mavrodes concedes, perhaps somewhat grudgingly,

> If God Himself is a temporal being, sustaining temporal relations, then indeed there can be time outside of the singularity. And God, the cause of the existence of the world, can temporally precede the existence of the world.[20]

Divine spatiality could also exist literally outside our world as all-embracing Divine Superspacetime; the temporal God of process theology is everlastingly embodied. Our finite universe may be and likely is only one part of God's body.

Divine causation could be non-physical or "incorporeal" as Classical Theology maintained; but if efficient causation is unintelligible without space, time, and embodied energy, then Process Theism attributes all of them everlastingly to God and thus can make theoretical sense of initial Divine transcendent but still spatiotemporal causation or creation from God's Superspacetime to our spacetime.

But our inquiry is not quite complete. Until both Divine and non-divine transcendent but contingent causes are ruled out, our Cosmological Argument from Contingency does not come to rest in God as a Necessary Being who always was and ever will be.

D. God Also Must Have Had a Cause

If everything has a cause, then God has a cause. So, who or what caused God? Naturalistic cosmologists really seem to savor this common objection to Theism. Carl Sagan expressed it nicely in his *Cosmos*.

> In many cultures it is customary to answer that God created the universe out of nothing. But this is mere temporizing. If we wish courageously to pursue the question, we must of course ask next where God comes from. If we say that God has always existed, why not save a step and conclude that the universe has always existed?[21]

Sagan's question ignores several important things. First, we know from Big Bang theory that the universe has not always existed. It is only fifteen billion or so years old. Just pronouncing that our universe is eternal cannot overcome that. Next, even if the universe were infinitely old, it would still be contingent because it depends on God in various ways at every moment, as previously explained. Sagan's premise–*Everything* (even God) has a cause–is inconsistent with his conclusion–The universe has no cause ("has always existed," as Sagan put it). If everything has a cause, then the universe has a cause.

Even Sagan does not really want to say that everything has a cause. Sagan's atheistic Naturalism supposes that our necessarily existing universe had no cause; but this is false. That our universe is an everlasting necessary being is decisively falsified by all the evidence supporting Big Bang Cosmology. Finally, Sagan misunderstands the concept of God. Nothing falsifies God's existence. God, unlike the universe, has always existed because God is that being who could not not exist. As St. Anselm discerned, if it were possible for God not to exist, God would not be God, the supremely worshipful reality than whom none greater can be conceived. Sagan's analogy between contingent and necessary being does not work!

The desire to "save a step" cannot change a world that every human experience shows to be contingent, temporal, finite, and capable of non-existence into an everlasting world that could not not exist. The very idea is self-contradictory. By contrast, neither experience nor logic conflicts with the concept of God as the one reality who could not not exist and who is causally self-sufficient, uncreated, everlasting, and indestructible. Experience clearly discloses that the world does not resemble God's everlasting and self-sufficient Primordial Nature in any of these ways. Nature does not manifest these divine attributes, and we cannot make an unworkable analogy go through merely to "save a step." Experience and inductive extrapolation show that our universe is a contingent entity, and no ingenious conceptual gerrymandering can alter that brute fact.

Poor philosophical reasoning is repeated interminably in cosmology. In *The Blind Watchmaker*, Richard Dawkins argues that it is futile to appeal to God to explain features of the world like DNA and proteins because this

> leaves unexplained the origin of the Designer. You have to say something like 'God was always there,' and if you allow yourself that kind of lazy way out, you might as well just say 'DNA was always there,' or 'Life was always there,' and be done with it.[22]

But we know for a fact that DNA and life were not always there, and saying so does not make it so. They are contingent wholes that require causes beyond themselves. Similarly, when Paul Davies considers the Theistic claim that God is the necessary being who has within himself the explanation of his own existence, he asks, "Why can't we use the same argument to explain the universe?"[23]

The answer is, the fifteen billion-year-old universe is composed entirely of contingent beings; and any whole composed entirely of contingent beings is a contingent whole, as experience clearly shows. We know that our world's existence is not everlasting, self-sufficient, or self-explanatory; and that is why we cannot use the same concepts to explain the universe that we use to explain God, whose Primordial Nature exists necessarily.

Stephen Hawking grew increasingly agnostic if not decisively atheistic over the years.[24] Hawking suggested in *A Brief History of Time* that there is really nothing for God to do if his "no boundary" proposal for cosmology is correct.

> So long as the universe had a beginning, we could suppose it had a creator. But if the universe is really completely self-contained, having no boundary or edge, it would have neither beginning nor end: it would simply be. What place, then, for a creator?[25]

We should now understand why Sagan, Dawkins, Davies, and Hawking are philosophically confused. A universe unlike ours with no beginning or end nevertheless would exist contingently if, like ours, it were composed entirely of contingently existing parts. The contingency of any universe like ours that had a beginning should be patently obvious. Our universe also contains no parts that exist necessarily–without any causal dependence of any kind on anything. A contingent universe with or without a beginning still needs God; and our universe had a beginning.

E. Atheism Is Simpler than Theism

Philosophers and scientists alike appeal to the Principle of Parsimony or Simplicity as a significant norm of rational explanation. Ockham's razor, as it is often called, says that we should not multiply explanatory entities unnecessarily, that our explanations should be as simple as possible. The lure of simplicity is as much aesthetic as rational. It is where the true and the beautiful come together. Yet, the ideal of simplicity must always be balanced by the norm of comprehensiveness, for we should not oversimplify. As Whitehead suggested, we should "Seek simplicity and distrust it."[26]

Naturalists may want to argue that the hypothesis of an infinite and self-sufficient nature or cosmos is simpler than the hypothesis of a contingent universe plus God. Pierre Simon Laplace, who believed in a self-sufficient and everlasting system of natural causes, told Napoleon that he had no need of the God hypothesis. Unfortunately, the naturalistic hypothesis is not scientifically verified or verifiable; in fact, it is falsified by Big Bang Cosmology. And neither an infinite single world nor infinitely many worlds is simpler than one infinite God. To avoid God, atheists have to resort to infinity–to something just as complex as God!

Simplicity is determined by counting either the number or the complexity of explanatory entities, or both. A theory with fewer or with less complex explanatory entities is simpler than one with more. Naturalist may reject theism on the grounds that one world alone is simpler than one world plus one God. Obviously, one is simpler than two. But things are not so simple. The universe of Naturalism is either one infinitely complex world, or infinitely many finite worlds. Is this really simpler than Theism's minimal postulate of one world and one God? The answer is complicated by the fact that this one God is said to be infinitely complex in Process Theism and infinitely simple in Classical Theism.

In arguing that Theism is not as simple as Naturalism, Naturalists may have in mind either the *number* of explanatory entities or their internal *complexity*. Complexity is a function of the number, kinds, and order of inner parts.

In *number*, one (finite) world plus one (finite) God is less simple than one (finite) world; so at this simplistic level, Naturalism wins the battle of simplicity; but, as already shown, Naturalists really do not believe in a finite world. The viable options for both Naturalism and Theism are much more complicated. Expressed in terms of contemporary cosmology, Naturalists are committed to either one infinite world (ours) or infinitely many finite worlds in a Superworld beyond and/or before ours. These options are very different; but which is the simplest? At the same order of infinity, a whole comprised of an infinite number of finite entities is just as complex as, indeed is numerically identical with, a single infinitely rich entity. Dennis Sciama argues that an infinite worlds metaphysics is simpler than a Theism in which God decides to create only one world because the it places as few constraints on reality as is compatible with observation.[27] Sciama obviously confuses simplicity with plenitude and observation with concoction. Victor J. Stenger remarks,

> Several commentators have argued that a cosmology of many universes violates Ockham's razor. I beg to differ. The entities that the law of parsimony forbids us from multiplying beyond necessity are theoretical hypotheses, not universes. The cosmology of many universes is more economical if it provides an explanation for the origin of our universe that does not require the highly nonparsimonious introduction of a supernatural element that has not heretofore been required to explain any observations.[28]

In reply, other universes *are* supernatural elements. And the standard objections are that theism multiplies the number of entities, not the number of hypotheses, and that it explains the existence of our world by appeal to an otherworldly being or beings. In both respects, however, there is no difference at all between God and infinitely many other worlds, except that God is numerically one, and the many worlds are numerically infinite. Stenger elsewhere characterizes the innumerable other worlds of many worlds metaphysics as "bizarre," "untestable,"[29] and a matter of "uneconomical speculation."[30] The Other Worlds of Quantum Cosmology are just as transcendent, supernatural, and inaccessible to direct sensory verification and warranted inductive inference as a transcendent God; and they outnumber God by infinity to one! So which hypothesis is the most "nonparsimonious"?

If we shift focus from extensional referents (things) to intensional meanings (hypothetical constructs), exactly the same point applies to hypotheses as to entities. Epistemologically, as explanations of the origin of our universe, God and infinitely many worlds are both theoretical hypotheses; and the claim that only the God-theory is really a theory is question-begging nonsense. Stenger's argument against God-as-a-theory is like dismissing evolution because it is just a theory.

The strongest and most appealing forms of Theism really do not conceive of God as infinite in every conceivable or verbalizable respect. God is infinite only in every logically consistent respect compatible with supreme goodness and worshipfulness. In number of explanatory entities, one infinite world alone seems at first to be simpler than one or more infinite worlds plus one infinite God, for one is simpler than two. We are habituated to finitistic thinking, but we are now dealing with infinities that do not sum up like finites. Infinity added to infinity just equals infinity. In complexity, it is not at all obvious that one infinite world is simpler than one infinite world plus one infinite God.

Richard Dawkins also failed to consider infinities when he argued that God is more complex, thus less simple, than the world. In his words,

> Any God capable of intelligently designing something as complex as the DNA/protein replicating machinery must have been at least as complex and organized as that machine itself. Far more so if we suppose him additionally capable of such advanced functions as listening to prayers and forgiving sins.[31]

Where infinities are involved, determining simplicity and multiplicity or complexity requires the use of transfinite mathematics. Most critics of theism fail to consider this, which is why Naturalists often seem to win the simplicity contest. In transfinite mathematics one infinity plus another (or plus any finite number) is equal to one infinity, assuming that the infinities are of the same order; so one infinite God plus an infinite world has no more members than an infinite world alone, or an infinite God alone. If God's infinity is of a higher order than the world's, then God plus one or more infinite worlds is equal to God's more complex order of infinity. The sum is still just one infinity. When infinities are totaled, no matter how many, they always sum up to one infinity, and one is the paradigm of numerical simplicity. All the foregoing options seem to be equally simple.

A further complication with respect to the *complexity* of explanatory entities must be considered. Classical Theists insist that God is absolutely simple in himself; only our thoughts about God are complex. In himself, God is somehow pure undifferentiated unity of being, and that constitutes God's simplicity. So what could be simpler than that? Well, we hardly know how to assess the meaning or the simplicity of such double-talk. By contrast, Process Theism drives no absolute wedge between God and our thoughts about God. We think of God as infinitely complex because God really is infinitely richer in properties than all lesser beings. God is more complex than an infinite world, for God's infinity is of a higher order than the world's. Theological comprehensiveness, explanatory power, and correctness sometimes demand complexity at the expense of inordinate simplicity.

So, is a naturalistic infinite-world-ensemble metaphysics really simpler than Theism? Where the order of infinity is the same, in number of explanatory entities, Naturalism's infinitely many worlds without Divinity are certainly no simpler than Creation *ex nihilo* Theism's one infinite God plus one finite world because God's infinity plus a finite unit (one finite world) just equals God's infinity.

Considering only the *number of worlds* involved, naturalistic infinite worlds metaphysics, not Theistic creation of our one world *ex nihilo*, egregiously violates Ockham's razor. One-world monotheism clearly wins the simplicity contest. One finite world is infinitely simpler (less numerous) than an infinity of worlds.

Process Theists postulate a loving God's creative involvement with an infinite number of worlds to love. This gives us an infinite God plus an infinity of worlds, but these sum up to one infinity, God's. In complexity of ultimate explanatory entities, infinitely many finite worlds considered as a totality would not rival God's own infinite complexity if God's richness belongs to a higher order of infinity. If their order is the same, they are equally complex.

No matter what, that Naturalism is simpler than Theism is by no means evident. To avoid God, non-theists must appeal to something almost if not entirely as rich and complex as God.

The most simple-minded naturalistic atheistic metaphysics would affirm the existence of a single completely self-contained, self-sufficient, and everlasting but totally *finite* world without Divinity; but this metaphysics is incoherent because infinite time, required for everlasting self-sufficiency, is incompatible with total finitude. Older Naturalists postulated the infinite duration of our one world, but the Big Bang came along. Naturalists are now driven to postulate infinitely many transcendent worlds–while still complaining about traditional religion's other worlds! The actualized infinity of infinite worlds metaphysics is not verified, not verifiable, lacks an Agent of Diversification and Selection, and fails to account for the remarkable fine-tuning of our universe for the production of complex valuable life. Creation *ex nihilo* Theism is the simplest theory that can account for our well-ordered world and satisfy rational explanatory norms of both simplicity and comprehensiveness, but even it must come to terms with the concept of God as infinitely creative.

F. There Is No Universe as a Whole

Kai Nielsen contends that the very notion of the universe as a whole is unintelligible. If true, this implies that the idea that anything, including God, created or caused the universe is also unintelligible. If he is right, the whole project of this book, the search for the cause of the origin of our universe in the Big Bang, is totally misguided; and its conclusion, that God did it, is utterly unintelligible. As Nielsen expresses the argument:

> It is only by thinking of the universe as some kind of gigantic thing or some kind of entity or totality that we can have a shot at intelligibly speaking of the universe as a whole. But "universe" may be just an umbrella term for the various things, events and processes there are. We cannot intelligibly speak of the sum of things so that we could intelligibly speak of them as a whole and ask if, and, if so, how, this universe was created and is sustained. Moreover, since the universe is not an object, event, process–not any kind of entity at all–there is no such thing as the universe for "universe" to stand for. Rather the term "universe" is an umbrella term standing for the objects, events, and processes there are. But there is no sense to trying to count the number that there are and to come to speak of the universe as a whole. "Universe" does not stand for some mysterious entity, but indifferently for those various discrete things. Things of which, since "object" is not a count word, we cannot count the sum. There is, that is, no way of summing them up and fixing the number of them that there are. This being so, we cannot coherently speak of the universe itself–that totality–being caused, created, sustained, and the like. To ask for the cause of the universe is to make what in the good old Ryleian days would have been called a category mistake.[32]

This argument suffers from many defects. First, Nielsen assumes that the only way to give meaning to the notion of the unity of nature, the universe as a whole, is to be able to *count* the number of its ingredients. This is itself a significant departure from the views of traditional Naturalists, who consistently identified nature or the universe with our system of space/time and all of its ingredients, no matter how many, and no matter whether they are practically countable. Naturalists traditionally claimed that this system of nature infinitely or eternally predated our existence. Traditional Naturalists would grant that we probably cannot count all natural things, events, and processes, even in principle (except perhaps by using transfinite mathematics, which really is a way of counting); but this would not persuade them that the concept of nature as a whole is unintelligible. Nielsen disagrees. Perhaps he thinks that the entities within the universe are not countable because they are infinite in number and cannot be counted finitistically, even in principle. Or he may think the rich natural order of things, whether finite or infinite, is so lacking in unity that thinking of it as "a whole" makes no sense. Either way, he is mistaken. Infinity is also a way of counting; but our universe is finite; and it has much more unity than Nielsen allows.

Our universe is not infinite in time, space, or number of real parts; and strategies are available for making the unity of the concept of the universe as a whole perfectly intelligible. Nielsen seems to ignore contemporary Big Bang Cosmology completely. It has developed and explicated these strategies for us in great detail, as by now we well know. We need not be capable in practice of counting the number of things in the universe in order to know many things that make the concept of "the universe as a whole" intelligible. (1) Everything within our system of spacetime had a common, intensely concentrated, and totally unified origin. (2) All cones of causation within our contemporary universe are derived from an original grandly unified Big Bang. (3) By extrapolating from pervasive laws of nature, we can trace the common evolutionary physical history of everything within our system of spacetime back to highly unified Big Bang origins. (4) This history goes back for approximately fifteen billion years and no further. For details, read again Chapter One of this book and subsequent discussions! Nielsen finds unintelligible what contemporary astrophysicists find exceedingly intelligible! And *they* show us exactly how to conceive meaningfully of our universe as a unified whole.

Nielsen has not learned an important lesson from contemporary "scientific" astrophysics and cosmology: asking about the duration of the totality of what he calls "the objects, events, and processes there are" makes perfectly good sense. Without being able in practice to count each of them individually, traditional Naturalists gave this answer: Their collective duration is infinite; our universe has existed throughout an infinite past. Nielsen may presuppose this without making his commitment explicit; perhaps he half-consciously hopes that no one will pry too deeply into the topic. Contemporary cosmologists both ask and effectively answer the question. The totality of "The objects, events, and processes there are" has endured for only fifteen billion years or so, no longer, and certainly not forever. All of the overwhelming evidence for the Big Bang given in Chapter One of this book attests to this conclusion.

Since the notion of our universe as a whole really does make sense, and since traditional Naturalists were clearly wrong in insisting upon its infinite duration, the question of a transcendent cause or creation of the universe is intelligible after all. Even atheistic Oscillationists and Quantum Cosmologists think so. Even cosmological atheists affirm a transcendent causes or causes of our universe, as previously explained. But was it God? That is the viable issue.

Consider one of Nielsen's principle arguments against the supernatural. "It isn't that we do not have to go beyond nature, but that we do not understand what such talk comes to. We have no idea of what it would be like to go beyond nature."[33] To find out what it would be like, all Nielsen has to do is ask any Oscillationist or Quantum Cosmologist. Atheistic supernaturalism is rampant in contemporary astrophysical cosmology. For Theism, transcendence is no problem because even a scientifically well informed Naturalism is unintelligible without it.

G. Transcendent Reality Is Unknowable

Immanuel Kant contended that all arguments for the existence of God fail because they apply conceptual categories of the understanding–like causation, purpose, and necessity, which legitimately pertain only to appearances–to real things in themselves. For Kant, all reality is experientially, causally, spatially, and temporally transcendent and theoretically unknowable; only appearances are knowable, spatial, causal, and temporal. Kant believed that appearance and reality in no way resemble one another with respect to spatial, temporal, causal, and all other metaphysical properties and relations.

The problem is, if all spatiotemporal and causal properties and relations are removed from our concept of reality, or from reality in itself, nothing is left. Since "being" is also a Kantian category of the understanding that applies only to appearances, not even being is left! Kant rejected all cosmological attempts to know the real world of nature, along with critically realistic theories of perception according to which appearances are like realities at least in being spatiotemporal and in being caused by the realities they resemble. Kant rejected philosophical and scientific realism in epistemology, cosmology, and theology largely because they apply the concept of "cause" to realities; but when ask why he must introduce real things in themselves at all, his answer was that they have to be there to *cause* us to have the perceptions or "appearances" that we have.

Critical Realism applied to cosmology does not conceive of transcendent reality as something merely beyond or before our sensory perceptions. Instead, transcendent reality is "beyond" and "before" our objectively existing world of nature, our universe, our cosmos, our very real system of space/time/mass/energy. This is what "transcendent" meant traditionally in both theistic and naturalistic metaphysics. Nature is our public, objective system of spacetime, as opposed to the private spacetimes of dreams and hallucinations and the inaccessible spacetimes of transcendent Other Worlds. Nature includes all things that exist within and have causal relations with at least some other entities within our public world; and anything that transcends this world is supernatural. Even if it has causal relations with it, transcendent reality either logically or temporally antedates our system of spacetime and has no fixed or ascertainable limited position within it. In Process Theism, the eternal Primordial Nature of God is always embodied in some concrete, contingent, spatiotemporal Consequent Nature; it is a necessary condition for every particular contingency in our world; yet it both transcends and is immanent in all actual worlds–if God, who alone knows for sure, has created more than one.

Scientific Cosmological Agnosticism, we saw in Chapter One, indicates that we can really know many features of our objective spacetime system scientifically, but we cannot have scientific knowledge of transcendent realities like Other Worlds, whether the otherness be supertemporal, superspatial, or both. Can we then have no rational or philosophical knowledge at all of the transcendent? If not, Teleological and Cosmological Arguments fail, for the God who allegedly caused our Big Bang and ordered our universe is a transcendent reality. Most contemporary "scientific" cosmology also fails if transcendence is beyond all human knowledge. Like God, antecedent or co-existent universes *are* transcendent realities, if real at all. The God hypothesis is a philosophical, not a scientific, explanation of our universe. So is any version of infinitely many Other Worlds. The world disclosed to us in natural science leads to the very edge of scientific knowledge and demands a rational philosophical explanation that natural science cannot give.

What caused the Big Bang itself to erupt? What caused a life-supporting universe instead of chaos or lifelessness to issue from the Big Bang? Physical science poses the questions, but it cannot answer them. Science periodically answers previously unanswered questions, but "What caused the loaded-for-life Big Bang?" is unanswerable in principle by natural science; its methodologies just do not extend that far. Can philosophical reflection do any better? Theistic meta-physics (after-physics) replies that some transcendent-imminent Ultimate Reality who knew what it was doing was responsible. Neither the God hypothesis nor that of infinitely many worlds is directly or inductively verified empirically; both postulate unseen transcendent or supernatural realities to explain the origin of our universe. But, all things considered, (as we have done!), the God hypothesis is the best explanation.

A life-supporting universe might happen by pure accident, given an infinite number of diversified universes. But we have no good reasons for giving that, and mere infinity requires an Agent of Diversification to yield a life-supporting universe. Logical possibility as such says nothing whatsoever about empirical probability. Mere infinity contains no Agent of Selection and Diversification. Infinitely many worlds metaphysics cannot explain why our ordered-for-life world exists or why lifeless universes are not repeated infinitely many times; it cannot guarantee an infinite diversity of individuals, their qualities, and their relational combinations. The Plenitudist notion of realizing all possibilities is not logically coherent. Pure infinity requires no diversity at all, much less the kind that selects for life. Plenitude is merely a ghost without a machine.

A teleological explanation of the life-supporting cosmic coincidences of our finite but well-ordered universe is needed; a personal, intelligent, benevolent, and life-loving Agent of Diversification and Selection best explains the origin and structure of our exquisitely designed contingent universe. That our world was designed for life-flourishing by a Divine, supercosmic, supercalculating, life-loving intellect is the most plausible hypothesis, the explanation best supported by a preponderance of the anthropic, cosmological, and philosophical evidence. Without a supercalculating God, our life-supporting universe would be infinitely improbable; and reason rejects infinite improbabilities. Reasonable persons certainly may, and perhaps must, come to this conclusion: God caused the Big Bang.

NOTES

Preface

1. From *A BRIEF HISTORY OF TIME*, by Stephen W. Hawking, copyright © 1988, 1996 by Stephen W. Hawking, p. 174 . Used by permission of Bantam Books, a division of Random House, Inc. and Writers House LLC.

2. George Smoot and Keay Davidson, *Wrinkles in Time* (New York: Avon Books, 1993), p. 289.

3. Hawking, p. 175.

4. George Lemaître, *The Primeval Atom* (New York: Van Nostrand, 1950).

5. Pope Pius XII, "Modern Science and the Existence of God," *The Catholic Mind*, (March 1952), pp. 182–192.

6. Eric Lerner, *The Big Bang Never Happened* (New York: Random House, 1991).

7. William L. Craig and Quentin Smith, *Theism, Atheism, and Big Bang Cosmology* (Oxford: Clarendon Press, 1993), p. 197. By permission of Oxford University Press.

8. M. A. Corey, *God and the New Cosmology: The Anthropic Design Argument* (Lanham, Md.: Rowman and Littlefield, 1993).

Chapter One

1. *Cf.* Alan Lightman and Roberta Brawer, *Origins: The Lives and Worlds of Modern Cosmologists* (Cambridge, Mass.: Harvard University Press, 1990); or *Cosmos, Bios, Theos: Scientists Reflect on Science, God, and the Origins of the Universe, Life, and Homo sapiens*, eds. Henry Margenau and Roy A. Varghese (La Salle, Ill.: Open Court, 1992).

2. Cheryl J. Beatty and Richard T. Fienberg, "Participatory Cosmology: The Big Bang Challenge," *Sky & Telescope* (March 1994) pp. 20–22.

3. George Gamow, *The Creation of the Universe* (New York: Mentor Books, 1960), p. 41.

4. Werner Heisenberg, *Physics and Philosophy: The Revolution in Modern Science* (New York: Harper & Row, 1958), p. 126.

5. Robert Jastrow, *God and the Astronomers* (New York: W. W. Norton & Co., 1978), p. 13n.

6. Frank J. Tipler, *The Physics of Immortality* (New York: Anchor Books, 1994), pp. x, 2, 78, 152, 214.

7. Wendy L. Freedman, *et al.*, "Distance to the Virgo Cluster Galaxy M100 from Hubble Telescope observations of Cepheids," *Nature* 371 (27 October 1994), p. 761.

8. See "Closing in on the Hubble Parameter," *Sky & Telescope*, 88 (December 1994), p. 10; and "A Young Universe with Old Stars," *Sky & Telescope*, 89 (January 1995), pp. 10–11.

9. Andrew Watson, "The Universe Shows Its Age," *Science*, 279 (13 February 1998), pp. 981–983.

10. Virginia Trimble, "Cosmic Discovery 1998," *Sky & Telescope*, 97:2 (February 1999), p. 38.

11. See, *e.g.*, Jastrow, *God and the Astronomers*. Also, Leon M. Lederman and David N. Schramm, *From Quarks to the Cosmos* (New York: Scientific American Library, 1989); and Owen Gingerich, "Perspectives on Science and Theology," in *How Large Is God?*, ed. John M. Templeton (Philadelphia: Templeton Foundation Press, 1997), pp. 23–46.

12. See Stephen W. Hawking, *A Brief History of Time* (New York: Bantam Books, 1988), pp. 8, 46, 49ff, 61, 88, 117, 133, 173.

13. *Ibid.*, pp. 49–51.

14. *Ibid.*, p. 50.

15. Lederman and Schramm, *From Quarks to the Cosmos*, p. 169. See below Chapter Four, endnote 74.

16. J. A. Wheeler, "Genesis and Observership," in *Foundational Problems in the Special Sciences*, eds. Robert E. Butts and Jaakko Hintikka (Dordrecht, Holland: D. Reidel Publishing Company, 1975), p. 15.

17. Alexander Friedmann, "On the Curvature of the World," *Zeitschrift für Physik*, 10 (1922), pp. 377ff.

18. *Cf.* Michael White and John Gribbin, *Stephen Hawking: A Life in Science* (New York: Penguin Books, 1992), p. 185.

19. Alan H. Guth, "Inflationary Universe: A Possible Solution to the Horizon and Flatness Problems,"*Physical Review D*, 23 (1981), pp. 347ff.

20. Alan H. Guth, "Starting the Universe: The Big Bang and Cosmic Inflation," in *Bubbles, Voids, and Bumps in Time: The New Cosmology*, ed. James Cornell (New York: Cambridge University Press, 1989), pp. 105–146, esp. pp. 132ff, reprinted with the permission of Cambridge University Press; Andrei Linde, Dmitri Linde, Arthur Mezhlumian, "From the Big Bang Theory to the Theory of a Stationary Universe," *Physical Review D*, 49:4 (15 February 1994), p. 1785. See George F. R. Ellis, *Before the Beginning: Cosmology Explained* (New York: Boyars/Bowerdean, 1997), section on "Testing the Nature of Fundamental Forces."

21. Alan H. Guth, *The Inflationary Universe: The Quest for a New Theory of Cosmic Origins* (Reading, Mass.: Perseus Books, 1997), p. 278.

22. Guth, "Starting the Universe: The Big Bang and Inflation," pp. 140ff.

23. *Ibid.*, pp. 133ff.

24. Andrew Watson, "Inflation Confronts an Open Universe," *Science* (6 March 1998), p. 1455.

25. Roger Penrose, "Difficulties with Inflationary Cosmology," in *Fourteenth Texas Symposium on Relativistic Astrophysics*, ed. E. J. Fenyves (New York: Academy of Science, 1989), pp. 249–264.

26. A. Karel Velan, *The Multi-Universe Cosmos: The First Complete Story of he Origin of the Universe* (New York: Plenum Press, 1992), pp. 145–147.

27. William L. Craig and Quentin Smith, *Theism, Atheism, and Big Bang Cosmology* (Oxford: Clarendon Press, 1993), pp. 68, 73. By permission of Oxford University Press.

28. In Lightman and Brawer, *Origins: The Lives and Worlds of Modern Cosmologists*, see pp. 62–63 (Fred Hoyle), pp. 147–148 (Dennis Sciama), p. 195 (Joseph Silk), pp. 315–317 (Edwin Turner). *Cf.* G. F. R. Ellis and Tony Rothman, "Has Cosmology Become Metaphysical?" *Astronomy*, 15:2 (February 1987), pp. 6–22. See also Peter Coles and George Ellis, "The Case for an Open Universe," *Nature*, 370 (25 August 1994), p. 610; Watson, "Inflation Confronts an Open Universe," p. 1455; and James Glanz, "Which Way to the Big Bang?" *Science* (28 May 1999), pp. 1448–1451.

29. Lightman and Brawer, *Origins*, p. 195.

30. Coles and Ellis, "The Case for an Open Universe," p. 610.

31. Alfred North Whitehead, *Adventures of Ideas* (New York: The Free Press, 1967), pp. 111–115, 136–137.

32. Quoted by Marcia Bartusiak in *Thursday's Universe: A Report from the Frontier on the Origin, Nature, and Destiny of the Universe* (Redmond, Wash.: Tempus Books, 1988), p. 49.

33. Edwin Hubble, *The Realm of the Nebulae* (New Haven: Yale University Press, 1936), p. 18.

34. *Ibid.*, p. 282.

35. David Russell, "Island Universes from Wright to Hubble," *Sky & Telescope*, 97:1 (January 1999), pp. 56–60.

36. David Burstein, "Unveiling a New Galaxy," *Nature*, 372 (3 November 1994), p. 38; see also pp. 77–79.

37. Ann Finkbeiner, "Hubble Telescope Settles Cosmic Distance Debate-or Does It?" *Science* (28 May 1999), p. 1438; "HST's Hubble Constant Announced–Again," *Sky & Telescope*, 100:5 (November 2000), p. 28.

38. See the quotations from Corliss Lamont at notes 20 and 21 in Chapter Two.

39. Charles W. Misner, Kip S. Thorne, and John A. Wheeler, *Gravitation* (New York: W. H. Freeman and Company, 1973), p. 1215.

40. Roger Penrose, "Singularities and Time-Asymmetry," in *General Relativity: An Einstein Centenary Survey*, eds. S. W. Hawking and W. Israel (Cambridge, England: Cambridge University Press, 1979), p. 588.

41. Roger Penrose, *The Emperor's New Mind: Concerning Computers, Minds, and the Laws of Physics* (New York: Penguin Books, 1991), p. 310.

42. Gamow, *The Creation of the Universe*, pp. 27–28.

43. Penrose, *The Emperor's New Mind*, p. 335.

44. *Ibid.*

45. From *A BRIEF HISTORY OF TIME*, by Stephen W. Hawking, copyright © 1988, 1996 by Stephen W. Hawking, p. 150. Used by permission of Bantam Books, a division of Random House, Inc. and Writers House LLC.

46. Don Page, "Will Entropy Decrease if the Universe Recollapses?" *Physical Review D*, 32:10 (15 November 1985), pp. 2496–2499.

47. Hawking, *A Brief History of Time*, pp. 150, 151.

48. Mark Israelit and Nathan Rosen, "A Singularity-Free Cosmological Model in General Relativity," *Astrophysical Journal*, 342 (15 July 1989), p. 634.

49. Brian Greene, *The Elegant Universe: Superstrings, Hidden Dimensions, and the Quest for the Ultimate Theory* (New York: W. W. Norton & Co., 1999), pp. 342–343.

50. Jastrow, *God and the Astronomers*, p. 28.

51. Quoted by Sten Odenwald in, "Spacetime: The Final Frontier," *Sky & Telescope*, 91:2 (February 1996), p. 29.

52. Robert Jastrow, "Science and the Creation," in *Creation*, ed. Thomas H Schattauer (New Haven: Yale Divinity School, 1980), p. 34.

53. Jastrow, *God and the Astronomers*, p. 28.

54. *Ibid.*, pp. 25–27.

55. Bartusiak, *Thursday's Universe*, pp. 281–282.

56. Hermann Bondi, *Cosmology* (Cambridge, England: The University Press, 1961), p. 21.

57. *Cf.* Margenau and Varghese, *Cosmos, Bios, Theos*, Ulrich J. Becker (pp. 28–29), John E. Fornaess (pp. 40–41), Robert Jastrow (pp. 46, 48), Vera Kistiakowsky (pp. 51–52), Louis Neal (p. 73), John A. Russell (p. 90), Arthur L. Schawlow (p. 106), Emilio Segre (pp. 108–109), Wolfgang Smith (p. 116), Charles H. Townes (p. 123), Eugene P. Wigner (p. 131), Frederick C. Robbins (pp. 194–196), and William Stoeger (pp. 265–266).

58. Jastrow, "Science and the Creation," p. 32.

59. *Ibid.*, pp. 34, 35.

60. Jastrow, *God and the Astronomers*, pp. 115–116.

61. Jastrow, "Science and the Creation," p. 35.

Chapter Two

1. Kai Nielsen, *Naturalism without Foundations* (Amherst, N.Y.: Prometheus Books, 1996), pp. 26, 29, 527.

2. David R. Griffin, *Religion and Scientific Naturalism* (Albany: State University of New York Press, 2000), pp. 11–17, 39–40.

3. See Rem B. Edwards, *Reason and Religion: An Introduction to the Philosophy of Religion* (New York: Harcourt Brace Jovanovich, 1972, and Lanham, Md.: The University Press of America, 1979), pp. 133–144. *Cf.* Holmes Rolston, III, *Science and Religion: A Critical Survey* (New York: Random House, 1987), p. 248.

4. Nielsen, *Naturalism Without Foundations*, pp. 33–34, 47, 54 n. 39, and elsewhere.

5. Corliss Lamont, "Naturalistic Humanism," in *The Humanist Alternative*, ed. Paul Kurtz (Buffalo, NY: Prometheus Books, 1973), p. 129.

6. John Dewey, *The Quest for Certainty* (New York: G. P. Putnam's Sons, 1960), p. 243.

7. Paul Kurtz, ed. *Humanist Manifestos I and II* (Buffalo, NY: Prometheus Books, 1973), p. 16.

8. Roy W. Sellars, *Religion Coming of Age* (New York: Macmillan, 1928), p. 141.

9. Carl Sagan, *Cosmos* (New York: Random House, 1980), p. 4.

10. Paul Kurtz, *In Defense of Secular Humanism* (Buffalo: Prometheus Books, 1983), p. 63.

11. Nielsen, *Naturalism Without Foundations*, p. 35.

12. Frank J. Tipler, *The Physics of Immortality* (New York: Anchor Books, 1994), pp. ix, 3, 116.

13. Lucretius, "The Nature of the Universe," in *Theories of the Universe from Babylonian Myth to Modern Science*, ed. Milton Munitz (New York: The Free Press, 1957), p. 55.

14. Roy W. Sellars, "Does Naturalism Need Ontology," *Journal of Philosophy*, 41: 25 (December 1944), p. 686.

15. Lamont, "Naturalistic Humanism," p. 129.

16. Sidney Hook, *The Quest for Being* (New York: St. Martin's Press, 1961), p. 237.

17. Kurtz, *Humanist Manifestos I and II*, p. 8.

18. Kurtz, *In Defense of Secular Humanism*, p. 65.

19. Quoted in: Joseph Priestley, *Letters to a Philosophical Unbeliever*, Part 1 (New York: Garland Publishing Co., 1983), pp. 382–383.

20. Corliss Lamont, *The Illusion of Immortality* (New York: Frederick Ungar Pub. Co., 1965), p. 186.

21. *Ibid.*, p. 271.

22. Sellars, *Religion Coming of Age*, p. 146.

23. *Ibid.*, p. 212.

24. Roy W. Sellers, *The Philosophy of Physical Realism* (New York: Russell & Russell, 1996), p. 15.

25. Kurtz, *Humanist Manifestos I and II*, p. 8.

26. Kurtz, *In Defense of Secular Humanism*, p. 214.

27. William R. Dennes, "The Categories of Naturalism," in *Naturalism and the Human Spirit*, ed. Y. H. Krikorian (New York: Columbia University Press, 1944), p. 288.

28. Hook, *The Quest for Being*, pp. 185–186.

29. Sterling Lamprecht, *The Metaphysics of Naturalism* (New York: Appleton-Century-Crofts, 1967), p. 202.

30. Hans Reichenbach, *The Rise of Scientific Philosophy* (Berkeley, Cal.: University of California Press, 1951), p. 208.

31. Kurtz, *In Defense of Secular Humanism*. p. 18.

32. Bertrand Russell, *Religion and Science* (New York: Oxford University Press, 1935), p. 243.

33. Hook, *The Quest for Being*, p. 78.

34. John Dewey, *A Common Faith* (New Haven: Yale University Press, 1934), pp. 32–33.

35. Sellars, *The Philosophy of Physical Realism*, p. 141.

36. Kurtz, *Humanist Manifestos I and II*, p. 16.

37. Stuart C. Dodd, "Hypotheses Defining Scientific Humanism: A Reformulation of Humanist Principles for Testing by Scientific Methods," *Journal of Human Relations*, 21:2 (Second Quarter 1973), p. 174.

38. Kurtz, *In Defense of Secular Humanism*, p. 117.

39. Nielsen, *Naturalism without Foundations*, p. 26.

40. Michael Devitt, *Coming to Our Senses: A Naturalistic Program for Semantic Localism* (Cambridge, England: Cambridge University Press, 1996), pp. 2, 49.

41. Richard Rorty, *Philosophy and the Mirror of Nature* (Princeton: Princeton University Press, 1979); and *Consequences of Pragmatism* (Minneapolis: University of Minnesota Press, 1982).

42. Paul Feyerabend, *Against Method*, rev. ed. (New York: Verso, 1988).

43. See Michael Friedman, "Philosophical Naturalism," *Proceedings and Addresses of the American Philosophical Association*, 74:2 (1997), pp. 7–21.

44. Arthur Peacock, "Science and God the Creator," in *Evidence of Purpose: Scientists Discover the Creator*, ed. John M. Templeton (New York: Continuum Pub. Co., 1994), p. 93.

45. Kurtz, *Humanist Manifestos I and II*, p. 14.

46. Lamont, *The Illusion of Immortality*, p. 262.

47. Kurtz, *In Defense of Secular Humanism*, pp. 68–69.

48. *Cf.* David Ehrenfeld, *The Arrogance of Humanism* (New York: Oxford University Press, 1978).

49. Michael Ruse, "The New Evolutionary Ethics," in *Evolutionary Ethics*, eds. Matthew H. Nitecki and Doris V. Nitecki (Albany: State University of New York Press, 1993), pp. 133–162; Michael Ruse, "Evolutionary Theory and Christian Ethics," *Zygon*,

29 (March 1994), pp. 5–24.

50. Neil O. Weiner, *The Harmony of the Soul* (Albany: State University of New York Press, 1993).

51. Patricia A. Williams, "Christianity and Evolutionary Ethics: Sketch Toward a Reconciliation," *Zygon*, 31:2 (June 1996), pp. 253–268.

52. Kurtz, *In Defense of Secular Humanism*, p. 67.

53. *Ibid.*, p. 110.

54. Nielsen, *Naturalism without Foundations*, pp. 26–27, 32, 64, and elsewhere.

55. *Ibid.*, pp. 12–19, Part Two, and elsewhere.

56. *Ibid.*, p. 13.

57. George Smoot and Keay Davidson, *Wrinkles in Time* (New York: Avon Books, 1993), pp. 95–97.

58.See Rem B. Edwards, *Reason and Religion: An Introduction to the Philosophy of Religion*, pp. 347–351.

59. See Norwood Russell Hanson, *Patterns of Discovery* (Cambridge, England: The University Press, 1969), Ch. 1.

60. See James F. Harris, *Against Relativism: A Philosophical Defense of Method* (La Salle, Ill.: Open Court, 1992), especially Ch. 7 and pp. 59–63. See also Nielsen, *Naturalism without Foundations*, pp. 27–29, 31–33, 64, 70, 309, 316–317.

61. See Timothy Ferris, *Coming of Age in the Milky Way* (New York: William Morrow and Co., 1988), Ch. 16 and pp. 298–299, 384–388. See also J. D. Barrow, *The Artful Universe* (New York: Oxford University Press, 1995).

62. I. L. Rozental, *Big Bang Big Bounce* (Berlin: Springer-Verlag, 1988), p. 127.

63. Stephen Hawking, *Black Holes and Baby Universes and Other Essays* (New York: Bantam Books, 1993), Ch. 10.

64. *Ibid.*, p. 120.

65. *Ibid.*, pp. 110–111.

66. For details see Leon M. Lederman and David N. Schramm, *From Quarks to the Cosmos: Tools of Discovery* (New York: Scientific American Library, 1989), pp. 163–165; Martin Rees, *Before the Beginning: Our Universe and Others* (Reading, Mass.: Perseus Books, 1997), pp. 155–158.

Chapter Three

1. See George Smoot and Keay Davidson, *Wrinkles in Time* (New York: Avon Books, 1993), pp. 246–247.

2. See Fred Hoyle, *The Nature of the Universe*, rev. ed. (New York: The New American Library, 1964), Ch. 6 and Hermann Bondi, *Cosmology*, 2nd ed. (Cambridge, England: The University Press, 1961), Ch. 12.

3. Hoyle, *The Nature of the Universe*, p. 108.

4. *Ibid.*, p. 110.

5. *Ibid.*

6. *Ibid.*

7. Bondi, *Cosmology*, p. 143.

8. *Ibid.*, p. 149.

9. *Ibid.*, p. 144.

10. A. Karel Velan, *The Multi-Universe Cosmos* (New York: Plenum Press, 1992), pp. 162–165, 183–184.

11. Bondi, *Cosmology*, pp. 23–26.

12. *Ibid.*, p. 144.

13. Hoyle, *The Nature of the Universe*, p. 113.

14. Bondi, *Cosmology*, p. 144.

15. Martin Rees, *Before the Beginning: Our Universe and Others* (Reading, Mass.: Perseus Books, 1997), p. 185.

16. Virginia Trimble, "Cosmic Discovery 1998," *Sky & Telescope*, 97:2 (February 1999), p. 37.

17. Hoyle, *The Nature of the Universe*, p. 111.

18. Bondi, *Cosmology*, p. 149.

19. Alan Lightman and Roberta Brawer, *Origins: The Lives and Worlds of Modern Cosmologists* (Cambridge, Mass.: Harvard University Press, 1990), p. 326.

20. *Ibid.*, p. 418.

21. Fred Hoyle, "The Steady-State Theory Revived?" *Comments on Astrophysics*, 13:2 (1989), pp. 81–86.

22. Lightman and Brawer, *Origins: The Lives and Worlds of Modern Cosmologists*, p. 57.

23. From *THE BIG BANG NEVER HAPPENED* by Eric Lerner, pp. 39–40. Copyright © 1991 by Eric J. Lerner. Reprinted by permission of Alfred A. Knopf, a Division of Random House Inc.

24. *Ibid.*, pp. 24–25.

25. *Ibid.*, p. 24.

26. *Ibid.*, pp. 31–32.

27. *Ibid.*, pp. 33–34.

28. Stephen Hawking, *Black Holes and Baby Universes and Other Essays* (New York: Random House, 1993), p. 151.

29. See Peter Coles and George Ellis, "The Case for an Open Universe," *Nature*, 370 (25 August 1994), pp. 610–612.

30. Lerner, *The Big Bang Never Happened*, p. 32.

31. See Smoot and Davidson, *Wrinkles in Time*, pp. 160–161.

32. Vera C. Rubin, "Weighing the Universe: Dark Matter and Missing Mass," in *Bubbles, Voids, and Bumps in Time: The New Cosmology*, ed. James Cornell (Cambridge, England: Cambridge University Press, 1991), pp. 102–103. Reprinted with the permission of Cambridge University Press.

33. Lerner, *The Big Bang Never Happened*, p. 34.

34. Rubin, "Weighing the Universe: Dark Matter and Missing Mass," p. 73.

35. Quoted by Don Page in Lightman and Brawer, *Origins: The Lives and Worlds of Modern Cosmologists*, p. 404.

36. Lerner, *The Big Bang Never Happened*, pp. 36ff.

37. *Ibid.*, pp. 38–39.

38. *Ibid.*

39. Rubin, "Weighing the Universe: Dark Matter and Missing Mass," p. 85.

40. *Ibid.*, pp. 86–87, 90.

41. *Ibid.*, p. 90.

42. Lerner, *The Big Bang Never Happened*, p. 35.

43. Rubin, "Weighing the Universe: Dark Matter and Missing Mass," pp. 98–99.

44. James Trefel, *The Dark Side of the Universe* (New York: Avon Books, 1988), p. 117.

45. H. Reeves, "Nuclear Processes in the Early Universe," in *Confrontation between Theories and Observations in Cosmology: Present Status and Future Programmes*, eds. J. Audouze and F. Melchiorri (Amsterdam: North Holland, 1990), p. 436.

46. Joshua Roth and Joel R. Primack, "Cosmology: All Sewn Up or Coming Apart at the Seams," *Sky & Telescope*, 91:1 (January 1996), p. 22.

47. James Glanz, "Exploding Stars Point to a Universal Repulsive Force," *Science*, 279 (30 January 1998), p. 651.

48. Roth and Primack, "Cosmology: All Sewn Up or Coming Apart at the Seams," pp. 23–26.

49. *Ibid.*, p. 25.

50. Coles and Ellis, "The Case for an Open Universe," p. 609.

51. Saul Perlmutter *et al.*, "Discovery of a Supermova Explosion at Half the Age of the Universe," *Nature*, 391 (1 January 1998), pp. 51–54; "Big Crunch Canceled," *The Economist*, 346 (17 January 1998); "Farthest Supernovae Favor Low-Density Universe," *Sky & Telescope* 95 (March 1998), p. 19.

52. See e.g., : J. Richard Gott, III, James E. Gunn, David N. Schramm, and Beatrice M. Tinsley, "An Unbounded Universe," *Astrophysical Journal*, 194 (1974), pp. 543–553; P. J. E. Peebles, "The Mean Mass Density of the Universe," *Nature*, 321 (1986), pp. 27–32; Alan R. Sandage and G. A. Tammann, "The Dynamical Parameters of the Universe," in *Large Scale Structure of the Universe, Cosmology and Fundamental Physics*, eds. G. Setti and L. Van Hove (Geneva: ESO and CERN, 1984); Martin Rees, "Concluding Comments: A Cosmologist's View," in G. Setti and L. Van Hove; Joseph Silk, *The Big Bang* (New York: W. H. Freeman and Co., 1980), pp. 388–393; Coles and Ellis, "The Case for an Open Universe," pp. 609–615. In Lightman and Brauer, *Origins: The Lives and Worlds of Modern Cosmologists*, see Fred Hoyle (p. 62), Alan Sandage (p. 79), Joseph Silk (p. 195), Vera Rubin (p. 299), and Edwin Turner (pp. 310–313). *Cf.* Roger Penrose in Stephen Hawking and Roger Penrose, *The Nature of Space and Time* (Princeton: Princeton University Press, 1996), pp. 119, 137.

53. Coles and Ellis, "The Case for an Open Universe," p. 614.

54. Glanz, "Exploding Stars Point to a Universal Repulsive Force." See also: James Glanz, "Astronomers See a Cosmic Antigravity Force at Work," *Nature*, 279 (27 February 1998), pp. 1298–1299; Ann K. Finkbeiner, "Cosmic Yardsticks: Supernovae and the Fate of the Universe," *Sky & Telescope*, 96:3 (1998), pp. 38–45; Andrew Watson, "Cosmology: Inflation Confronts an Open Universe," *Science*, 279: 5356 (6 March 1998), p. 1455; James Glanz, "Exploding Stars Flash New Bulletins from Distant Universe," *Science*, 280:5366 (May 1998), pp. 1008–1009; Trimble, p. 38; Neta A. Bahcall, Jeremiah P. Ostriker, Saul Perlmutter, and Paul J. Steinhardt, "The Cosmic Triangle: Revealing the State of the Universe," *Science*, 284 (May 28, 1999), pp. 1481–1488. Donald Goldsmith, *The Runaway Universe: The Race to Find the Future of the Cosmos* (New York: Perseus Pub. Co., 2000); Joshua Roth, "Farthest Supernova Clinches Case for Accelerating Universe," *Sky & Telescope* 102:1 (July 2001), pp. 20–21.

55. James Glanz, "Breakthrough of the Year: Astronomy: Cosmic Motion Revealed," *Science*, 282:5397 (18 December 1998), pp. 2156–2157.

56. James Glanz, "American Physical Society: Celebrating a Century of Physics, *en Masse*," *Science* (2 April 1999), p. 34.

57. "An Accelerating Universe? Not So Fast," *Sky & Telescope*, 98:6 (December 1999), p. 18; James Glanz, "No Backing Off From the Accelerating Universe," *Science* (13 November 1998); James Glanz, "Calculating Weight of Emptiness is No. 1

Cosmological Mystery," *Minneapolis Star Tribune* (5 December 1999), p. 19 A; Kathy Sawyer, "Cosmic Driving Force?; Scientists' Work on 'Dark Energy' Mystery could yield a New View of the Universe," *The Washington Post* (19 February 2000), p. A 1.

58. Wayne Hu, "Ringing in the New Cosmology," *Nature*, 404 (27 April 2000), pp. 939–940. See also P. de Bernardis, *et al.*, "A Flat Universe From High-Resolution Maps of the Cosmic Microwave Background Radiation," *Nature*, 404 (27 April 2000), pp. 955–959.

59. Lerner, *The Big Bang Never Happened*, pp. 15, 41.

60. See Hannes Alfvén, *Worlds-Antiworlds* (San Francisco: W. H. Freeman, Co., 1966).

61. Lerner, *The Big Bang Never Happened*, p. 41.

62. *Ibid.*, p. 14.

63. *Ibid.*

64. *Ibid.*, p. 42.

65. *Ibid.*, pp. 42–43.

66. *Ibid.*, pp. 45–49.

67. *Ibid.*, p. 44.

68. *Ibid.*, p. 49.

69. *Ibid.*, p. 50.

70. *Ibid.*

71. *Ibid.*, p. 31.

72. *Ibid.*, pp. 425–430.

73. *Ibid.*, p. 217.

74. *Ibid.*, p. 52.

75. *Ibid.*

76. *Ibid.*, pp. 425–428.

77. *Ibid.*, p. 278.

78. James Glanz, "Sky Survey Racks up Record-Setting Quasars," *Science*, 282 (11 December 1998), pp. 1969–1970.

79. F. Duccio Macchetto and Mark Dickinson, "Galaxies in the Young Universe," *Scientific American*, 276: 5 (May 1997), pp. 92–99. See also "Hubble Revisits Its Deep Field," *Sky & Telescope*, 97:1 (1999), p. 24.

80. Kazuyuki Omukai and Ryoichi Nishi, "Formation of Primordial Protostars," *Astrophysical Journal*, 501:1 (20 November 1998), pp. 1ff; Samantha Parker, "The Eagle's Nest," *Sky & Telescope*, 91:2 (February 1996), pp. 32–34.

81. Alan M. MacRobert and Joshua Roth, "The Planet of 51 Pegasi," *Sky & Telescope*, 91:1 (January 1996), pp. 38–40; David C. Black, "Other Suns, Other Planets," *Sky & Telescope*, 92:2 (August 1996), pp. 20–27; William Cochran, "Extrasolar Planets," *Physics World*, 10:7 (1 July 1997), pp. 31ff; "New Extrasolar Planets Hint at More to Come," *Science*, 288:5463 (7 April 2000), pp. 23–25; "Extrasolar Planets Aplenty," *Sky & Telescope*, 100:5 (November 2000), pp. 24–25.

82. Alan H. Guth, *The Inflationary Universe: The Quest for a New Theory of Cosmic Origins* (Reading, Mass.: Persus Books, 1997), p. 186.

83. Milton Munitz, *Cosmic Understanding: Philosophy and Science of the Universe* (Princeton: Princeton University Press, 1986), pp. 165–166.

84. Gary H. Sanders and David Beckett, "LIGO: An Antenna Tuned to the Songs of Gravity," *Sky & Telescope*, 100:4 (October 2000), p. 44.

85. Lerner, *The Big Bang Never Happened*, p. 389.

86. *Ibid.*, p. 403.

87. *Ibid.*, pp. 5, 12, 39, 41, 54, 91, 127, 214, 227, 333, 360, 383, 388, 403, 406, 416.

88. *Ibid.*, Ch. 10.

89. *Ibid.*, pp. 64, 108.

90. *Ibid.*, pp. 355, 360–362.

91. *Ibid.*, p. 108.

92. *Ibid.*, p. 388.

93. Sidney Hook, *The Quest for Being* (New York: St. Martin's Press, 1961), p. 78.

94. Lerner, *The Big Bang Never Happened*, p. 41.

95. Hoyle, *The Nature of the Universe*, p. 109.

96. See Gary Steigman, "Observational Tests of Antimatter Cosmologies," *Annual Reviews of Astronomy and Astrophysics*, 14 (1976), pp. 339–372.

97. Lerner, *The Big Bang Never Happened*, pp. 225–227.

98. I. L. Rozental, *Big Bang Big Bounce* (Berlin: Springer-Verlag, 1988), p. 40.

99. *Ibid.*, pp. 40–41; and Trimble, "Cosmic Discovery 1998," p. 37.

100. Lerner, *The Big Bang Never Happened*, p. 215.

101. *Ibid.*, pp. 216–217.

102. *Ibid.*, p. 218.

103. See Robert P. Kirshner "Measuring the Universe: Redshifts and Standard Candles," in *Bubbles, Voids and Bumps in Time: The New Cosmology*, ed. James Cornell (New York: Cambridge University Press, 1989), pp. 26–49. See also the summary in Wendy L. Freedman, *et al.*, "Distance to the Virgo Cluster Galaxy M100 from Hubble Telescope Observation of Cepheids," *Nature*, 371 (27 October 1994), pp. 757–758.

104. Michael Perryman, "Hipparcos: The Stars," *Sky & Telescope*, 97:6 (June 1999), pp. 40–50.

105. Eugene F. Mallove, "Review of Eric J. Lerner, *The Big Bang Never Happened*," in *Sky & Telescope*, 82:5 (November 1991), p. 492.

106. Smoot and Davidson, *Wrinkles in Time*, p. 286.

107. Kim A. McDonald, "Discovery of Largest, Most Ancient Structures in Cosmos Said to Confirm 'Big Bang' Model of Universe's Creation," *The Chronicle of Higher Education*, 38 (29 April 1992).

108. Lerner, *The Big Bang Never Happened*, p. 52.

109. "The Great Attractor's Hidden Heart," *Sky & Telescope*, 91:6 (June 1996), p. 12.

110. Smoot and Davidson, *Wrinkles in Time*, p. 242.

111. *Ibid.*, Chs. 11 and 12.

112. McDonald, p. A 9. Also see Sally Stephens, "New Image of the Universe Soon After Creation: What the COBE Results Mean," *Mercury*, 21 (May/June 1992), pp. 91–92 and Michael D. Lemonick, "Echoes of the Big Bang," *Time* (4 May 1992), pp. 62ff.

113. Smoot and Davidson, *Wrinkles in Time*, p. 282.

114. "The Primordial Soup: A Recipe for Life," *Sky & Telescope*, 95:2 (February 1998), p. 20.

Chapter Four

1. Alan M. MacRobert, "Beyond the Big Bang," *Sky & Telescope* (March 1983), p. 211.

2. I. L. Rozental, *Big Bang Big Bounce* (New York: Springer-Verlag, 1985), p. ix.

3. George Gamow, *One Two Three...Infinity* (New York: The American Library, 1954), pp. 314–315. See also Gamow, *The Creation of the Universe* (New York: The New American Library, 1957), pp. 36–37, 134.

4. *Ibid.*, p. 42.

5. George Gamow, "Modern Cosmology," *Scientific American*, 190:3 (March 1954), p. 63.

6. Gamow, *The Creation of the Universe*, p. 37.

7. Gamow, *One Two Three...Infinity*, p. 314.

8. A. Karel Velan, *The Multi-Universe Cosmos: The First Complete Story of the Origin of the Universe* (New York: Plenum Press, 1992), p. 336.

9. Mark Israelit and Nathan Rosen, "A Singularity-Free Cosmological Model in General Relativity," *The Astrophysical Journal*, 342 (15 July 1989), p. 634.

10. See Abner Shimony, "Quantum Physics and the Philosophy of Whitehead," *Boston Studies in the Philosophy of Science*, 2 (1965), pp. 307–330.

11. Alfred North Whitehead, *Process and Reality*, corrected ed. (New York: The Free Press, 1978), p. 91.

12. *Ibid.*, p. 95.

13. *Ibid.*, p. 97.

14. William L. Craig and Quentin Smith, *Theism, Atheism, and Big Bang Cosmology* (Oxford: Clarendon Press, 1993), pp. 43–44, 120.

15. *Ibid.*, p. 43.

16. *Ibid.*, p. 208.

17. *Ibid.*, p. 211.

18. *Ibid.*, p. 209.

19. *Ibid.*, p. 210.

20. *Ibid.*

21. Richard C. Tolman, *Relativity, Thermodynamics, and Cosmology* (Oxford: Clarendon Press, 1934).

22. *E.g.*, see *ibid.*, Ch. 10. See also Charles W. Misner, Kip S. Thorne and John A. Wheeler, *Gravitation* (New York: W. H. Freeman and Co., 1973), pp. 935, 940, 1196–1217.

23. John Gribbin, "Oscillating Universes Bounce Back," *Nature*, 259 (1976), p. 15.

24. John Gribbin, *In the Beginning: After COBE and Before the Big Bang* (Boston: Little Brown and Co., 1993), Ch. 9.

25. Stephen W. Hawking and Roger Penrose, "The Singularities of Gravitational Collapse and Cosmology," *Proceedings of the Royal Society of London*, A314 (1970), pp. 529–548.

26. Stephen W. Hawking, *A Brief History of Time* (New York: Bantam Books, 1988), pp. 50–51 and Ch. 8.

27. Roger Penrose, "Some Remarks on Gravity and Quantum Mechanics," in *Quantum Structure of Space and Time*, eds. M. J. Duff and C. J. Isham (Cambridge, England: Cambridge University Press, 1982), p. 4.

28. See John Boslough, *Stephen Hawking's Universe* (New York: Avon Books, 1989), p. 40. See also Michael White and John Gribbin, *Stephen Hawking: A Life in Science* (New York: Penguin Books, 1992), pp. 178–179; L. P. Grishchuk and Ya. B. Zeldovich, "Complete Cosmological Theories," in *Quantum Structure of Space and Time*, eds. M. J. Duff and C. J. Isham (Cambridge, England: Cambridge University Press, 1982), p. 414; George Smoot and Keay Davidson, *Wrinkles in Time* (New York: Avon

Books, 1993), p. 186; Chris Isham, "Quantum Gravity," in *The New Physics*, ed. Paul Davies (Cambridge, England: Cambridge University Press, 1989), p. 71.

29. Stephen Hawking, *Black Holes and Baby Universes and Other Essays* (New York: Bantam Books, 1993), p. 91.

30. *Ibid.*, p. 61.

31. Frank J. Tipler, *The Physics of Immortality* (New York: Anchor Books, 1994), pp. 64–65, 71, 117, 136.

32. See Werner Heisenberg, *Physics and Philosophy: The Revolution in Modern Science* (New York: Harper & Row, 1958), pp. 31–35; Heinz R. Pagels, *Perfect Symmetry: The Search for the Beginning of Time* (New York: Bantam Books, 1986), pp. 304 ff; George Smoot and Keay Davidson, *Wrinkles in Time* (New York: Avon Books, 1993), p. 84 n; Victor J. Stenger, *The Unconscious Quantum: Metaphysics in Modern Physics and Cosmology* (Amherst, N.Y.: Prometheus Books, 1995), pp. 189–190, 201–202, 208–211.

33. See *e.g.*, Leon M. Lederman and David N. Schramm, *From Quarks to the Cosmos* (New York: Scientific American Library, 1989), p. 161.

34. Timothy Ferris, *Coming of Age in the Milky Way* (New York: William Morrow and Company, 1988), p. 328. See also Brian Greene, *The Elegant Universe: Superstrings, Hidden Dimensions, and the Quest for the Ultimate Theory* (New York: W. W. Norton & Co., 1999).

35. Stephen Hawking and Roger Penrose, *The Nature of Space and Time*, (Princeton, N.J.: Princeton University Press, 1996), p. 38.

36. Michio Kaku, *Visions: How Science Will Revolutionize the 21st Century* (New York: Anchor Books, 1997), p. 342.

37. From *A BRIEF HISTORY OF TIME*, by Stephen W. Hawking, copyright © 1988, 1996 by Stephen W. Hawking, p. 50. Used by permission of Bantam Books, a division of Random House, Inc. and Writers House LLC.

38. Hawking and Penrose, *The Nature of Space and Time*, p. 4.

39. *Ibid.*, p. 103.

40. Hawking, *A Brief History of Time*, pp. 88–92; Roger Penrose, "Singularities and Time-Asymmetry," in, *General Relativity*, eds. S. W. Hawking and W. Israel (Cambridge, England: Cambridge University Press, 1979), pp. 617–629. See also Stephen Hawking, "Theoretical Advances in Relativity," in *Some Strangeness in the Proportion*, ed. Harry Woolf (Reading, Mass.: Addison-Wesley, 1980), pp. 145–152.

41. Hawking and Penrose, *The Nature of Space and Time*. p. 103.

42. *Ibid.*, pp. 99–103.

43. Roger Penrose, "What Does the Big Bang Tell Us about Quantum Gravity?" *Memorie della Societa Astronomica Italiana*, 62:3 (1991), p. 612.

44. *Ibid.*, p. 612.

45. *Ibid.*, pp. 610–612.

46. *Ibid.*, p. 614.

47. Hawking, *A Brief History of Time*, p. 50.

48. *Ibid.*, p. 138.

49. *Ibid.*, p. 139.

50. *Ibid.*

51. Hawking, *Black Holes and Baby Universes and Other Essays*, pp. 81–83, 92, 96.

52. Hawking, *A Brief History of Time*, p. 136.

53. Carl Sagan, "Introduction," Hawking, *A Brief History of Time*, p. x.

54. Hawking, *A Brief History of Time*, p. 139.

55. Robert Jastrow, "What Forces Filled the Universe with Energy Fifteen Billion Years Ago?" in *Cosmos, Bios, Theos*, eds. Henry Margenau and Abraham Varghese (La Salle, Ill.: Open Court, 1992), p. 49.

56. Hawking, *Black Holes and Baby Universes and Other Essays*, p. 111.

57. Hawking and Penrose, *The Nature of Space and Time*, pp. 98, 101, 106, 119, 131, 137.

58. Hawking, *A Brief History of Time*, pp. 110–113.

59. *Ibid.*, pp. 135–141.

60. *Ibid.*, pp. 8, 166.

61. See the many essays in David R. Griffin, ed. *Physics and the Ultimate Significance of Time* (Albany: State University of New York Press, 1986), esp. David Griffin's "Preface," pp. vii–xv, and his "Introduction: Time and the Fallacy of Misplaced Concreteness," pp. 1–48.

62. John D. Barrow and Mariusz P. Dabrowski, "Oscillating Universes," *Monthly Notices of the Royal Astronomical Society*, 275 (1995), p. 855.

63. Robert J. Russell, "Cosmology, Creation, and Contingency," in *Cosmos as Creation: Theology and Science in Consonance*, ed. Ted Peters (Nashville: Abingdon Press, 1989), pp. 188–189.

64. James Trefel, *The Dark Side of the Universe* (New York: Avon Books, 1988), p. 189.

65. See: Mark Israelit and Nathan Rosen, "A Singularity-Free Cosmological Model in General Relativity," *The Astrophysical Journal*, 342 (15 July 1989), pp. 627–634; Mark Israelit, "Cosmological Perturbations in the Early Prematter Universe," *The Astrophysical Journal*, 435 (1 November 1994), pp. 8–15.

66. Rozental, p. 105.

67. *Ibid.*

68. Alan H. Guth, *The Inflationary Universe: The Quest for a New Theory of Cosmic Origins* (Reading, Mass.: Persus Books, 1998), note * on p. 26.

69. Martin Rees, *Before the Beginning: Our Universe and Others* (Reading, Mass.: Perseus Books, 1997), p. 195.

70. See Timothy Ferris, *The Red Limit: The Search for the Edge of the Universe* (New York: Quill, 1983), p. 119.

71. Gamow, *The Creation of the Universe*, p. 35.

72. From *THE BIG BANG NEVER HAPPENED* by Eric Lerner, p. 217. Copyright © 1991 by Eric J. Lerner. Reprinted by permission of Alfred A. Knopf, a Division of Random House Inc.

73. Velan, *The Multi-Universe Cosmos*, p. 223.

74. See Roger Penrose, "Singularities of Spacetime," in *Theoretical Principles in Astrophysics and Relativity*, eds. Norman R. Lebovitz, *et al.* (Chicago: The University of Chicago Press, 1978), p. 218; P. C. W. Davies, *The Accidental Universe* (Cambridge, England: Cambridge University Press, 1982), p. 29; Heinz R. Pagels, *Perfect Symmetry: The Search for the Beginning of Time* (New York: Bantam Books, 1986), p. 338; Andrei Linde, "Inflation and Quantum Cosmology," in *Three Hundred Years of Gravitation*, eds. Stephen W. Hawking and Werner Israel (Cambridge, England: Cambridge University Press, 1987), pp. 611, 620; Joseph Silk, *The Big Bang*, rev. and updated ed. (New York: W. H. Freeman and Co., 1989), pp. 112–113, 115, 118–119; Mark Israelit and Nathan Rosen, "A Singularity-Free Cosmological Model in General Relativity," *The Astrophysical Journal*, 342 (15 July 1989), pp. 627–628, 633; John Gribbin and Martin

Rees, *Cosmic Coincidences* (New York: Bantam Books. 1989), pp. 68, 256, 280; Paul Davies, *The Mind of God: The Scientific Basis for a Rational World* (New York: Simon & Schuster, 1992), pp. 62–66; George Smoot and Keay Davidson, *Wrinkles in Time* (New York: Avon Books, 1993), pp. 161, 283; David Bohm and B. J. Hiley, *The Undivided Universe: An Ontological Interpretation of Quantum Theory* (London: Routledge, 1993), pp. 348, 374; Andrei Linde, Dmitri Linde, and Arthur Mezhlumian, "From the Big Bang Theory to the Theory of a Stationary Universe," *Physical Review D*, 49:4 (15 February 1994), pp. 1788–1791, 1808; Gribbin, *In the Beginning, After COBE and Before the Big Bang*, p. 166; Mark Israelit, "Cosmological Perturbations in the Early Prematter Universe," *The Astrophysical Journal*, 435 (1 November 1994), pp. 8–9; Victor Stenger, *The Unconscious Quantum*, 1995, pp. 208–209, 219–220; Sten Odenwald, "Space-Time: The Final Frontier," *Sky & Telescope*, 91:2 (February 1996), pp. 27–29; Martin Rees, *Before the Beginning: Our Universe and Others*, pp. 172, 210; Brian R. Greene, *The Elegant Universe: Superstrings, Hidden Dimensions, and the Quest for the Ultimate Theory* (New York: W. W. Norton, Co., 1999), pp. 236, 239, 252–254, 357–358.

75. Milton Munitz, *The Mystery of Existence* (New York: Appleton-Century-Crofts, 1965), p. 141.

76. *Ibid.*, p. 139.

77. *Ibid.*, p. 141.

78. Robert Jastrow, "Science and the Creation," in *Creation*, ed. Thomas H. Schattauer (New Haven: Yale Divinity School, 1980), p. 35.

79. Hawking, *A Brief History of Time*, p. 51.

80. Kaku, *Visions: How Science Will Revolutionize the 21st Century*, p. 351.

81. Roger Penrose, *Shadows of the Mind* (New York: Oxford University Press, 1994), p. 257.

82. Pagels, *Perfect Symmetry*, p. 167.

83. Rees, *Before the Beginning: Our Universe and Others*, p. 169; see also p. 195.

84. John Gribbin and Martin Rees, *Cosmic Coincidences: Dark Matter, Mankind, and Anthropic Cosmology* (New York: Bantam Books, 1989), p. 68.

85. Israelit and Rosen, "A Singularity-Free Cosmological Model in General Relativity," 1989, p. 628.

86. Jonathan J. Halliwell, "Quantum Cosmology and the Creation of the Universe," *Scientific American*, 265:6 (1 December 1991), p. 82.

87. Paul Davies, "The First One Second of the Universe," in *Images of the Universe*, ed. Carole Stott (Cambridge, England: Cambridge University Press, 1991), p. 209.

88. Alan Lightman and Roberta Brawer, *Origins: The Lives and Worlds of Modern Cosmologists* (Cambridge, Mass.: Harvard University Press, 1990), pp. 18, 19.

89. MacRobert, "Beyond the Big Bang," p. 213.

90. Steven Weinberg, *The First Three Minutes: A Modern View of the Origin of the Universe*, updated ed. (New York: Basic Books, 1988), p. 154.

91. Ya. B. Zel'Dovich and I. D. Novikov, *The Structure and Evolution of the Universe* (Chicago: The University of Chicago Press, 1983), pp. 658–663.

92. *Ibid.*, pp. 220, 221.

93. *Ibid.*, pp. 468, 474.

94. S. A. Bludman, "Thermodynamics and the End of a Closed Universe," *Nature*, 308 (March 1984), p. 322.

95. Charles W. Misner, Kip S. Thorne, John A. Wheeler, *Gravitation* (New York: W. H. Freeman & Co., 1973), pp. 1214 ff.

96. M. A. Markov, "Problems of a Perpetually Oscillating Universe," *Annals of Physics*, 155 (1984), p. 490.

97. *Ibid.*, p. 335. See also pp. 344 and 354.

98. R. H. Dicke and P. J. E. Peebles, "The Big Bang Cosmology–Enigmas and Nostrums," in *General Relativity*, eds. S. W. Hawking and W. Israel (Cambridge, England: Cambridge University Press, 1979), pp. 511–512.

99. Stephen Toulmin, *The Return to Cosmology: Postmodern Science and the Theology of Nature* (Berkeley: University of California Press, 1982), pp. 35–52.

100. *Ibid.*, p. 42.

101. *Ibid.*, pp. 42–43.

102. *Ibid.*, p. 39.

103. Stenger, *The Unconscious Quantum*, p. 265.

Chapter Five

1. See: B. S. DeWitt and N. Graham, eds. *The Many-Worlds Interpretation of Quantum Mechanics. A Fundamental Exposition by Hugh Everett, III, with Papers by J. A. Wheeler, B. S. DeWitt, L. N. Cooper and D. Van Vechten, and N. Graham* (Princeton: Princeton University Press, 1973).

2. Michael D. Lemonick, "Visions 21/Science & Space: Will We Discover Another Universe?" *Time* (10 April 2000), pp. 82ff.

3. A. A. Starobinsky, "A New Type of Isotropic Cosmological Models Without Singularity," *Physics Letters*, 91B (1980), pp. 99–102.

4. Andrei Linde, "Inflation and Quantum Cosmology," in *Three Hundred Years of Gravitation*, eds. Stephen Hawking and Werner Israel (Cambridge, England: Cambridge University Press, 1987), pp. 604–630.

5. M. A. Markov, "Problems of a Perpetually Oscillating Universe," *Annals of Physics*, 155 (1987), pp. 333–357. J. M. Aman and M. A. Markov, "Oscillating Universe in the State p 0," Preprint P-0290 (1983).

6. I. L. Rozental, *Big Bang, Big Bounce* (Berlin: Springer-Verlag, 1988), pp. 123–124.

7. *Ibid.*, p. 124.

8. John Gribbin, *In the Beginning: After COBE and Before the Big Bang* (Boston: Little, Brown and Co., 1993, p. 244.

9. Katsuhiko Sato, Hideo Kodama, Misao Sasaki, and Kei-ichi Madea, "Multi-Production of Universes by First-Order Phase Transition of a Vacuum," *Physics Letters*, 108B:2 (14 January 1982), p. 105. See also John D. Barrow, *Theories of Everything: The Quest for Ultimate Explanation* (Oxford: The Clarendon Press, 1991), p. 107.

10. Willem B. Drees, *Beyond the Big Bang: Quantum Cosmologies and God* (La Salle, Ill.: Open Court, 1990), p. 47, fig. 2.

11. John Hick, *An Interpretation of Religion* (New Haven: Yale University Press, 1989), p. 87.

12. See Sato *et al.*, "Multi-Production of Universes by First-Order Phase Transition of a Vacuum," pp. 103–107; Rozental, *Big Bang, Big Bounce*, pp. 107–110; Charles W. Misner, Kip S. Thorne, and John A. Wheeler, *Gravitation* (New York: W. H. Freeman and Co., 1973), pp. 1202–1203; Milton Munitz, *Cosmic Understanding: Philosophy and Science of the Universe* (Princeton: Princeton University Press, 1986), pp. 131–137; Paul

Davies and John Gribbin, *The Matter Myth* (New York: Simon & Schuster, 1992), pp. 142–148, 162–164; A. Karel Velan, *The Multi-Universe Cosmos: The First Complete Story of the Origin of the Universe* (New York: Plenum Press, 1992), Ch. 14; Robert Matthews, "Nothing Like a Vacuum," *New Scientist*, (February 1995), pp. 30ff; Sten Odenwald, "Space-Time: The Final Frontier," *Sky & Telescope*, 91:2 (February 1996), pp. 24–29.

13. Alan Guth, "Starting the Universe," in *Bubbles, Voids, and Bumps in Time: The New Cosmology*, ed. James Cornell (Cambridge, England: Cambridge University Press, 1989), p. 128.

14. Rozental, *Big Bang, Big Bounce*, p. 109; Victor J. Stenger, *Not by Design: The Origin of the Universe* (Buffalo: Prometheus Books, 1988), p. 168.

15. Rozental, *Big Bang, Big Bounce*, p. 107.

16. Odenwald, "Space-Time: The Final Frontier," pp., 26–27.

17. Misner, Thorne, and Wheeler, *Gravitation*, p. 1202.

18. George Smoot and Keay Davidson, *Wrinkles in Time* (New York: Avon Books, 1993), p. 188.

19. See Steven K. Blau and Alan Guth, "Inflationary Cosmology," in *Three Hundred Years of Gravitation*, eds. Stephen Hawking and Werner Israel (Cambridge, England: Cambridge University Press, 1987), pp. 542, 550–555; John Gribbin and Martin Rees, *Cosmic Coincidences: Dark Matter, Mankind, and Anthropic Cosmology* (New York: Bantam Books, 1989), pp. 280–283; Leon M. Lederman and David N. Schramm, *From Quarks to the Cosmos: Tools of Discovery* (New York: Scientific American Library, 1989), pp. 176–180; Heinz Pagels, *Perfect Symmetry: The Search for the Beginning of Time* (New York: Bantam Books, 1986), pp. 339 ff.

20. Andrei Linde, Dmitri Linde, Arthur Mezhlumian, "From the Big Bang Theory to the Theory of a Stationary Universe," *Physical Review D*, 49:4 (15 February 1994), p. 1784.

21. Blau and Guth, "Inflationary Cosmology," pp. 592, 593.

22. Linde, "Inflation and Quantum Cosmology," p. 625.

23. *Ibid.*, p. 625, fig. 13.2.

24. See Barrow, *Theories of Everything*, pp. 106–111; Davies and Gribbin, *The Matter Myth*, pp. 273–279; Paul Halpern, *Cosmic Wormholes: The Search for Interstellar Shortcuts* (New York: NAL-Dutton, 1992).

25. Linde, Linde, and Mezhlumian, "From the Big Bang Theory to the Theory of a Stationary Universe," p. 1803.

26. Linde, "Inflation and Quantum Cosmology;" p. 618.

27. Velan, *The Multi-Universe Cosmos*, Ch. 16 and pp. 148–150, 193, 200.

28. *Ibid.*, p. 187.

29. See Hugh Everett, III in DeWitt and Graham, eds. *The Many-Worlds Interpretation of Quantum Mechanics*, 1973.

30. See all the articles in the special edition of *Noûs* devoted to the "Foundations of Quantum Mechanics," 18:4 (November 1984).

31. Frank J. Tipler, "The Many-Worlds Interpretation of Quantum Mechanics in Quantum Cosmology," in *Quantum Concepts in Space and Time*, eds. R. Penrose and C. J. Isham (Oxford: Clarendon Press, 1986), p. 206.

32. *Ibid.*, p. 210.

33. *Ibid.*, p. 208.

34. Richard A. Healey, "How Many Worlds?" in *Noûs*, 18:4 (November 1984), p. 591.

35. Blau and Guth, "Inflationary Cosmology," p. 541.

36. Rozental, *Big Bang, Big Bounce*, p. 124.

37. Michael White and John Gribbin, *Stephen Hawking; A Life in Science* (New York: Penguin Books, 1992), p. 188.

38. Hugh Everett, III, "'Relative State' Formulation of Quantum Mechanics," *Reviews of Modern Physics*, 29:3 (July 1957), p. 460 n.

39. Bryce S. DeWitt, "Quantum Mechanics and Reality," *Physics Today*, 23:9 (September 1970), p. 30.

40. From *THE BIG BANG NEVER HAPPENED* by Eric Lerner, p. 5. Copyright © 1991 by Eric J. Lerner. Reprinted by permission of Alfred A. Knopf, a Division of Random House Inc.

41. Heinz R. Pagels, *Perfect Symmetry, The Search for the Beginning of Time* (New York: Bantam Books, 1986), pp. 264–265.

42. Abner Shimony, "Role of the Observer in Quantum Theory," *American Journal of Physics*, 36 (1963), p. 773, n. 33.

43. Healey, "How Many Worlds?" p. 599.

44. Victor J. Stenger, *The Unconscious Quantum: Metaphysics in Modern Physics and Cosmology* (Amherst, N.Y.: Prometheus Books, 1995), p. 176.

45. *Ibid.*, p. 198.

46. *Ibid.*, p. 144.

47. Arthur O. Lovejoy, *The Great Chain of Being* (Cambridge, Mass.: Harvard University Press, 1961, pp. 52ff.

48. Dennis W. Sciama, "The Anthropic Principle and the Non-Uniqueness of the Universe," in *The Anthropic Principle: Proceedings of the Second Venice Conference on Cosmology and Philosophy*, eds. F. Bertola and U. Curi (Cambridge, England: Cambridge University Press, 1993), p. 107.

49. *Ibid.*, p. 108.

50. Stephen Hawking, *Black Holes and Baby Universes and Other Essays* (New York: Bantam Books, 1993), p. 95.

51. See Steven Hawking, *A Brief History of Time* (New York: Bantam Books, 1988), p. 134. See also his *Black Holes and Baby Universes and Other Essays*, pp. 45, 78–80, 92–93.

52. Thanks to my student Rob Dalton for pointing this out to me.

53. Rozental, *Big Bang, Big Bounce*, p. 8.

54. Linde, "Inflation and Quantum Cosmology," p. 607.

55. *Ibid.*, p. 621. See also Rozental, *Big Bang, Big Bounce*, pp. ix, 91.

56. DeWitt, "Quantum Mechanics and Reality," p. 33.

57. Lerner, *The Big Bang Never Happened*, p. 55.

58. Frank J. Tipler, *The Physics of Immortality* (New York: Anchor Books, 1994), pp. 220–225.

59. *Ibid.*, 167–173.

60. *Ibid.*, p. 213.

61. Timothy Ferris, *Coming of Age in the Milky Way* (New York: William Morrow and Co., 1988), p. 369.

Chapter Six

1. John A. Wheeler, "Genesis and Observership," in *Foundational Problems in the Special Sciences*, eds. Robert E. Butts and Jaakko Hintikka (Dordrecht: D. Reidel, 1977),

pp. 3–33. John A. Wheeler, "Beyond the Black Hole," in *Some Strangeness in the Proportion*, ed. Harry Woolf (Reading, Mass.: Addison-Wesley, 1980), pp. 341–375.

2. See Bernard d'Espagnat, "Quantum Theory and Reality," *Scientific American*, 241 (November 1979), pp. 158–181; David N. Mermin, "Is the Moon There When Nobody Looks? Reality and the Quantum Theory," *Physics Today*, 38 (April 1985), pp. 38–47; Paul Davies, *God and the New Physics* (New York: Simon & Schuster, 1983), Ch. 8; Victor J. Stenger, *The Unconscious Quantum* (Amherst, N.Y.: Prometheus Books, 1995), Chs. 5 and 6; David Bohm and B. J. Hiley, *The Undivided Universe: An Ontological Interpretation of Quantum* Theory (London and New York: Routledge, 1993), Ch. 7 and elsewhere.

3. See Wheeler, "Genesis and Observership," pp. 22–25; Brian Greene, *The Elegant Universe: Superstrings, Hidden Dimensions, and the Quest for the Ultimate Theory* (New York: W. W. Norton, & Co., 1999), pp. 97–108; Bohm and Hiley, pp. 32–35, 41–42, and elsewhere.

4. John A. Wheeler, "Bohr, Einstein, and the Strange Lesson of the Quantum," in *Mind in Nature*, ed. Richard Q. Elvee (San Francisco: Harper & Row, 1981), p. 29.

5. Wheeler, "Genesis and Observership," p. 6.

6. Eugene Wigner, *Symmetries and Reflections* (Bloomington: Indiana University Press, 1967), p. 172.

7. Wheeler, "Genesis and Observership," p. 32.

8. *Ibid.*, p. 21.

9. *Ibid.*, p. 25.

10. *Ibid.*

11. *Ibid.*, p. 28.

12. Fred W. Hallberg, "Barrow and Tipler's Anthropic Cosmological Principle," *Zygon*, 23:2 (June 1988), p. 144.

13. See Wheeler's "Genesis and Observership," p. 26.

14. Wheeler, "Beyond the Black Hole," p. 362.

15. Richard Rorty, "Does Academic Freedom Have Philosophical Presuppositions? Academic Freedom and the Future of the University," *Academe*, 80:6 (November/December 1994), pp. 52–63.

16. See J. Leplin, ed. *Scientific Realism* (Berkeley: University of California Press, 1984); Arthur Peacocke, *Intimations of Reality: Critical Realism in Science and Religion* (Notre Dame: University of Notre Dame Press), 1984.

17. Werner Heisenberg, *Philosophical Problems of Quantum Physics* (Woodbridge, Conn.: Ox Bow Press, 1979), pp. 38, 86.

18. Quoted by Mermin, "Is the Moon There When Nobody Looks?" p. 40.

19. Greene, *The Elegant Universe*, pp. 143–148.

20. Alfred North Whitehead, *Science and the Modern World* (New York: The Free Press, 1967), p. 34.

21. Sten Oldenwald, "Space-Time: The Final Frontier," *Sky & Telescope*, 91:2 (February 1996), p. 28.

22. John S. Bell, *Speakable and Unspeakable in Quantum Mechanics* (Cambridge, England: Cambridge University Press, 1986), p. 191.

23. Bohm and Hiley, *The Undivided Universe*, pp. 104–106, 125, and elsewhere.

24. *Ibid.*, pp. 32, 41.

25. See David R. Griffin, *Religion and Scientific Naturalism* (Albany: State University of New York Press, 2000), pp. 120–122.

26. George Smoot and Keay Davidson, *Wrinkles in Time* (New York: Avon Books, 1993), pp. 116–118.

27. Paul Davies and John Gribbin, *The Matter Myth* (New York: Simon & Schuster, 1992), p. 309.

28. Victor J. Stenger, *The Unconscious Quantum: Metaphysics in Modern Physics and Cosmology* (Amherst, N.Y.: Prometheus Books, 1995), pp. 79, 80.

29. Holmes Rolston, III, *Science and Religion* (New York: Random House, 1987), p. 58.

30. Stephen Hawking and Roger Penrose, *The Nature of Space and Time* (Princeton: Princeton University Press, 1996), p. 134.

31. Roger Penrose, "Newton, Quantum Theory, and Reality," in *Three Hundred Years of Gravitation*, eds. Stephen Hawking and Werner Israel (Cambridge, England: Cambridge University Press, 1987), p. 46.

32. Davies, *God and the New Physics*, p. 107.

33. See Mermin, "Is the Moon There When Nobody Looks?" pp. 38–39; d'Espagnat, "Quantum Theory and Reality," pp. 165, 172; 177–179; and Davies and Gribbin, *The Matter Myth*, pp. 221–226.

34. Penrose, "Newton, Quantum Theory, and Reality," pp. 26, 27.

35. d'Espagnat, "Quantum Theory and Reality," p. 158.

36. *Ibid.*, p. 177.

37. Bell, *Speakable and Unspeakable in Quantum Mechanics*, p. 171.

38. d'Espagnat, "Quantum Theory and Reality," pp. 179–180.

39. Penrose, "Newton, Quantum Theory, and Reality," pp. 29–31. See also Roger Penrose, *Shadows of the Mind* (New York: Oxford University Press, 1994), pp. 290–296.

40. See the contributions of David Bohm, Basil Hiley, and John S. Bell to *The Ghost in the Atom*, eds. P. C. W. Davies and J. R. Brown (Cambridge, England: Cambridge University Press, 1986). See also Bohm and Hiley, *The Undivided Universe* Ch. 7.

41. John A. Jungerman, *World in Process: Creativity and Interconnectedness in the New Physics* (Albany: State University of New York Press, 2000), pp. 86–89.

42. Henry Krips, *The Metaphysics of Quantum Theory* (Oxford: Clarendon Press, 1987).

43. Stenger, *The Unconscious Quantum*, p. 39. See also his Chs. 5 and 6.

44. Werner Heisenberg, *Physics and Philosophy: The Revolution in Modern Science* (New York: Harper & Row, 1958), p. 137.

45. Bell, *Speakable and Unspeakable in Quantum Mechanics*, p. 170.

46. Stenger, *The Unconscious Quantum*, p. 185.

47. Abner Shimony, "Events and Processes in the Quantum World," in *Quantum Concepts in Space and Time*, eds. R. Penrose and C. J. Isham (Oxford: Clarendon Press, 1986), p. 184.

48. Paul Davies, *The Cosmic Blueprint: New Discoveries in Nature's Creative Ability to Order the Universe* (New York: Simon & Schuster, 1989), pp. 170–174.

49. Bell, *Speakable and Unspeakable in Quantum Mechanics*, pp. 117–127, 173–195.

50. John Wheeler, "Not Consciousness but the Distinction Between the Probe and the Probed as Central to the Elemental Quantum Act of Observation," in *The Role of Consciousness in the Physical World*, ed. Robert G. Jahn (Boulder, Colo.: Westview Press, 1981), p. 97.

Chapter Seven

1. Alan Guth, "Starting the Universe," in *Bubbles, Voids, and Bumps in Time: The New Cosmology*, ed. James Cornell (Cambridge, England: Cambridge University Press, 1989), p. 136.

2. Edward P. Tryon, "Is the Universe a Vacuum Fluctuation?" *Nature*, 246 (14 December 1973), pp. 396–397.

3. Peter Atkins, *The Creation* (Oxford: W. H. Freeman), 1981.

4. A. Vilenkin, "Creation of Universes from Nothing," *Physical Letters*, 117B (1982), pp. 25–28.

5. Victor J. Stenger, *Not by Design: The Origin of the Universe* (Buffalo: Prometheus Books, 1988) and *The Unconscious Quantum: Metaphysics in Modern Physics and Cosmology* (Amherst, N.Y.: Prometheus Books, 1995).

6. See Quentin Smith and William L. Craig, *Theism, Atheism, and Big Bang Cosmology* (Oxford: Clarendon Press, 1993). By permission of Oxford University Press.

7. See Steven K. Blau and Alan Guth, "Inflationary Cosmology," in *Three Hundred Years of Gravitation*, eds. Stephen Hawking and Werner Israel (Cambridge, England: Cambridge University Press, 1987), pp. 546, 556.

8. Tryon, "Is the Universe a Vacuum Fluctuation?" p. 397.

9. *Ibid.*

10. Atkins, *The Creation*, p. 119.

11. Stenger, *Not by Design*, p. 173. *Cf.* Stenger, *The Unconscious Quantum*, pp. 21, 229.

12. Craig and Smith, *Theism, Atheism, and Big Bang Cosmology*, p. 127.

13. *Ibid.*, p. 135.

14. Stenger, *Not by Design*, p. 174.

15. Andrei Linde, Dmitri Linde, Arthur Mezhlumian, "From the Big Bang Theory to the Theory of a Stationary Universe," *Physical Review D*, 49:4 (15 February 1994), p. 1792.

16. *E.g.*, in *Cosmos, Bios, Theos*, eds. Henry Margenau and Roy Abraham Varghese (La Salle, Ill.: Open Court, 1992), see Robert Jastrow (p. 46), Arno Penzias (p. 83), and Abdus Salam (p. 99). See also endnotes 44–59 in Chapter Three of this book.

17. See Kevin Krisciunas, "Science with the Keck Telescope," *Sky & Telescope* (September 1994), p. 26.

18. Stenger, *The Unconscious Quantum*, p. 240.

19. *Ibid.*, p. 219.

20. *Ibid.*, p. 220.

21. Stenger, *Not By Design*, p. 174.

22. Stenger, *The Unconscious Quantum*, p. 240.

23. Abdus Salam, "Science and Religion: Reflections on Transcendence and Secularization," in *Cosmos, Bios, Theos*, eds. Henry Margenau and Roy Abraham Varghese (La Salle, Ill.: Open Court, 1992), p. 99.

24. Stenger, *Not By Design*, pp. 175–177.

25. A. Zee, "Nonconservation of Baryon Number," in *Gamow Cosmology*, eds. LF. Melchiorri and R. Ruffini (Amsterdam: North-Holland, 1986), p. 420.

26. Gary Steigman, "Observational Tests of Antimatter Cosmologies," *Annual Reviews of Astronomy and Astrophysics* 14 (1976), p. 355.

27. John D. Barrow, *Theories of Everything: The Quest for Ultimate Explanation* (Oxford: Clarendon Press, 1991), pp. 133–134.

28. George Smoot and Keay Davidson, *Wrinkles in Time* (New York: Avon Books, 1993), p. 109.

29. Heinz R. Pagels, *The Cosmic Code: Quantum Physics as the Language of Nature* (New York: Bantam Books, 1983), p. 284.

30. See Leon M. Lederman and David N. Schramm, *From Quarks to the Cosmos: Tools of Discovery* (New York: Scientific American Library, 1989), pp. 162–165; see also Barrow, *Theories of Everything*, pp. 134–135.

31. A. G. Cohen, A. De RúJula, and S. L. Glashow, "A Matter-Antimatter Universe?" *Astrophysical Journal*, 495:2 (10 March 1998). See also Gary Taubes, "Astrophysics: Theorists Nix Distant Antimatter Galaxies," *Science*, 539 (10 October 1997, pp. 1 ff.

32. Lederman and Schramm, *From Quarks to Cosmos*, p. 163; Alan H. Guth, *The Inflationary Universe: The Quest for a New Theory of Cosmic Origins* (Reading, Mass.: Perseus Books, 1997), p. 107.

33. A. Karel Velan, *The Multi-Universe Cosmos: The First Complete Story of the Origin of the Universe* (New York: Plenum Press, 1992), pp. 350–351.

34. Michael White and John Gribbin, *Stephen Hawking, A Life in Science* (New York: Penguin Books, 1992), p. 211.

35. Stephen Hawking, *A Brief History of Time* (New York: Bantam Books, 1988), p. 46.

36. Stephen Hawking, ed. *Stephen Hawking's A Brief History of Time: A Reader's Companion* (New York: Bantam Books, 1992), p. 168.

37. Stephen Hawking, *Black Holes and Baby Universes and Other Essays* (New York: Bantam Books, 1993), pp. 147–155.

38. *Ibid.*, pp. 147–148.

39. Peter Coles and George Ellis, "The Case for an Open Universe," *Nature*, 370 (25 August 1994), p. 609.

40. *Ibid.*, p. 612.

41. *Ibid.*, p. 614.

42. See "Old Galaxy in a Young Universe," *Sky & Telescope*, 93:2 (September 1996), p. 11.

43. Hawking, *A Brief History of Time*, p. 133.

44. Craig and Smith, *Theism, Atheism, and Big Bang Cosmology*, p. 301.

45. *Ibid.*, pp. 125–129, esp. p. 128.

46. Heinz R. Pagels, *Perfect Symmetry: The Search for the Beginning of Time* (New York: Bantam Books), p. 347.

47. Stenger, *Not by Design*, p. 12.

48. Paul Davies, *God and the New Physics* (New York: Simon & Schuster, 1983), p. 215.

49. *Ibid.*

50. *Ibid.*

51. Roger Penrose, "What Does the Big Bang Tell Us about Quantum Gravity?" *Memorie della Società Astronomica Italiana*, 62:3 (1991), p. 608.

52. Craig and Smith, *Theism, Atheism, and Big Bang Cosmology*, pp. 61–63.

53. *Ibid.*, pp. 178–191.

54. *Ibid.*, p. 123.

55. *Ibid.*, pp. 122–123.

56. David Atkatz, "Quantum Cosmology for Pedestrians," *American Journal of Physics*, 62:7 (1 July 1994), p. 662.

57. Craig and Smith, *Theism, Atheism, and Big Bang Cosmology*, pp. 127–128.
58. Hawking, *A Brief History of Time*, p. 133.
59. Stenger, *The Unconscious Quantum*, pp. 142, 150.
60. *Ibid.*, pp. 205–206.
61. *Ibid.*, p. 159.
62. *Ibid.*, p. 21.

Chapter Eight

1. See Frank J. Tipler, "The Omega Point Theory: A Model of an Evolving God," in *Physics, Philosophy, and Theology*, eds. Robert J. Russell, William R. Stoeger, and George V. Coyne (Vatican City State: Vatican Observatory, 1988), pp. 323–324; George Smoot and Keay Davidson, *Wrinkles in Time* (New York: Avon Books, 1993), pp. 293–294.

2. Brandon Carter, "Large Number Coincidences and the Anthropic Principle in Cosmology," in *Confrontation of Cosmological Theories with Observation*, ed. M. S. Longair (Dordrecht: Reidel, 1974), pp. 291–298.

3. John D. Barrow and Frank J. Tipler, *The Anthropic Cosmological Principle* (Oxford: Oxford University Press, 1988). By permission of Oxford University Press.

4. P. C. W. Davies, *The Accidental Universe* (Cambridge, England: Cambridge University Press, 1982).

5. Robert H. Dicke, "Dirac's Cosmology and Mach's Principle," *Nature*, 192 (1961), pp. 170–171.

6. Freeman Dyson, "Time without End: Physics and Biology in an Open Universe," *Reviews of Modern Physics*, 51, pp. 447 ff.

7. Stephen W. Hawking, *A Brief History of Time: From the Big Bang to Black Holes* (New York: Bantam Books, 1988).

8. B. J. Carr and M. J. Rees, "The Anthropic Principle and the Structure of the Physical World," *Nature*, 278 (1979), p. 605.

9. John Leslie, *Universes* (London: Routledge, 1989).

10. John A. Wheeler, "Genesis and Observership," in *Foundational Problems in the Special Sciences*, eds. Robert E. Butts and Jaakko Hintikka (Dordrecht-Holland: D. Reidel Publishing Company, 1975). See also John A. Wheeler, "The Universe as Home for Man," in *The Nature of Scientific Discovery*, ed. Owen Gingerich (Washington: Smithsonian Institution Press, 1975), pp. 261–296.

11. Barrow and Tipler, *The Anthropic Cosmological Principle*, p. 1.

12. Carter, "Large Number Coincidences," p. 293.

13. See Steven J. Dick, *Life on Other Worlds: The 20th-Century Extraterrestrial Life Debate* (Cambridge, England: Cambridge University Press, 1998); Paul Davies, *The Fifth Miracle: The Search for the Origin and Meaning of Life* (New York: Simon & Schuster, 1999); and Bruce Jakosky, *The Search for Life on Other Planets* (Cambridge, England: Cambridge University Press, 1999).

14. *Cf.* Rem B. Edwards, *Reason and Religion* (New York: Harcourt Brace Jovanovich, 1972, and Washington, DC: University Press of America, 1979), p. 285.

15. See Ch. 8 of Barrow and Tipler, *The Anthropic Cosmological Principle*.

16. *Ibid.*, p. 13.

17. See Carter, "Large Number Coincidences," pp. 291–293, and Barrow and Tipler, *The Anthropic Cosmological Principle*, pp. 16–21 and elsewhere.

18. Carter, "Large Number Coincidences," p. 291.

19. From *A BRIEF HISTORY OF TIME*, by Stephen W. Hawking, copyright © 1988, 1996 by Stephen W. Hawking, p. 124. Used by permission of Bantam Books, a division of Random House, Inc. and Writers House LLC.

20. Carter, "Large Number Coincidences," p. 294.

21. Barrow and Tipler, *The Anthropic Cosmological Principle*, p. 22.

22. Leslie, *Universes*, pp. 89–91.

23. Barrow and Tipler, *The Anthropic Cosmological Principle*, p. 249; for the writings of G. F. R. Ellis, see their footnote 96 on p. 283.

24. *Ibid.*, p. 249.

25. *Ibid.*, p. 248.

26. *Ibid.*, pp. 248–249.

27. Robert Nozick, *Philosophical Explanations* (Cambridge, Mass.: The Belknap Press, 1981), pp. 128 ff.

28. Arthur O. Lovejoy, *The Great Chain of Being* (Cambridge, Mass.: Harvard University Press, 1961), p. 315.

29. *Ibid.*, pp. 52 ff.

30. Anton C. Pegis, ed. *Basic Writings of St. Thomas Aquinas* (New York: Random House, 1945), Vol. 1, pp. 462–463.

31. St. Thomas Aquinas, *On the Truth of the Catholic Faith: Summa Contra Gentiles*, Book 1 (New York: Doubleday & Co., 1955), pp. 258–259.

32. *Ibid.*, pp. 258–259, 266, and Book 2, pp. 68–69, 77–79, 117, 135–139.

33. Leslie, *Universes*, p. 2.

34. *Ibid.*, p. 171.

35. Alfred North Whitehead, *Process and Reality*, corrected ed . (New York: The Free Press, 1978), p. 19.

36. M. A. Corey, *God and the New Cosmology: The Anthropic Design Argument* (Lanham, Md.: Rowman & Littlefield, 1993), p. 250.

37. David Hume, *Dialogues Concerning Natural Religion*, Part 8.

38. See J. B. Hartle, "The Quantum Mechanics of Cosmology," in *Quantum Cosmology and Baby Universes*, eds. S. Coleman, J. B. Hartle, T. Piran, and S. Weinberg (Singapore: World Scientific, 1991), pp. 65–151.

39. Alfred North Whitehead, *Science and the Modern World* (New York: The Free Press, 1967), pp. 174, 178.

40. Alan Guth, *The Inflationary Universe: The Quest for a New Theory of Cosmic Origins* (Reading, Mass.: Perseus Books, 1997), p. 250.

41. Ian Hacking, "The Inverse Gambler's Fallacy: The Argument from Design. The Anthropic Principle Applied to Wheeler Universes," *Mind*, 96 (1987), pp. 331–340.

42. *Ibid.*, p. 135.

43. *Ibid.*, p. 136.

44. *Ibid.*

45. *Ibid.*, p. 137.

46. Fred W. Hallberg, "Barrow and Tipler's Anthropic Cosmological Principle," *Zygon*, 23:2 (June 1988), pp. 146–147.

47. *Ibid.*, p. 147.

Chapter Nine

1. John D. Barrow and Frank J. Tipler, *The Anthropic Cosmological Principle* (Oxford: Oxford University Press, 1989), p. 23 and Ch. 10. By permission of Oxford

University Press.

2. Frank J. Tipler, "The Omega Point as *Eschaton*: Answers to Pannenberg's Questions for Scientists," *Zygon*, 24:2 (June 1989), p. 250.

3. Frank J. Tipler, *The Physics of Immortality* (New York: Anchor Books, 1994).

4. Barrow and Tipler, *The Anthropic Cosmological Principle*, p. 615.

5. *Ibid.*, p. 667.

6. *Ibid.*

7. *Ibid.*, p. 677.

8. *Ibid.*, p. 682, n. 123, and pp. 676–677.

9. Frank J. Tipler, "The Omega Point Theory: A Model of an Evolving God," in *Physics, Philosophy, and Theology*, eds. Robert J. Russell, William R. Stoeger, and George V. Coyne (Vatican City State: Vatical Observatory, 1988), pp. 312–331; and Tipler, *The Physics of Immortality*, pp. 153–158 and throughout.

10. Tipler, *The Physics of Immortality*, p. 5.

11. Marvin Gardner, "WAP, SAP, PAP, and FAP," *The New York Review of Books*, 33 (8 May 1986), p. 25. Cited by Fred W. Hallberg, "Barrow and Tipler's Anthropic Cosmological Principle," *Zygon*, 23:2 (June 1988), p. 147.

12. Tipler, *The Physics of Immortality*, pp. 124–124.

13. Timothy Ferris, *Coming of Age in the Milky Way* (New York: William Morrow and Co., 1988), p. 371.

14. Tipler, *The Physics of Immortality*, pp. 48–49.

15. John A. Jungerman, *World in Process: Creativity and Interconnection in the New Physics* (Albany: State University of New York Press, 2000), pp. 27–28.

16. *Ibid.*, p. 28.

17. See Rem B. Edwards, "Systemic Value and Valuation," in *Forms of Value and Valuation: Theory and Applications*, eds. Rem B. Edwards and John W. Davis (Lanham, Md.: University Press of America, 1991), pp. 37–55.

18. Barrow and Tipler, *The Anthropic Cosmological Principle*, p. 614.

19. *Ibid.*, p. 315.

20. *Ibid.*, p. 316.

21. Tipler, "The Omega Point Theory: A Model of an Evolving God," pp. 316, 327.

22. *Ibid.*, p. 327.

23. *Ibid.*

24. Tipler, *The Physics of Immortality*, Ch. 9.

25. Steven Weinberg, *The First Three Minutes: A Modern View of the Origin of the Universe, Updated Edition* (New York: Basic Books, 1988), p. 154.

26. Alan Lightman and Roberta Brawer, *Origins: The Lives and Worlds of Modern Cosmologists* (Cambridge, Mass.: Harvard University Press, 1990).

27. Steven Weinberg, *Dreams of a Final Theory* (New York: Pantheon Books, 1992), pp. 255 ff.

28. From A BRIEF HISTORY OF TIME, by Stephen W. Hawking, copyright © 1988, 1996 by Stephen W. Hawking, p. 140. Used by permission of Bantam Books, a division of Random House, Inc. and Writers House LLC.

29. Stephen Hawking, in the television program *Master of the Universe*, broadcast on BBC in 1990.

30. John Hick, *The Interpretation of Religion* (New Haven: Yale University Press, 1989), p. 121.

31. Victor J. Stenger, *The Unconscious Quantum: Metaphysics in Modern Physics and Cosmology* (Amherst, N.Y.: Prometheus Books, 1996), p. 17.

32. *Ibid.*, p. 193.

33. Tipler, "The Omega Point Theory: A Model of an Evolving God," p. 315.

34. See Fred C. Adams and Gregory Laughlin, "The Future of the Universe," *Sky & Telescope*, 96:2 (August 1998), pp. 32–39.

35. See M. A. Corey, *God and the New Cosmology: The Anthropic Design Argument* (Lanham, Md.: Rowman & Littlefield, 1993), pp. 56–58.

36. Pierre Teilhard de Chardin, *The Phenomenon of Man* (New York: Harper & Row, 1959), pp. 226–228.

37. See Rem B. Edwards, "Universals, Individuals, and Intrinsic Good," in *Forms of Value and Valuation*, eds. Edwards and Davis, pp. 81–104.

38. Freeman J. Dyson, "Time Without End: Physics and Biology in an Open Universe," *Review of Modern Physics*, 51 (1979), pp. 447–460.

39. See Rem B. Edwards, *Pleasures and Pains: A Theory of Qualitative Hedonism* (Ithaca, N.Y.: Cornell University Press, 1979).

40. Ralph Waldo Emerson, "Self-Reliance" in *The Complete Works of Ralph Waldo Emerson*, ed. Edward Waldo Emerson (Boston: Houghton Mifflin Co., 1903), Vol. 2, p. 53.

41. "Top Scientist Pessimistic for our Survival," *Knoxville News Sentinel* (3 October 2000), p. A 2.

Chapter Ten

1. See Rem B. Edwards, *Reason and Religion* (Washington, D. C.: University Press of America, 1979), Ch. 9.

2. See Rem B. Edwards, "The Pagan Dogma of the Absolute Unchangeableness of God," *Religious Studies*, 14 (1978), pp. 305–314.

3. See Edwards, *Reason and Religion*, Ch. 8.

4. "The Westminster Confession of Faith," in *The Constitution of the Presbyterian Church (U. S. A.):* Part 1, *Book of Confessions* (New York: The Office of the General Assembly, 1983), section 6.001.

5. Mark W. Worthing, *God, Creation, and Contemporary Physics*, (Minneapolis: Fortress Press, 1996), pp. 70, 78, 195–196.

6. Joel Friedman, "The Natural God: A God Even an Atheist Can Believe In," *Zygon*, 21:3 (September 1986), pp. 369–388.

7. Bertrand Russell, *Why I Am Not a Christian* (New York: Simon & Schuster, 1957), p. 107.

8. See Charles Hartshorne, "The Ethics of Contributionism," in *Responsibilities to Future Generations: Environmental Ethics*, ed. Ernest Partridge (Buffalo: Prometheus Books, 1981), pp 103–107. See also Charles Hartshorne, *The Divine Relativity* (New Haven: Yale University Press, 1948), pp. 46, 58, 124, 128, 130, 131, 133, 141.

9. Thomas M. Dicken and Rem B. Edwards, *Dialogues on Values and Centers of Value: Old Friends, New Thoughts* (Amsterdam—New York: Editions Rodopi, in press), Ch. 2, Section 5.

10. See Rem B. Edwards, "Process Thought and the Spaciness of Mind," *Process Studies*, 19 (1990), pp. 156–166.

11. See Lewis Ford, *Transforming Process Theism* (Albany: State University of New York Press, 2000).

12. Sidney Norton Deane, ed. *St. Anselm* (La Salle, Ill.: Open Court Publishing Company, 1954), p. 190.

13. Anton C. Pegis, ed. *Basic Writings of Saint Thomas Aquinas* (New York: Random House, 1945), Vol. 1, pp. 216 and 226.

14. Dean, *St. Anselm*, pp. 13–14.

15. "The Westminster Confession of Faith," sect. 6.011.

16. "The Shorter Catechism," in *The Constitution of the Presbyterian Church (U.S.A): Part 1, Book of Confessions* (New York: The Office of the General Assembly, 1983), sect. 7.001.

17. Hartshorne, *The Divine Relativity*, pp. 130, 131.

18. "The Westminster Confession," sect. 6.012.

19. David Bohm and B. J. Hiley, *The Undivided Universe: An Ontological Interpretation of Quantum Theory* (London and New York: Routledge, 1993), pp. 218, 220–221, 230.

20. John A. Jungerman, *World in Process: Creativity and Interconnection in the New Physics* (Albany: State University of New York Press, 2000), pp. 102–103, 113.

21. Bohm and Hiley, *The Undivided Universe*, p. 375.

22. Paul Tillich, *Systematic Theology* (London: Nisbit & Co., 1955), Vol. 1, pp. 263, 227.

23. See note 11 above.

24. Robert C. Neville, *Creativity and God: A Challenge to Process Theology* (New York: Seabury Press, 1980), pp. 51–52, 90–93, 95, 140–141.

25. *Ibid.*, p. 40.

26. See Rem B. Edwards, "The Human Self: An Actual Entity or a Society?" *Process Studies*, 5:3 (Fall 1975), pp. 195–203; Rem B. Edwards, "Kraus's Boethian Interpretation of Whitehead's God," *Process Studies*, 11 (Spring 1981), pp. 30–34; Rem B. Edwards, "God and Process, " in *Logic, God, and Metaphysics*, ed. James F. Harris (The Hague: Kluwer Academic Publishers, 1992), pp. 45–57.

27. Ford, *Transforming Process Theism*, *passim*, but esp. pp. 233–326.

28. Jungerman, *World in Process*, p. 114.

29. See Alfred North Whitehead, *Adventures of Ideas* (New York: The Free Press, pp. 192–195, 210–211, 230–332; Alfred North Whitehead, *Process and Reality* (New York: The Free Press, 1978), pp. 84–87, 212–215.

30. Charles Hartshorne, *Creative Synthesis and Philosophical Method* (La Salle, Ill.: Open Court Publishing Co., 1970), p. 192.

31. See David R. Griffin, *God, Power, and Evil: A Process Theodicy* (Philadelphia: The Westminster Press, 1976); and David R. Griffin, *Evil Revisited: Responses and Reconsiderations* (Albany: State University of New York Press, 1991). *Cf.* David Basinger, *Divine Power in Process Theism* (Albany: State University of New York Press, 1988).

32. See Griffin, *Evil Revisited: Responses and Reconsiderations*, p. 101.

33. *Ibid.*, p. 107.

34. *Ibid.*, p. 119. See also David R. Griffin, *Religion and Scientific Naturalism* (Albany: State University of New York Press, 2000), pp. 292–293.

35. Ford, *Transforming Process Theism*, pp. 270–271.

36. See J. Gerald Janzen, "Modes of Power and the Divine Relativity," *Encounter*, 36:4 (Autumn 1975), pp. 379–406; Arthur Peacocke, *Theology for a Scientific Age: Being and Becoming–Natural, Divine, and Human* (Minneapolis: Fortress Press, 1993), pp. 121–126; and Edwards, "God and Process," pp. 41–57.

37. Griffin, *God, Power, and Evil: A Process Theodicy*, pp. 174–195, 226, 241.

38. Arthur Peacocke, *Theology for a Scientific Age* (Minneapolis: Fortress Press, 1993), pp. 121–123, 308–311.

39. John Polkinghorne, *The Faith of a Physicist* (Minneapolis: Fortress Press, 1994), pp. 61–71, 85, 151.

40. Nancey Murphy and George F. R. Ellis, *On the Moral Nature of the Universe: Theology, Cosmology, and Ethics* (Minneapolis: Fortress Press, 1996), p. 246.

41. Peacocke, *Theology for a Scientific Age*, pp. 370–372, n. 75.

42. Ian Barbour, *Religion in an Age of Science* (San Francisco: Harper & Row, 1990), pp. 268–270.

43. David Griffin, ed. *Physics and the Ultimate Significance of Time* (Albany: State University of New York Press, 1986), p. 139.

44. Robert C. Neville, *God as Creator* (Chicago: University of Chicago Press, 1968); Robert C. Neville, *Creativity and God: A Challenge to Process Theology* (New York: Seabury Press, 1980), pp. 33–35, 44–46; Robert C. Neville, *A Theology Primer* (Albany, N.Y.: State University of New York Press, 1991), pp. 28–48.

45. John B. Cobb, Jr., "Is God Creator *Ex Nihilo*?" *Process and Faith* website, (July-August, 1999).

46. Charles Hartshorne, *Creative Synthesis and Philosophical Method*, p. 125.

47. Charles Hartshorne, "Response to Alston," *Existence and Actuality: Conversations with Charles Hartshorne*, eds. John B. Cobb, Jr., and Franklin I. Gamwell (Chicago: University of Chicago Press, 1987), p. 100.

48. Lewis Ford, "An Alternative to *Creatio Ex Nihilo*," *Religious Studies*, 19 (1983), 205–213.

49. John B. Cobb, Jr. and David R. Griffin, *Process Theology: An Introductory Exposition*, Philadelphia: Westminster Press, 1976, p. 65.

50. Griffin, *Evil Reconsidered*, p. 23.

51. Alan M. MacRobert, "Beyond the Big Bang," *Sky & Telescope* (March 1983), p. 211.

52. Alfred North Whitehead, *Science and the Modern World* (New York: The Free Press, 1967, pp. 53–54; *Process and Reality* (New York: The Free Press, 1978), pp. 177, 199.

53. Charles Hartshorne, "Response to Martin." *Existence and Actuality: Conversations with Charles Hartshorne*, eds. John B. Cobb, Jr., and Franklin Gamwell (Chicago: University of Chicago Press, 1987), p. 74.,

54. Hartshorne, *Creative Synthesis and Philosophic Method*, pp. 30, 125, 126.

55. Whitehead, *Process and Reality*, pp. 253, 308.

56. Alfred North Whitehead, *Religion in the Making* (New York: The Macmillan Co., 1926), p. 153.

57. Charles Hartshorne, *Man's Vision of God and the Logic of Theism* (Hamden, Conn.: Archon Books, 1964), pp. 94, 234, 239.

58. Alfred North Whitehead, *Science and Philosophy* (Patterson, N.J.: Littlefield, & Adams, 1964), p.144.

59. Hartshorne, *Creative Synthesis and Philosophic Method*, pp. 53–54.

60. *Ibid.*, pp. 63, 65, 125, 126.

61. Whitehead, *Process and Reality*, p. 91.

62. *Ibid*, p. 95.

63. *Ibid.*, p. 97.

64. *Ibid.*, pp. 66, 97, 288–289.

65. Jorge Luis Nobo, *Whitehead's Metaphysics of Extension and Solidarity* (Albany, N.Y.: State University of New York Press, 1986), pp. 205–218.

66. *Whitehead, Process and Reality*, p. 98.

67. *Ibid.*, p. 97.

68. Edwards, *Reason and Religion*, p. 172; Lewis S. Ford, *The Lure of God: A Biblical Background for Process Theism* (Philadelphia: Fortress Press, 1978), p. 21.

69. Cobb and Griffin, *Process Theology*, p. 64.

70. Edwards, *Reason and Religion*, p. 175

71. Griffin, *Evil Revisited*, pp. 24–25, 99–100.

72. Ann K. Finkbeiner, "Cosmic Yardsticks: Supernovae and the Fate of the Universe," *Sky & Telescope*, 96 (1998), pp. 38–45; James Glanz, "Breakthrough of the Year: Astronomy: Cosmic Motion Revealed," *Science*, 282 (18 December 1998), pp. 2156–2157; James Glanz, "American Physical Society: Celebrating a Century of Physics, en Masse," *Science*, 284 (2 April 1999), p. 34. See Chapter Three above, notes 47–59.

73. Whitehead, *Process and Reality*, p. 19.

74. Charles Hartshorne, *Wisdom as Moderation* (Albany, N.Y.: State University of New York Press, 1987), p. 96.

75. Whitehead, *Process and Reality*, p. 343.

Chapter Eleven

1. John Hick, *An Interpretation of Religion* (New Haven: Yale University Press, 1989), p. 129.

2. From *A BRIEF HISTORY OF TIME*, by Stephen W. Hawking, copyright © 1988, 1996 by Stephen W. Hawking, p. 125. Used by permission of Bantam Books, a division of Random House, Inc. and Writers House LLC.

3. *Ibid.*, p. 121.

4. B. J. Carr and M. J. Rees, "The Anthropic Principle and the Structure of the Physical World," *Nature*, 278 (1979), p. 609.

5. John D. Barrow and Frank J. Tipler, *The Anthropic Cosmological Principle* (Oxford: Oxford University Press, 1988), p. 124. By permission of Oxford University Press.

6. See P. C. W. Davies, *The Accidental Universe* (New York: Cambridge University Press), pp. 117–118. See also Barrow and Tipler, *The Anthropic Cosmological Principle*, pp. 250–255.

7. Edwin A. Abbott, *Flatland: A Romance of Many Dimensions* (New York: Dover Publications, 1952).

8. Hawking, *A Brief History of Time*, p. 164.

9. *Ibid.*, p. 165.

10. John Leslie, *Universes* (London: Routledge, 1989).

11. M. A. Corey, *God and the New Cosmology: The Anthropic Design Argument* (Lanham, Md.: Rowman & Littlefield, 1993), esp. Chapter 4.

12. See Roger Penrose, "Big Bangs, Black Holes and 'Times Arrow'," in *The Nature of Time*, eds. Raymond Flood and Michael Lockwood (Oxford: Basil Blackwell, 1986), esp. pp. 49 ff.

13. See Jaakko Hintikka, "Gaps in the Great Chain of Being," *Proceedings and Addresses of the American Philosophical Association*, 49 (November 1976), pp. 22–38.

14. Hawking, *A Brief History of Time*, p. 125.

15. Steven Weinberg, *Dreams of a Final Theory* (New York: Pantheon Books, 1992), p. 223.

16. *Ibid.*, p. 245.

17. Roger Penrose, *The Emperor's New Mind* (New York: Penguin Books, 1989), p. 340.

18. *Ibid.*, p. 344.

19. Fred Hoyle, "The Universe: Past and Present Reflections," *Engineering and Science*, 45:2 (November 1981), p. 12.

20. *Ibid.*

21. *Ibid.*

22. Victor J. Stenger, *Not by Design: The Origin of the Universe* (Buffalo, N.Y.: Prometheus Books, 1988), Ch. I.

23. From *THE BIG BANG NEVER HAPPENED* by Eric Lerner, p. 400. Copyright © 1991 by Eric J. Lerner. Reprinted by permission of Alfred A. Knopf, a Division of Random House Inc.

24. Holmes Rolston, III, *Science and Religion: A Critical Survey* (New York: Random House, 1987), pp. 104–109, 115–119. See also M. A. Corey, *Back to Darwin: The Scientific Case for Deistic Evolution* (Lanham, Md.: Rowman and Littlefield, 1994); M. A. Corey, *The Natural History of Creation: Biblical Evolutionism and the Return of Natural Theology* (Lanham, Md.: Rowman and Littlefield, 1995); and David R. Griffin, *Religion and Scientific Naturalism* (Albany: State University of New York Press, 2000), Ch. 8.

25. Philip Hefner, "The Evolution of the Created Co-Creator," in *Cosmos as Creation*, ed. Ted Peters (Nashville: Abingdon Press, 1989), p. 213.

26. See Steven J. Dick, *The Biological Universe* (Cambridge, England: Cambridge University Press, 1996) and other references in Chapter Eight above, note 13.

27. Robert Jastrow, "Science and the Creation," in *Creation*, ed. Thomas H. Schattauer (New Haven: Yale Divinity School, 1980), p. 35.

28. Robert Jastrow, "What Are the Chances for Life?" *Sky & Telescope*, 93:6 (June 1997), p. 63.

29. Alan M. MacRobert, "A Tour of Extrasolar Planets," *Sky & Telescope*, 96:3 (March 1997), pp. 80–81; Geoffrey Marcy and R. Paul Butler, "The Diversity of Planetary Systems," *Sky & Telescope*, 97:3 (March 1998), pp. 30–37; "Extrasolar Planets Aplenty," *Sky & Telescope*, 100:4 (November 2000), pp. 24–25. See Chapter Three above, note 81.

30. "Extrasolar Planet Seen Transiting Its Star," *Sky & Telescope*, 99:2 (February 2000), p. 16.

31. Barrow and Tipler, *The Anthropic Cosmological Principle*, p. 92.

32. John Hick, *Evil and the God of Love* (New York: Harper & Row, 1966), parts 3 and 4.

33. See Frans de Waal, *Good Natured: The Origins of Right and Wrong in Humans and Other Animals* (Cambridge, Mass.: Harvard University Press, 1996).

34. Mary Midgley, *Evolution as a Religion* (London: Methuen, 1985), p. 120.

35. See Richard Restak, *The Brain* (New York: Bantam Books, 1984), pp. 156, 162.

36. Steven Weinberg, *Dreams of a Final Theory: The Search for the Fundamental Laws of Nature* (New York: Pantheon Books), p. 250.

37. See Rolston, *Science and Religion*, Ch. 3.

38. See M. A. Corey, *Back to Darwin*, 1994, and *The Natural History of Creation*, 1995.

39. Alfred North Whitehead, *Process and Reality*, corrected ed. (New York: The Free Press, 1978), p. 351.

40. John Hick, *Death and Eternal Life* (New York: Harper & Row, 1976).

Twelve

1. See William L. Craig and Quentin Smith, *Theism, Atheism, and Big Bang Cosmology* (Oxford: Clarendon Press, 1993), Chs. 1–3. By permission of Oxford University Press.

2. For the texts from Bonaventure and Aquinas, and from Siger of Brabant's *On the Eternity of the World*, see *Medieval Philosophical Texts in Translation*, No. 16, eds. Cyril Vollert, Lottie H. Kendierski, and Paul M. Byrne (Milwaukee: Markette University Press, 1964).

3. See, *e.g.*, Heinz R. Pagels, *Perfect Symmetry: The Search for the Beginning of Time* (New York: Bantam Books, 1985), pp. 183, 224; Chris Isham, "Quantum Gravity," in Paul Davies, ed. *The New Physics* (Cambridge, England: Cambridge University Press, 1992), p. 83; John Gribbin, *In the Beginning: After Cobe and Before the Big Bang* (Boston: Little, Brown and Co., 1993), p. 247. Stephen Hawking, *Black Holes and Baby Universes and Other Essays* (New York: Bantam Books, 1993), p. 56.

4. Wolfhart Pannenberg, "The Doctrine of Creation in Modern Science," in *Cosmos as Creation: Theology and Science in Consonance*, ed. Ted Peters (Nashville: Abingdon Press: 1989), p. 161.

5. For arguments I. and II. I am indebted to my student, John R. Fitzpatrick.

6. See Rem B. Edwards, *Reason and Religion: An Introduction to the Philosophy of Religion* (New York: Harcourt, Brace, Jovanovich, 1972 and Washington, D.C.: University Press of America, 1979), pp. 264–267.

7. Kai Nielsen, *Naturalism without Foundations* (Amherst, N.Y.: Prometheus Books, 1996), p. 108.

8. See A. Mishra, *et al.*, "Higgs-Particle Production through Vacuum Excitations," *Physical Review, D*, 44:1 (1 July 1991); John W. Norbury, "Higgs-Boson Production in Nucleus-Nucleus Collisions," *Physical Review, D*, 42:11 (1 December 1990); and A. Dobrovolskaya and V. Novikov, "On Heavy Higgs Boson Production," *Zeitschrift für Physik, C*, 52:3 (1991).

9. Werner Heisenberg, *Physics and Philosophy: The Revolution in Modern Science* (New York: Harper & Row, 1958), p. 71. See also pp. 70 ff. and 117 ff.

10. See Ch. 9 of Edwards, *Reason and Religion*.

11. See Ian G. Barbour, *Religion in an Age of Science* (New York: Harper & Row, 1990), pp. 129–135.

12. John R. Bartlett, *The First and Second Book of the Maccabees* (Cambridge, England: Cambridge University Press, 1973), p. 274.

13. From *THE BIG BANG NEVER HAPPENED* by Eric Lerner, p. 79. Copyright © 1991 by Eric J. Lerner. Reprinted by permission of Alfred A. Knopf, a Division of Random House Inc.

14. Charles Hartshorne, *Omnipotence and Other Theological Mistakes* (Albany: State University of New York Press, 1984), pp. 18, 118.

15. Charles Hartshorne, *Man's Vision of God* (Hamden, Conn.: Archon Books, 1964), pp. 263, 278.

16. Holmes Rolston, III, *Science and Religion* (New York: Random House, 1987), pp. 321–322.

17. Charles Hartshorne, *The Divine Relativity* (New Haven: Yale University Press, 1948), pp. xi–xii.

18. George Mavrodes, "Comments on Rem Edwards' 'Antecedent Universe Cosmologies.'" presented at a meeting of The Society for Philosophy of Religion in Savannah, Ga. (25 February 1994).

19. J. A. Wheeler, "Genesis and Observership, in *Foundational Problems in the Special Sciences*, eds. Robert E. Butts and Jaakko Hintikka (Dordrecht-Holland: D. Reidel Publishing Company, 1975) p. 15.

20. *Ibid.*

21. Carl Sagan, *Cosmos* (New York: Random House, 1980), p. 212.

22. Richard Dawkins, *The Blind Watchmaker* (New York: W. W. Norton, 1987), p. 141.

23. Paul Davies, *God & the New Physics* (New York: Simon & Schuster, 1983), p. 47.

24. Michael White and John Gribbin, *Stephen Hawking: A Life in Science* (New York: Penguin Books, 1992), pp. 165–171.

25. From *A BRIEF HISTORY OF TIME*, by Stephen W. Hawking, copyright © 1988, 1996 by Stephen W. Hawking, pp. 140–141. Used by permission of Bantam Books, a division of Random House, Inc. and Writers House LLC.

26. Alfred North Whitehead, *The Concept of Nature* (Cambridge, England: Cambridge University Press, 1971), p. 163.

27. Dennis W. Sciama, "The Anthropic Principle and the Non-Uniqueness of the Universe," in *The Anthropic Principle: Proceedings of the Second Venice Conference on Cosmology and Philosophy*, eds. F. Bertola and U. Curi (Cambridge, England: Cambridge University Press, 1993), p. 108.

28. Victor J. Stenger, *The Unconscious Quantum: Metaphysics in Modern Physics and Cosmology* (Amherst, N.Y.: Prometheus Books, 1995), p. 236.

29. *Ibid.*, p. 198.

30 *Ibid.*, p. 144.

31. Dawkins, *The Blind Watchmaker*, p. 141.

32. Nielsen, *Naturalism without Foundations*, p. 518.

33. *Ibid.*, p. 49.

BIBLIOGRAPHY

Abbott, Edwin A. *Flatland: A Romance of Many Dimensions*. New York: Dover Publications, 1952.

"An Accelerating Universe? Not So Fast," *Sky & Telescope*, 98:6 (December 1999), p. 18.

Adams, Fred C., and Gregory Laughlin. "The Future of the Universe," *Sky & Telescope*, 96:2, pp. 32–39.

Alfvén, Hannes. *Worlds-Antiworlds*. San Francisco: W. H. Freeman, Co., 1966.

Aman, J. M., and M. A. Markov. "Oscillating Universe in the State p 0," Preprint P-0290, (1983).

Atkatz, David. "Quantum Cosmology for Pedestrians," *American Journal of Physics*, 62:7 (1 July 1994), pp. 619–627.

Atkins, Peter. *The Creation*. Oxford: W. H. Freeman, 1981.

Bahcall, Neta A., Jeremiah P. Ostriker, Saul Perlmutter, and Paul J. Steinhardt. "The Cosmic Triangle: Revealing the State of the Universe," *Science*, 284 (May 28 1999), pp. 1481–1488.

Barbour, Ian. *Religion in an Age of Science*. San Francisco: Harper & Row, 1990.

Barrow, John D. *The Artful Universe*. New York: Oxford University Press, 1995.

———. *Theories of Everything: The Quest for Ultimate Explanation*. Oxford: The Clarendon Press, 1991.

Barrow, John D., and Mariusz P. Dabrowski. "Oscillating Universes," *Monthly Notices of the Royal Astronomical Society*, 275 (1995), pp. 850–862.

Barrow, John D., and Frank J. Tipler. *The Anthropic Cosmological Principle*. Oxford: Oxford University Press, 1988.

Bartlett, John R. *The First and Second Book of the Maccabees*. Cambridge, England: Cambridge University Press, 1973.

Bartusiak, Marcia. *Thursday's Universe: A Report from the Frontier on the Origin, Nature, and Destiny of the Universe*. Redmond, Wash.: Tempus Books, 1988.

Basinger, David. *Divine Power in Process Theism*. Albany: State University of New York Press, 1988.

Beatty, Cheryl J., and Richard T. Fienberg. "Participatory Cosmology: The Big Bang Challenge," *Sky & Telescope* (March 1994), pp. 20–22.

Bell, John S. *Speakable and Unspeakable in Quantum Mechanics*. Cambridge, England: Cambridge University Press, 1987.

Bernstein, Jeremy. "The Einstein-Besso Letters," *The New Yorker* (27 February 1989), pp. 92–93.

"Big Crunch Canceled," *The Economist*, 346 (17 January 1998).

Black, David C. "Other Suns. Other Planets," *Sky & Telescope*, 92:2 (August 1996), pp. 20–27.

Blau, Steven K., and Alan Guth. "Inflationary Cosmology," in *Three Hundred Years of Gravitation*, ed. Stephen Hawking and Werner Israel. Cambridge, England: Cambridge University Press, 1987, pp. 524–603.

Bludman. S. A. "Thermodynamics and the End of a Closed Universe," *Nature*, 308 (March 1984), pp. 319–322.

Bohm, David, and B. J. Hiley. *The Undivided Universe: An Ontological Interpretation of Quantum Theory*. London: Routledge, 1993.

Bondi, Hermann. *Cosmology*. Cambridge, England: The University Press, 1961.

Boslough, John. *Stephen Hawking's Universe*. New York: Avon Books, 1989.

Burstein, David. "Unveiling a New Galaxy," *Nature*, 372 (3 November 1994), p. 38.

Carr, B. J., and M. J. Rees. "The Anthropic Principle and the Structure of the Physical World," *Nature*, 278 (1979), pp. 605–612.

Carter, Brandon. "Large Number Coincidences and the Anthropic Principle in Cosmology," in *Confrontation of Cosmological Theories with Observation*, ed. M. S. Longair. Dordrecht: Reidel, 1974, pp. 291–298.

"Closing in on the Hubble Parameter," *Sky & Telescope*, 88 (December 1994), p. 10.

Cobb, Jr., John B. "Is God Creator *Ex Nihilo?*" *Process and Faith* website, (July/August, 1999).

Cobb, Jr., John B., and David R. Griffin. *Process Theology: An Introductory Exposition*. Philadelphia: Westminster Press, 1976.

Cohen, A. G., A. De RúJula, and S. L. Glashow. "A Matter-Antimatter Universe?" *Astrophysical Journal*, 495:2 (10 March 1998), pp. 539–549.

Coles, Peter, and George Ellis. "The Case for an Open Universe," *Nature*, 370 (25 August 1994), pp. 610–612.

Corey, M. A. *Back to Darwin: The Scientific Case for Deistic Evolution*. Lanham, Md.: Rowman and Littlefield, 1994.

———. *God and the New Cosmology: The Anthropic Design Argument*. Lanham, Md.: Rowman and Littlefield, 1993.

———. *The Natural History of Creation: Biblical Evolutionism and the Return of Natural Theology*. Lanham, Md.: Rowman and Littlefield, 1995.

Craig, William L., and Quentin Smith. *Theism, Atheism, and Big Bang Cosmology*. Oxford: Clarendon Press, 1993.

Davies, Paul. *The Cosmic Blueprint–New Discoveries in Nature's Creative Ability to Order the Universe*. New York: Simon & Schuster, 1989.

———. *The Fifth Miracle: The Search for the Origin and Meaning of Life*. New York: Simon & Schuster, 1999.

———. "The First One Second of the Universe," in *Images of the Universe*, ed. Carole Stott. Cambridge, England: Cambridge University Press, 1991, pp. 199–214.

———. *God and the New Physics*. New York: Simon & Schuster. 1983.

———. *The Mind of God: The Scientific Basis for a Rational World*. New York: Simon & Schuster, 1992.

Davies, Paul, and John Gribbin. *The Matter Myth*. New York: Simon & Schuster, 1992.

Davies, P. C. W. *The Accidental Universe*. Cambridge, England: Cambridge University Press, 1982.

Davies, P. C. W., and Brown, J. R., eds. *The Ghost in the Atom*. Cambridge, England: Cambridge University Press, 1986.

Dawkins, Richard. *The Blind Watchmaker*. New York: W. W. Norton, 1987.

Deane, Sidney Norton, ed. *St. Anselm*. La Salle, Ill.: Open Court Publishing Company, 1954.

de Bernardis, P. *et al*. "A Flat Universe From High-Resolution Maps of the Cosmic Microwave Background Radiation," *Nature*, 404 (27 April 2000), pp. 955–959.

Dennes, William R. "The Categories of Naturalism," in *Naturalism and the Human Spirit*, ed. Y. H. Krikorian. New York: Columbia University Press, 1944, pp. 270–204.

d'Espagnat, Bernard. "Quantum Theory and Reality," *Scientific American*, 241 (November 1979), pp. 158–181.

Devitt, Michael. *Coming to Our Senses: A Naturalistic Program for Semantic Localism.* Cambridge, England: Cambridge University Press, 1966.

de Waal, Frans. *Good Natured: The Origins of Right and Wrong in Humans and Other Animals*. Cambridge, Mass.: Harvard University Press, 1996.

Dewey, John. *A Common Faith.* New Haven: Yale University Press, 1934.

———. *The Quest for Certainty.* New York: G. P. Putnam's Sons, 1960.

DeWitt, B. S. "Quantum Mechanics and Reality," *Physics Today*, 23:9, (September 1970), pp. 30–35.

DeWitt, B. S., and N. Graham, eds. *The Many-Worlds Interpretation of Quantum Mechanics. A Fundamental Exposition by Hugh Everett, III with Papers by J. A. Wheeler, B. S. DeWitt, L. N. Cooper and D. Van Vechten, and N. Graham.* Princeton: Princeton University Press, 1973.

Dick, Steven J. *The Biological Universe*. Cambridge, England: Cambridge University Press, 1996.

———. *Life on Other Worlds: The 20th-Century Extraterrestrial Life Debate*. Cambridge, England: Cambridge University Press, 1998.

Dicke, Robert H. "Dirac's Cosmology and Mach's Principle," *Nature*, 192 (1961), pp. 170–171.

Dicke, R. H., and P. J. E. Peebles. "The Big Bang Cosmology–Enigmas and Nostrums," in *General Relativity*, eds. S. W. Hawking and W. Israel. Cambridge, England: Cambridge University Press, 1979, pp. 504–517.

Dicken, Thomas M., and Rem B. Edwards, *Dialogues on Values and Centers of Value: Old Friends, New Thoughts*. Amsterdam—New York: Editions Rodopi, in press.

Dobrovolskaya, A., and V. Novikov. "On Heavy Higgs Boson Production," *Zeitschrift für Physik C*, 52:3, (1991), pp. 427–436.

Dodd, Stuart C. "Hypotheses Defining Scientific Humanism: A Reformulation of Humanist Principles for Testing by Scientific Methods," *Journal of Human Relations*, 21:2 (Second Quarter 1973).

Drees, Willem B. *Beyond the Big Bang: Quantum Cosmologies and God.* La Salle, Ill.: Open Court, 1990.

Dyson, Freeman. "Time without End: Physics and Biology in an Open Universe," *Reviews of Modern Physics* 51:3 (July 1979), pp. 447–460.

Edwards, Rem B. "God and Process," in *Logic, God, and Metaphysics*, ed. James F. Harris. The Hague: Kluwer Academic Publishers, 1992, pp. 45–57.

———. "The Human Self: An Actual Entity or a Society?" *Process Studies*, 5:3 (Fall 1975), pp. 195–203.

———. "Kraus's Boethian Interpretation of Whitehead's God," *Process Studies*, 11:1 (Spring 1981), pp. 30–34.

———. "The Pagan Dogma of the Absolute Unchangeableness of God," *Religious Studies*, 14 (1978), pp. 305–314.

———. *Pleasures and Pains: A Theory of Qualitative Hedonism*. Ithaca, N. Y.: Cornell University Press, 1979.

———. "Process Thought and the Spaciness of Mind," *Process Studies*, 19:3 (Fall 1990), pp. 156–166.

———. *Reason and Religion: An Introduction to the Philosophy of Religion*. New York: Harcourt Brace Jovanovich, 1972; and Lanham, Md.: The University Press of America, 1979.

———. "Systemic Value and Valuation," in *Forms of Value and Valuation: Theory and Applications*, eds. Rem B. Edwards and John W. Davis. Lanham, Md.: University Press of America, 1991, pp. 37–55.

———. "Universals, Individuals, and Intrinsic Good," in *Forms of Value and Valuation: Theory and Applications*, eds. Rem B. Edwards and John W. Davis. Lanham, Md.: University Press of America, 1991, pp. 81–104.

Ehrenfeld, David, *The Arrogance of Humanism*. New York: Oxford University Press, 1978.

Ellis, George F. R. *Before the Beginning: Cosmology Explained*. New York: Boyars/Bowerdean, 1997.

Emerson, Ralph Waldo. "Self-Reliance," in *The Complete Works of Ralph Waldo Emerson*, ed. Edward Waldo Emerson. Boston: Houghton Mifflin Co., 1903, Vol. 2.

Everett, III, Hugh. "'Relative State' Formulation of Quantum Mechanics," *Reviews of Modern Physics*, 29:3 (July 1957), pp. 454 ff.

"Extrasolar Planet Seen Transiting Its Star," *Sky & Telescope*, 99:2 (February 2000), pp. 16–17.

"Extrasolar Planets Aplenty," *Sky & Telescope*, 100:5 (November 2000), pp. 24–25.

"Farthest Supernovae Favor Low-Density Universe," *Sky & Telescope*, 95 (March 1998), p. 19.

Ferris, Timothy. *Coming of Age in the Milky Way*. New York: William Morrow & Co., 1988.

———. *The Red Limit: The Search for the Edge of the Universe*. New York: Quill, 1983.

Feyerabend, Paul. *Against Method*, rev. ed. New York: Verso, 1988.

Finkbeiner, Ann K. "Cosmic Yardsticks: Supernovae and the Fate of the Universe," *Sky & Telescope*, 96 (1998), pp. 38–45.

———."Hubble Telescope Settles Cosmic Distance Debate–or Does It?" *Science*, 284:5419 (28 May 1999), p. 1438.

Ford, Lewis. "An Alternative to *Creatio Ex Nihilo*," *Religious Studies*, 19 (1983), pp. 205–213.

———. *The Lure of God: A Biblical Background for Process Theism*. Philadelphia: Fortress Press, 1978.

Freedman, Wendy L., *et al.* "Distance to the Virgo Cluster Galaxy M100 from Hubble Telescope Observations of Cepheids," *Nature*, 371 (27 October 1994), pp. 757–762.

"Foundations of Quantum Mechanics," *Noûs*, 18:4 (November 1984)–entire issue.

Friedmann, Alexander. "On the Curvature of the World." *Zeitschrift für Physik* 10 (1922), pp. 377ff.

Friedman, Joel. "The Natural God: A God Even an Atheist Can Believe In," *Zygon*, 21:3 (September 1986), pp. 369–388.

Friedman, Michael. "Philosophical Naturalism," *Proceedings and Addresses of the American Philosophical Association*, 74:2 (1997), pp. 7–21.

Gamow, George. *One Two Three ... Infinity*. New York: The American Library, 1954.

———. "Modem Cosmology," *Scientific American*, 190:3 (March 1954), pp. 55–63.

———. *The Creation of the Universe*. New York: Mentor Books, 1960.

Gardner, Marvin. "WAP, SAP, PAP, and FAP," *The New York Review of Books*, 33 (8 May 1986), p. 25.

Gingerich, Owen. "Perspectives on Science and Theology," in *How Large Is God?* ed. John M. Templeton. Philadelphia: Templeton Foundation Press, 1997, pp. 23–46.

Glanz, James. "American Physical Society: Celebrating a Century of Physics *en Masse*," *Science*, 283 (2 April 1999), pp. 34–36.

———. "Astronomers See a Cosmic Antigravity Force at Work," *Nature*, 279 (27 February 1998), pp. 1298–1299.

———. "Breakthrough of the Year: Astronomy: Cosmic Motion Revealed," *Science*, 282: 5397 (18 December 1998), pp. 2156–2157.

———. "Calculating Weight of Emptiness is No. 1 Cosmological Mystery," *Minneapolis Star Tribune*, (5 December 1999), p. 19A.

———. "Exploding Stars Flash New Bulletins from Distant Universe," *Science*, 280: 5366 (May 1998), pp. 1008–1009.

———. "Exploding Stars Point to a Universal Repulsive Force," *Science*, 279 (30 January 1998), p. 651–652.

———. "No Backing off From the Accelerating Universe," *Science* (13 November 1998), pp. 1249–1251.

———. "Sky Survey Racks up Record-Setting Quasars," *Science*, 282 (11 December 1998) , pp. 1969–1970.

———. "Which Way to the Big Bang?" *Science*, 284:5419 (28 May 1999), pp. 1448–1451.

Goldsmith, Donald. *The Runaway Universe: The Race to Find the Future of the Cosmos*. New York: Perseus Pub. Co., 2000.

"The Great Attractor's Hidden Heart," *Sky & Telescope*, 91:6 (December 1999), p. 12.

Greene, Brian. *The Elegant Universe: Superstrings, Hidden Dimensions, and the Quest for the Ultimate Theory*. New York: W. W. Norton & Co., 1999.

Gribbin, John. "Oscillating Universes Bounce Back," *Nature*, 259 (1976), pp. 15–16.

———. *In the Beginning: After COBE and Before the Big Bang*. Boston: Little Brown and Co., 1993.

Gribbin, John, and Martin Rees. *Cosmic Coincidences: Dark Matter, Mankind, and Anthropic Cosmology*. New York: Bantam Books, 1989.

Griffin, David R. *Evil Revisited: Responses and Reconsiderations*. Albany: State University of New York Press, 1991.

———. *God., Power, and Evil: A Process Theodicy*. Philadelphia: The Westminster Press, 1976.

———. *Religion and Scientific Naturalism*. Albany: State University of New York Press, 2000.

Griffin, David R., ed. *Physics and the Ultimate Significance of Time*. Albany: State University of New York Press, 1986.

Grishchuk, L. P., and Ya. B Zeldovich. "Complete Cosmological Theories," in *Quantum Structure of Space and Time*, ed. M. J. Duff and C. J. Isham. Cambridge, England: Cambridge University Press, 1982, pp. 409–422.

Gunn, James E., David N. Schramm, and Beatrice M. Tinsley. "An Unbounded Universe," *Astrophysical Journal*, 194 (1994), pp. 543–553.

Guth, Alan H. "Inflationary Universe: A Possible Solution to the Horizon and Flatness Problems," *Physical Review, D*, 23 (1981), pp. 347–356.

———. "Starting the Universe: The Big Bang and Cosmic Inflation," in *Bubbles, Voids, and Bumps in Time: The New Cosmology*, ed. James Cornell. New York: Cam-bridge University Press, 1989, pp. 105–146.

———. *The Inflationary Universe: The Quest for a New Theory of Cosmic Origins*. Reading, Mass.: Perseus Books, 1997.

Hacking, Ian. "The Inverse Gambler's Fallacy: The Argument from Design. The Anthropic Principle Applied to Wheeler Universes." *Mind*, 96 (1987), pp. 331–340.

Hallberg, Fred W. "Barrow and Tipler's Anthropic Cosmological Principle," *Zygon*, 23:2 (June 1988), pp. 139–157.

Halliwell, Jonathan J. "Quantum Cosmology and the Creation of the Universe," *Scientific American*, 265:6 (1 December 1991), pp. 76–82.

Halpern, Paul. *Cosmic Wormholes: The Search for Interstellar Shortcuts*. New York: NAL-Dutton, 1992.

Hanson, Norwood Russell. *Patterns of Discovery*. Cambridge, England: Cambridge University Press, 1969.

Harris, James F. *Against Relativism: A Philosophical Defense of Method*. La Salle, Ill: Open Court, 1992.

Hartle, J. B. "The Quantum Mechanics of Cosmology," in *Quantum Cosmology and Baby Universes*, eds. S. Coleman, J. B. Hartle, T. Piran, and S. Weinberg. Singapore: World Scientific, 1991, pp. 65–151.

Hartshorne, Charles. *Creative Synthesis and Philosophical Method.* La Salle. Ill.: Open Court Publishing Co., 1970.

———. *The Divine Relativity.* New Haven: Yale University Press, 1948.

———. "The Ethics of Contributionism," in *Responsibilities to Future Generations: Environmental Ethics*, ed. Ernest Partridge. Buffalo: Prometheus Books, 1981, pp. 103–107.

———. *Man's Vision of God and the Logic of Theism.* Hamden, Conn.: Archon Books, 1964.

———. *Omnipotence and Other Theological Mistakes.* Albany: State University of New York Press, 1984.

———. "Response to Alston," in *Existence and Actuality: Conversations with Charles Hartshorne*, eds. John B. Cobb, Jr. and Franklin I. Gamwell. Chicago: University of Chicago Press, 1987, pp. 98–102.

———. "Response to Martin," in *Existence and Actuality: Conversations with Charles Hartshorne*, eds. John B. Cobb. Jr. and Franklin Gamwell. Chicago: University of Chicago Press, 1987, pp. 66–77.

———. *Wisdom as Moderation.* Albany, N.Y.: State University of New York Press, 1987.

Hawking, Stephen W. *A Brief History of Time: From the Big Bang to Black Holes.* New York: Bantam Books, 1988.

———. *Black Holes and Baby Universes and Other Essays.* New York: Bantam Books, 1993.

———. *Master of the Universe*, broadcast on BBC in 1990.

———. "Theoretical Advances in Relativity," in *Some Strangeness in the Proportion*, ed. Harry Woolf. Reading, Mass.: Addison-Wesley, 1980.

Hawking, Stephen, W., Editor. *Stephen Hawking's A Brief History of Time: A Reader's Companion.* New York: Bantam Books, 1992.

Hawking, Stephen, and Roger Penrose. "The Singularities of Gravitational Collapse and Cosmology," *Proceedings of the Royal Society of London*, A314 (1970), pp. 529–548.

———. *The Nature of Space and Time.* Princeton: Princeton University Press, 1996.

Healey, Richard A. "How Many Worlds?" *Noûs*, 18:4 (November 1984), pp. 591–616.

Hefner, Philip. "The Evolution of the Created Co-Creator," in *Cosmos as Creation*, ed. Ted Peters. Nashville: Abingdon Press, 1989, pp. 211–233.

Heisenberg, Werner. *Physics and Philosophy: The Revolution in Modern Science*. New York: Harper & Row, 1958.

———. *Philosophical Problems of Quantum Physics*. Woodbridge, Conn.: Ox Bow Press, 1979.

Hick, John. *Death and Eternal Life*. New York: Harper & Row, 1976.

———. *Evil and the God of Love*. New York: Harper & Row, 1966.

———. *An Interpretation of Religion*. New Haven: Yale University Press, 1989.

Hintikka, Jaakko. "Gaps in the Great Chain of Being," *Proceedings and Addresses of the American Philosophical Association*, 49 (November 1976), pp. 22–38.

Hook, Sidney. *The Quest for Being*. New York: St. Martin's Press, 1961.

Hoyle, Fred. *The Nature of the Universe*, rev. ed. New York: The New American Library, 1964.

———. "The Steady-State Theory Revived?" *Comments on Astrophysics*, 13:2 (1989), pp. 81–86.

———. "The Universe: Past and Present Reflections," *Engineering and Science*, 45:2 (November 1981), pp. 8–12.

"HST's Hubble Constant Announced–Again," *Sky & Telescope*, 100:5 (November 2000), p. 28.

Hu, Wayne. "Ringing in the New Cosmology," *Nature*, 404 (27 April 2000), pp. 939–940.

Hubble, Edwin. *The Realm of the Nebulae*. New Haven: Yale University Press, 1936.

"Hubble Revisits Its Deep Field," *Sky & Telescope*, 97:1 (1999), p. 24.

Isham, Chris. "Quantum Gravity," in *The New Physics*, ed. Paul Davies. Cambridge, England: Cambridge University Press, 1989, pp. 70–93.

Israelit, Mark. "Cosmological Perturbations in the Early Prematter Universe," *The Astrophysical Journal*, 435 (1 November 1994), pp. 8–15.

Israelit, Mark, and Rosen, Nathan. "A Singularity-free Cosmological Model in General Relativity," *Astrophysical Journal*, 342 (15 July 1989), pp. 627–634.

Jakosky, Bruce. *The Search for Life on Other Planets*. Cambridge, England: Cambridge University Press, 1999.

Janzen, J. Gerald. "Modes of Power and the Divine Relativity." *Encounter*, 36:4 (Autumn 1975), pp. 379–406.

Jastrow, Robert. *God and the Astronomers*. New York: W. W. Norton & Co., 1978.

———. "Science and the Creation," in *Creation* ed. Thomas H Schattauer. New Haven: Yale Divinity School, 1980, pp. 32–35.

———. "What Are the Chances for Life?" *Sky & Telescope*, 93:6 (June 1997), p. 63.

———. "What Forces Filled the Universe with Energy Fifteen Billion Years Ago?" in *Cosmos, Bios, Theos*, eds. Henry Margenau and Abraham Varghese. La Salle, Ill.: Open Court. 1992, pp. 45–49.

Jungerman, John A. *World in Process: Creativity and Interconnection in the New Physics*. Albany: State University of New York Press, 2000.

Kaku, Michio. *Visions: How Science will Revolutionize the 21st Century*. New York: Anchor Books, 1997.

Kirshner, Robert P. "Measuring the Universe: Redshifts and Standard Candles," in *Bubbles, Voids, and Bumps in Time: The New Cosmology*, ed. James Cornell. New York: Cambridge University Press, 1989, pp. 26–49.

Krips, Henry. *The Metaphysics of Quantum Theory*. Oxford: Clarendon Press, 1987.

Krisciunas, Kevin. "Science with the Keck Telescope," *Sky & Telescope* (September 1994), p. 26.

Kurtz, Paul, ed. *Humanist Manifestos I and II*, Buffalo, N.Y.: Prometheus Books, 1973.

———. *In Defense of Secular Humanism*. Buffalo, N.Y.: Prometheus Books, 1983.

Lamont, Corliss. *The Illusion of Immortality*. New York: Frederick Ungar Pub. Co., 1965.

———. "Naturalistic Humanism," in *The Humanist Alternative*, ed. Paul Kurtz. Buffalo, N.Y.: Prometheus Books, 1973, pp. 129–132.

Lamprecht, Sterling. *The Metaphysics of Naturalism*. New York: Appleton-Century-Crofts, 1967.

Lederman, Leon, M., and David N. Schramm. *From Quarks to the Cosmos*. New York: Scientific American Library, 1989.

Lemaître, George. *The Primeval Atom.* New York: Van Nostrand, 1950.

Lemonick, Michael D., "Echoes of the Big Bang," *Time* (4 May 1992), pp. 62 ff.

———. "Visions 21/Science & Space: Will We Discover Another Universe?" *Time* (10 April 2000), pp. 82 ff.

Leplin, J, Editor. *Scientific Realism.* Berkeley: University of California Press, 1984.

Lerner, Eric. *The Big Bang Never Happened.* New York: Random House, 1991.

Leslie, John. *Universes.* London: Routledge, 1989.

Lightman, Alan, and Roberta Brawer. *Origins: The Lives and Worlds of Modern Cosmologists.* Cambridge, Mass.: Harvard University Press, 1990.

Linde, Andrei. "Inflation and Quantum Cosmology," in *Three Hundred Years of Gravitation*, eds. Stephen W. Hawking and Werner Israel. Cambridge, England: Cambridge University Press, 1987, pp. 631–651.

Linde, Andrei, Dimitri Linde, and Arthur Mezhlumian. "From the Big Bang Theory to the Theory of a Stationary Universe," *Physical Review, D*, 49:4 (15 February 1994), pp. 1783–1826.

Lovejoy, Arthur O. *The Great Chain of Being.* Cambridge, Mass.: Harvard University Press, 1961.

Lucretius. "The Nature of the Universe," in *Theories of the Universe from Babylonian Myth to Modern Science*, ed. Milton Munitz. New York: The Free Press, 1957, pp. 41–57.

Macchetto, F. Duccio, and Mark Dickinson. "Galaxies in the Young Universe," *Scientific American*, 276: 5 (May 1997), pp. 92–99.

MacRobert, Alan M. "Beyond the Big Bang," *Sky & Telescope* (March 1983), pp. 211–213.

———. "A Tour of Extrasolar Planets," *Sky & Telescope*, 96:3 (March 1997), pp. 80–81.

MacRobert, Alan M., and Joshua Roth. "The Planet of 51 Pegasi," *Sky & Telescope*, 91:1 (January 1996), pp. 38–40.

Mallove, Eugene F. "Review of Eric J. Lerner, *The Big Bang Never Happened*," in *Sky & Telescope*, 82:5 (November 1991), pp. 492–493.

Marcy, Geoffrey, and Paul R. Butler. "The Diversity of Planetary Systems," *Sky & Telescope*, 97:3 (March 1998), pp. 30–37.

Margenau, Henry, and Roy A. Varghese, eds. *Cosmos, Bios, Theos: Scientists Reflect on Science, God, and the Origins of the Universe, Life, and Homo sapiens.* La Salle, Ill.: Open Court, 1992.

Markov, M. A. "Problems of a Perpetually Oscillating Universe," *Annals of Physics*, 155 (1984), pp. 333–357.

Matthews, Robert. "Nothing Like a Vacuum," *New Scientist* (February 1995), pp. 30–33.

Mavrodes, George. "Comments on Rem Edwards' 'Antecedent Universe Cosmologies,'" presented at a meeting of *The Society for Philosophy of Religion* in Savannah, Ga. (25 February 1994).

McDonald, Kim A. "Discovery of Largest, Most Ancient Structures in Cosmos Said to Confirm 'Big Bang' Model of Universe's Creation." *The Chronicle of Higher Education*, 38 (29 April 1992), p. A 9.

Mermin, David N. "Is the Moon There When Nobody Looks? Reality and the Quantum Theory," *Physics Today*, 38 (April 1985), pp. 38–47.

Midgley, Mary. *Evolution as a Religion*. London: Methuen, 1985.

Mishra, A., *et al.* "Higgs-particle Production through Vacuum Excitations," *Physical Review, D*, 44:1 (1 July 1991), pp. 110–117.

Misner, Charles W., Kip S. Thorne, and John A. Wheeler. *Gravitation*. New York: W. H. Freeman and Company, 1973.

Munitz, Milton. *Cosmic Understanding: Philosophy and Science of the Universe.* Princeton: Princeton University Press, 1986.

———. *The Mystery of Existence*. New York: Appleton-Century-Crofts, 1965.

Murphy, Nancey, and George F. R. Ellis. *On the Moral Nature of the Universe: Theology, Cosmology, and Ethics*. Minneapolis: Fortress Press, 1996.

Neville, Robert C. *Creativity and God: A Challenge to Process Theology*. New York: Seabury Press, 1980.

———. *God as Creator*. Chicago: University of Chicago Press, 1968.

———. *A Theology Primer*. Albany: N.Y.: State University of New York Press, 1991.

Nielsen, Kai. *Naturalism without Foundations*. Amherst, N.Y.: Prometheus Books, 1996.

Nobo, Jorge Luis. *Whitehead's Metaphysics of Extension and Solidarity*. Albany, N.Y.: State University of New York Press, 1986.

Norbury, John W. "Higgsboson Production in Nucleus-Nucleus Collisions," *Physical Review, D*, 42:11 (1 December 1990), pp. 3696–3698.

Nozick, Robert. *Philosophical Explanations*. Cambridge, Mass.: The Belknap Press. 1981.

Odenwald, Sten. "Space-time: The Final Frontier," *Sky & Telescope*. 91:2 (February 1996), pp. 24–29.

"Old Galaxy in a Young Universe," *Sky & Telescope*, 93:2 (September 1996), p. 11.

Page, Don. "Will Entropy Decrease if the Universe Recollapses?" *Physical Review, D* 32:10 (15 November 1985), pp. 2496–2499.

Pagels, Heinz R. *The Cosmic Code: Quantum Physics as the Language of Nature*. New York: Bantam Books, 1983.

———. *Perfect Symmetry: The Search for the Beginning of Time*. New York: Bantam Books, 1986.

Pannenberg, Wolfhart. "The Doctrine of Creation in Modern Science," in *Cosmos as Creation: Theology and Science in Consonance*, ed. Ted Peters. Nashville: Abingdon Press: 1989, pp. 152–176.

Parker, Samantha. "The Eagle's Nest," *Sky & Telescope*, 91:2 (February 1996), pp. 32–34.

Peacocke, Arthur. *Intimations of Reality: Critical Realism in Science and Religion*. Notre Dame, Ind.: University of Notre Dame Press, 1984.

———. "Science and God the Creator," in *Evidence of Purpose: Scientists Discover the Creator*, ed. John M. Templeton. New York: Continuum Publishing Co., 1994, pp. 91–104.

———. *Theology for a Scientific Age: Being and Becoming–Natural, Divine, and Human*. Minneapolis: Fortress Press, 1993.

Peebles, P. J. E. "The Mean Mass Density of the Universe," *Nature*, 321 (1986), pp. 27–32.

Pegis, Anton C., Editor. *Basic Writings of St. Thomas Aquinas*. New York: Random House, 1945), Vol. I.

Penrose, Roger. "Big Bangs, Black Holes and 'Times Arrow'," in *The Nature of Time*, ed. Raymond Flood and Michael Lockwood. Oxford: Basil Blackwell, 1986, pp. 36–79.

———."Difficulties with Inflationary Cosmology," in *Fourteenth Texas Symposium on Relativistic Astrophysics*, ed. E. J. Fenyves. New York: Academy of Science, 1989, pp. 249–264.

———. *The Emperor's New Mind & Concerning Computers, Minds, and the Laws of Physics.* New York: Penguin Books, 1991.

———. "Newton, Quantum Theory and Reality," in *Three Hundred Years of Gravitation*, eds. Stephen Hawking and Werner Israel. Cambridge, England: Cambridge University Press, 1987, pp. 17–49.

———. *Shadows of the Mind.* New York: Oxford University Press, 1994.

———."Singularities and Time-Asymmetry," in *General Relativity: An Einstein Centenary Survey*, eds. S. W. Hawking and W. Israel. Cambridge, England: Cambridge University Press, 1979, pp. 581–638.

———. "Singularities of Spacetime," in *Theoretical Principles in Astrophysics and Relativity*, ed. Norman R. Lebovitz, *et al.* Chicago: The University of Chicago Press, 1978, pp. 217–243.

———. "Some Remarks on Gravity and Quantum Mechanics," in *Quantum Structure of Space and Time*, eds. M. J. Duff and C. J. Isham. Cambridge, England: Cambridge University Press, 1982, pp. 3–10.

———. "What Does the Big Bang Tell Us about Quantum Gravity?" *Memorie della Società Astronomica Italiana*, 62:3 (1991), pp. 607–614.

Perlmutter, Saul, *et al.*, "Discovery of a Supernova Explosion at Half the Age of the Universe," *Nature*, 391 (1 January 1998), pp. 51–54.

Perryman, Michael. "Hipparcos: The Stars," *Sky & Telescope*, 97:6 (June 1999), pp. 40–50.

Pius XII, Pope. "Modem Science and the Existence of God," *The Catholic Mind* (March 1952), pp. 182–192.

Polkinghorne, John. *The Faith of a Physicist.* Minneapolis: Fortress Press, 1994.

Priestley, Joseph. *Letters to a Philosophical Unbeliever*, Part 1. New York: Garland Publishing Co., 1983.

"The Primordial Soup: A Recipe for Life," *Sky & Telescope*, 95:2 (February 1998), p. 20.

Rees, Martin. *Before the Beginning: Our Universe and Others.* Reading, Mass.: Perseus Books, 1997.

———. "Concluding Comments: A Cosmologist's View," in *Large Scale Structure of the Universe, Cosmology and Fundamental Physics*, eds. G. Setti and L. Van Hove. Geneva: ESO and CERN, 1984.

Reeves, H. "Nuclear Processes in the Early Universe," in *Confrontation between Theories and Observations in Cosmology: Present Status and Future Programmes*, eds. J. Audouze and F. Melchiorri. Amsterdam: North Holland, 1990, pp. 425–455.

Reichenbach, Hans. *The Rise of Scientific Philosophy*. Berkeley, Cal.: University of California Press, 1951.

Restak, Richard. *The Brain*. New York: Bantam Books, 1984.

Rolston, III, Holmes. *Science and Religion: A Critical Survey*. New York: Random House, 1987.

Rorty, Richard. *Consequences of Pragmatism*. Minneapolis: University of Minnesota Press, 1982.

———. "Does Academic Freedom Have Philosophical Presuppositions? Academic Freedom and the Future of the University," *Academe*, 80:6 (November/December 1994), pp. 52–63.

———. *Philosophy and the Mirror of Nature*. Princeton: Princeton University Press, 1979.

Roth, Joshua. "Farthest Supernova Clinches Case for Accelerating Universe," *Sky & Telescope* 102:1 (July 2001), pp. 20–21.

Roth, Joshua, and Joel R. Primack. "Cosmology: All Sewn Up or Coming Apart at the Seams," *Sky & Telescope*, 91:1 (January 1996), pp. 20–26.

Rozental, I. L. *Big Bang Big Bounce*. Berlin: Springer-Verlag, 1988.

Rubin, Vera C. "Weighing the Universe: Dark Matter and Missing Mass," in *Bubbles, Voids, and Bumps in Time: The New Cosmology*, ed. James Cornell. Cambridge, England: Cambridge University Press, 1991, pp. 73–104.

Ruse, Michael. "The New Evolutionary Ethics," in *Evolutionary Ethics*, eds. Matthew H. Nitecki and Doris V. Nitecki. Albany: State University of New York Press, 1993, pp. 133–162.

———. "Evolutionary Theory and Christian Ethics," *Zygon*, 29, (March 1994), pp. 5–24.

Russell, Bertrand. *Religion and Science*. New York: Oxford University Press, 1935.

———. *Why I Am Not a Christian*. New York: Simon & Schuster, 1957.

Russell, David. "Island Universes from Wright to Hubble" *Sky & Telescope*, 97:1 (January 1999), pp. 56–60.

Russell, Robert J. "Cosmology, Creation, and Contingency," in *Cosmos as Creation: Theology and Science in Consonance*, ed. Ted Peters. Nashville: Abingdon Press, 1989, pp. 177–209.

Sagan, Carl. *Cosmos*. New York: Random House, 1980.

Salam, Abdus. "Science and Religion: Reflections on Transcendence and Secularization," in *Cosmos, Bios, Theos*, eds. Henry Margenau and Roy Abraham Varghese. La Salle, Ill.: Open Court, 1992, pp. 93–104.

Sanders, Gary H., and David Beckett. "LIGO: An Antenna Tuned to the Songs of Gravity," *Sky & Telescope*, 100:4 (October 2000), pp. 41–48.

Sandage, Alan R. and G. A. Tammann. "The Dynamical Parameters of the Universe," in *Large Scale Structure of the Universe, Cosmology and Fundamental Physics*, eds. G. Setti and L. Van Hove. Geneva: ESO and CERN, 1984.

Sato, Katsuhiko, Hideo Kodama, Misao Sasaki, and Iei-ichi Madea. "Multiproduction of Universes by First-Order Phase Transition of a Vacuum," *Physics Letters*, 108B:2 (14 January 1982), pp. 103–107.

Sawyer, Kathy. "Cosmic Driving Force?; Scientists' Work on 'Dark Energy' Mystery Could Yield a New View of the Universe," *The Washington Post* (19 February 2000), p. A1.

Sciama, Dennis W. "The Anthropic Principle and the Non-Uniqueness of the Universe," in *The Anthropic Principle: Proceedings of the Second Venice Conference on Cosmology and Philosophy*, eds. F. Bertola and U. Curi. Cambridge, England: Cambridge University Press, 1993, pp. 107–109

Sellars, Roy W. "Does Naturalism Need Ontology," *Journal of Philosophy*, 41:25 (December 1944) pp. 686–693.

———. *The Philosophy of Physical Realism*. New York: Russell & Russell, 1996.

———. *Religion Coming of Age*. New York: Macmillan, 1928.

Shimony, Abner. "Role of the Observer in Quantum Theory," *American Journal of Physics*, 36 (1963).

———. "Quantum Physics and the Philosophy of Whitehead," *Boston Studies in the Philosophy of Science*, 2 (1965), pp. 307–330.

———. "Events and Processes in the Quantum World," in *Quantum Concepts in Space and Time*, eds. R. Penrose and C. J. Isham. Oxford: Clarendon Press, 1986, pp. 182–203.

"The Shorter Catechism," in *The Constitution of the Presbyterian Church (U.S.A): Part 1, Book of Confessions*. New York: The Office of the General Assembly, 1983, Section 7.001.

Silk, Joseph. *The Big Bang*. New York: W. H. Freeman and Co., 1980.

———. *The Big Bang: Revised and Updated Edition*. New York: W. H. Freeman and Co., 1989.

Smoot, George, and Keay Davidson. *Wrinkles in Time*. New York: Avon Books, 1993.

Starobinsky, A. A. "A New Type of Isotropic Cosmological Models Without Singularity," *Physics Letters*, 91B (24 March 1980), pp. 99–102.

Steigman, Gary "Observational Tests of Antimatter Cosmologies," *Annual Reviews of Astronomy and Astrophysics*, 14 (1976), pp. 339–372.

Stephens, Sally. "New Image of the Universe Soon After Creation: What the COBE Results Mean," *Mercury*, 21 (May/June 1992), pp. 91–92.

Stenger, Victor J. *Not by Design: The Origin of the Universe*. Buffalo: Prometheus Books, 1988.

———. *The Unconscious Quantum: Metaphysics in Modern Physics and Cosmology*. Amherst, N.Y.: Prometheus Books, 1995.

Taubes, Gary. "Theorists Nix Distant Antimatter Galaxies," *Science*, 278 (10 October 1997), p. 226.

Teilhard de Chardin, Pierre. *The Phenomenon of Man*. New York: Harper & Row, 1959.

Thomas Aquinas, St. *On the Truth of the Catholic Faith: Summa Contra Gentiles*, Book 1. New York: Doubleday & Co., 1955.

Tillich, Paul. *Systematic Theology*. London: Nisbit & Co., 1955, Vol. 1.

Tipler, Frank J. "The Many-Worlds Interpretation of Quantum Mechanics in Quantum Cosmology," in *Quantum Concepts in Space and Time*, eds. R. Penrose and C. J. Isham. Oxford: Clarendon Press, 1986, pp. 204–214.

———. "The Omega Point Theory: A Model of an Evolving God," in *Physics, Philosophy, and Theology*, eds. Robert J. Russell, William R. Stoeger, and George V. Coyne. Vatican City State: Vatican Observatory, 1988, pp. 323–324.

———. "The Omega Point as *Eschaton:* Answers to Pannenberg's Questions for Scientists," *Zygon*, 24:2 (June 1989), pp. 217–253.

———. *The Physics of Immortality*. New York: Anchor Books, 1994.

Tolman, Richard C. *Relativity, Thermodynamics and Cosmology*. Oxford: Clarendon Press, 1934.

"Top Scientist Pessimistic for Our Survival," *Knoxville News Sentinel* (3 October 2000), p. A2.

Toulmin, Stephen. *The Return to Cosmology: Postmodern Science and the Theology of Nature*. Berkeley: University of California Press, 1982.

Trefel, James. *The Dark Side of the Universe*. New York: Avon Books, 1988.

Trimble, Virginia. "Cosmic Discovery 1998," *Sky & Telescope*, 97:2 (February 1999), pp. 32–40.

Tryon, Edward P. "Is the Universe a Vacuum Fluctuation," *Nature*, 246 (14 December 1973), pp. 396–397.

Velan, A. Karel. *The Multi-Universe Cosmos: The First Complete Story of the Origin of the Universe*. New York: Plenum Press, 1992.

Vilenkin., A. "Creation of Universes from Nothing," *Physical Letters*, 117B (1982), pp. 25–28.

Vollert, Cyril, Lottie H. Kendierski, and Paul M. Byrne, eds. *Medieval Philosophical Texts in Translation*, No. 16. Milwaukee: Markette University Press, 1964.

Watson, Andrew. "The Universe Shows its Age," *Science*, 279 (13 February 1998), pp. 981–983.

———. "Inflation Confronts an Open Universe," *Science*, 279:5356 (6 March 1998), p. 1455.

Weinberg, Steven. *Dreams of a Final Theory*. New York: Pantheon Books, 1992.

———. *The First Three Minutes: A Modern View of the Origin of the Universe*. Updated ed. New York: Basic Books, 1988.

Weiner, Neil O. *The Harmony of the Soul*. Albany: State University of New York Press, 1993.

"The Westminster Confession of Faith," in *The Constitution of the Presbyterian Church (U. S. A)*: Part 1, *Book of Confessions*. New York: The Office of the General Assembly, 1983, section 6.001.

Wheeler. John A. "Genesis and Observership," in *Foundational Problems in the Special Sciences*, eds. Robert E. Butts and Jaakko Hintikka. Dordrecht, Holland: D. Reidel Publishing Company, 1975, pp. 3–33.

———. "Beyond the Black Hole," in *Some Strangeness in the Proportion*, ed. Harry Woolf. Reading, Mass.: Addison-Wesley, 1980.

———. "Bohr, Einstein, and the Strange Lesson of the Quantum," in *Mind in Nature*, ed. Richard Q. Elvee. San Francisco: Harper & Row, 1981, pp. 1–30.

———. "Not Consciousness but the Distinction Between the Probe and the Probed as Central to the Elemental Quantum Act of Observation," in *The Role of Consciousness in the Physical World*, ed. Robert G. Jahn. Boulder, Colo.: Westview Press, 1981.

———. "The Universe as Home for Man," in *The Nature of Scientific Discovery*, ed. Owen Gingerich. Washington: Smithsonian Institution Press, 1975, pp. 261–296.

White, Michael, and John Gribbin. *Stephen Hawking: A Life in Science*. New York: Penguin Books, 1992.

Whitehead, Alfred North. *The Concept of Nature*. Cambridge, England: Cambridge University Press, 1971.

———. *Process and Reality*, corrected ed. New York: The Free Press, 1978.

———. *Religion in the Making*. New York: The Macmillan Co., 1926.

———. *Science and the Modern World*. New York: The Free Press, 1967.

———. *Science and Philosophy*. Patterson, N.J.: Littlefield, & Adams, 1964.

Wigner, Eugene. *Symmetries and Reflections*. Bloomington: Indiana University Press, 1967.

Williams, Patricia A. "Christianity and Evolutionary Ethics: Sketch Toward a Reconciliation," *Zygon*, 31:2 (June 1996), pp. 253–268.

Worthing, Mark W. *God, Creation, and Contemporary Physics*. Minneapolis: Fortress Press, 1996.

"A Young Universe with Old Stars," *Sky & Telescope*, 89 (January 1995) pp. 10–11.

Zee, A. "Nonconservation of Baryon Number," in *Gamow Cosmology*, ed. LF. Melchiorri and R. Ruffini. Amsterdam: North-Holland, 1986, pp. 415– 430.

Zel'Dovich, Ya. B., and D. Novikov. *The Structure and Evolution of the Universe*. Chicago: The University of Chicago Press, 1983.

ABOUT THE AUTHOR

REM B. EDWARDS received his A.B. degree from Emory University in 1956, where he was elected to Phi Beta Kappa. During graduate school, he was a Danforth Graduate Fellow. He received a B.D. degree from Yale University Divinity School in 1959 and a Ph.D. from Emory University in 1962. He taught for four years at Jacksonville University in Florida, moved from there to the University of Tennessee, Knoxville in 1966, and retired from there partly in 1997 and partly in 1998. He continues to be professionally active and kept an office on the University campus until the end of May 2000. He was a U. T. Chancellor's Research Scholar in 1985 and a Lindsay Young Professor from 1987 to 1998.

His areas of specialization are Philosophy of Religion, American Philosophy, Ethical Theory, Medical Ethics with a special interest in Mental Health Care, Ethics and Animals, and Formal Axiology.

He is the author and/or editor of sixteen books, including *Reason and Religion* (New York: Harcourt, 1972 and Lanham, Md.: University Press of America, 1979); *Pleasures and Pains: A Theory of Qualitative Hedonism* (Ithaca, N. Y.: Cornell University Press, 1979); with Glenn Graber, *BioEthics* (San Diego: Harcourt, 1988); with John W. Davis, *Forms of Value and Valuation: Theory and Applications* (Lanham, Md.: University Press of America, 1991); *Formal Axiology and Its Critics* (Amsterdam—New York: Editions Rodopi, 1995); *Violence, Neglect, and the Elderly*, co-edited with Roy Cebik, Glenn Graber, and Frank H. Marsh (Greenwich, Conn.: JAI Press, 1996); *New Essays on Abortion and Bioethics* (Greenwich, Conn.: JAI Press, 1997); *Ethics of Psychiatry: Insanity, Rational Autonomy, and Mental Health Care* (Buffalo, NY: Prometheus Books, 1997); *Values, Ethics, and Alcoholism*, co-edited with Wayne Shelton (Greenwich, Conn.: JAI Press, 1997); *Bioethics for Medical Education*, co-edited with Dr. Edward Bittar (Stamford, Conn.: JAI Press, 1999), *Religious Values and Valuations* (Signal Mountain, Tenn.: Paidia Publishing Co, 2000); and *Dialogues on Values and Centers of Value: Old Friends, New Thoughts*, co-authored with Thomas M. Dicken (Amsterdam—New York: Editions Rodopi, in press). Edwards is also the author of over sixty articles and reviews, including "How Process Theology Can Affirm Creation *Ex Nihilo*," *Process Studies*, 29:1 (2000), pp. 77–96.

He is an Associate Editor with the Value Inquiry Book Series, published by Editions Rodopi, where he is responsible for the Hartman Institute Axiology Studies special series. For a number of years he was co-editor of the Advances in Bioethics book series published by JAI Press.

Edwards has been the President of the Tennessee Philosophical Association (1973-1974), the Society for Philosophy of Religion (1981-1982), and the Southern Society for Philosophy and Psychology (1984-1985). He is a Charter Member and Fellow of the Robert S. Hartman Institute for Formal and Applied Axiology, has served on its Board of Directors since 1987, and since 1989 has been its Secretary-Treasurer. He chaired the committee that established the website for the R. S. Hartman Institute at: http://www.hartmaninstitute.org

INDEX

VIBS

The **Value Inquiry Book Series** is co-sponsored by:

Adler School of Professional Psychology
American Indian Philosophy Association
American Maritain Association
American Society for Value Inquiry
Association for Process Philosophy of Education
Canadian Society for Philosophical Practice
Center for Bioethics, University of Turku
Center for International Partnerships, Rochester Institute of Technology
Center for Professional and Applied Ethics, University of North Carolina at Charlotte
Centre for Applied Ethics, Hong Kong Baptist University
Centre for Cultural Research, Aarhus University
Centre for the Study of Philosophy and Religion, College of Cape Breton
College of Education and Allied Professions, Bowling Green State University
Concerned Philosophers for Peace
Conference of Philosophical Societies
Department of Moral and Social Philosophy, University of Helsinki
Gannon University
Gilson Society
Global Association for the Study of Persons
Ikeda University
Institute of Philosophy of the High Council of Scientific Research, Spain
International Academy of Philosophy of the Principality of Liechtenstein
International Center for the Arts, Humanities, and Value Inquiry
International Society for Universal Dialogue
Natural Law Society

Philosophical Society of Finland
Philosophy Born of Struggle Association
Philosophy Seminar, University of Mainz
Pragmatism Archive
R.S. Hartman Institute for Formal and Applied Axiology
Research Institute, Lakeridge Health Corporation
Russian Philosophical Society
Society for Iberian and Latin-American Thought
Society for the Philosophic Study of Genocide and the Holocaust
Society for the Philosophy of Sex and Love
Yves R. Simon Institute.

Titles Published

1. Noel Balzer, *The Human Being as a Logical Thinker.*

2. Archie J. Bahm, *Axiology: The Science of Values.*

3. H. P. P. (Hennie) Lötter, *Justice for an Unjust Society.*

4. H. G. Callaway, *Context for Meaning and Analysis: A Critical Study in the Philosophy of Language.*

5. Benjamin S. Llamzon, *A Humane Case for Moral Intuition.*

6. James R. Watson, *Between Auschwitz and Tradition: Postmodern Reflections on the Task of Thinking.* A volume in **Holocaust and Genocide Studies.**

7. Robert S. Hartman, *Freedom to Live: The Robert Hartman Story,* edited by Arthur R. Ellis. A volume in **Hartman Institute Axiology Studies.**

8. Archie J. Bahm, *Ethics: The Science of Oughtness.*

9. George David Miller, *An Idiosyncratic Ethics; Or, the Lauramachean Ethics.*

10. Joseph P. DeMarco, *A Coherence Theory in Ethics.*

11. Frank G. Forrest, *Valuemetrics*$^{\aleph}$*: The Science of Personal and Professional Ethics.* A volume in **Hartman Institute Axiology Studies.**

12. William Gerber, *The Meaning of Life: Insights of the World's Great Thinkers.*

13. Richard T. Hull, Editor, *A Quarter Century of Value Inquiry: Presidential Addresses of the American Society for Value Inquiry.* A volume in **Histories and Addresses of Philosophical Societies.**

14. William Gerber, *Nuggets of Wisdom from Great Jewish Thinkers: From Biblical Times to the Present.*

15. Sidney Axinn, *The Logic of Hope: Extensions of Kant's View of Religion.*

16. Messay Kebede, *Meaning and Development.*

17. Amihud Gilead, *The Platonic Odyssey: A Philosophical-Literary Inquiry into the* Phaedo.

18. Necip Fikri Alican, *Mill's Principle of Utility: A Defense of John Stuart Mill's Notorious Proof.* A volume in **Universal Justice.**

19. Michael H. Mitias, Editor, *Philosophy and Architecture.*

20. Roger T. Simonds, *Rational Individualism: The Perennial Philosophy of Legal Interpretation.* A volume in **Natural Law Studies.**

21. William Pencak, *The Conflict of Law and Justice in the Icelandic Sagas.*

22. Samuel M. Natale and Brian M. Rothschild, Editors, *Values, Work, Education: The Meanings of Work.*

23. N. Georgopoulos and Michael Heim, Editors, *Being Human in the Ultimate: Studies in the Thought of John M. Anderson.*

24. Robert Wesson and Patricia A. Williams, Editors, *Evolution and Human Values.*

25. Wim J. van der Steen, *Facts, Values, and Methodology: A New Approach to Ethics.*

26. Avi Sagi and Daniel Statman, *Religion and Morality.*

27. Albert William Levi, *The High Road of Humanity: The Seven Ethical Ages of Western Man,* edited by Donald Phillip Verene and Molly Black Verene.

28. Samuel M. Natale and Brian M. Rothschild, Editors, *Work Values: Education, Organization, and Religious Concerns.*

29. Laurence F. Bove and Laura Duhan Kaplan, Editors, *From the Eye of the Storm: Regional Conflicts and the Philosophy of Peace.* A volume in **Philosophy of Peace.**

30. Robin Attfield, *Value, Obligation, and Meta-Ethics.*

31. William Gerber, *The Deepest Questions You Can Ask About God: As Answered by the World's Great Thinkers.*

32. Daniel Statman, *Moral Dilemmas.*

33. Rem B. Edwards, Editor, *Formal Axiology and Its Critics.* A volume in **Hartman Institute Axiology Studies.**

34. George David Miller and Conrad P. Pritscher, *On Education and Values: In Praise of Pariahs and Nomads.* A volume in **Philosophy of Education.**

35. Paul S. Penner, *Altruistic Behavior: An Inquiry into Motivation.*

36. Corbin Fowler, *Morality for Moderns.*

37. Giambattista Vico, *The Art of Rhetoric* (*Institutiones Oratoriae,* 1711-1741), from the definitive Latin text and notes, Italian commentary and introduction by Giuliano Crifò, translated and edited by Giorgio A. Pinton and Arthur W. Shippee. A volume in **Values in Italian Philosophy.**

38. W. H. Werkmeister, *Martin Heidegger on the Way,* edited by Richard T. Hull. A volume in **Werkmeister Studies.**

39. Phillip Stambovsky, *Myth and the Limits of Reason.*

40. Samantha Brennan, Tracy Isaacs, and Michael Milde, Editors, *A Question of Values: New Canadian Perspectives in Ethics and Political Philosophy.*

41. Peter A. Redpath, *Cartesian Nightmare: An Introduction to Transcendental Sophistry.* A volume in **Studies in the History of Western Philosophy.**

42. Clark Butler, *History as the Story of Freedom: Philosophy in Intercultural Context,* with Responses by sixteen scholars.

43. Dennis Rohatyn, *Philosophy History Sophistry.*

44. Leon Shaskolsky Sheleff, *Social Cohesion and Legal Coercion: A Critique of Weber, Durkheim, and Marx.* Afterword by Virginia Black.

45. Alan Soble, Editor, *Sex, Love, and Friendship: Studies of the Society for the Philosophy of Sex and Love, 1977-1992.* A volume in **Histories and Addresses of Philosophical Societies.**

46. Peter A. Redpath, *Wisdom's Odyssey: From Philosophy to Transcendental Sophistry.* A volume in **Studies in the History of Western Philosophy.**

47. Albert A. Anderson, *Universal Justice: A Dialectical Approach.* A volume in **Universal Justice.**

48. Pio Colonnello, *The Philosophy of José Gaos.* Translated from Italian by Peter Cocozzella. Edited by Myra Moss. Introduction by Giovanni Gullace. A volume in **Values in Italian Philosophy.**

49. Laura Duhan Kaplan and Laurence F. Bove, Editors, *Philosophical Perspectives on Power and Domination: Theories and Practices.* A volume in **Philosophy of Peace.**

50. Gregory F. Mellema, *Collective Responsibility.*

51. Josef Seifert, *What Is Life? The Originality, Irreducibility, and Value of Life.* A volume in **Central-European Value Studies.**

52. William Gerber, *Anatomy of What We Value Most.*

53. Armando Molina, *Our Ways: Values and Character,* edited by Rem B. Edwards. A volume in **Hartman Institute Axiology Studies.**

54. Kathleen J. Wininger, *Nietzsche's Reclamation of Philosophy.* A volume in **Central-European Value Studies.**

55. Thomas Magnell, Editor, *Explorations of Value.*

56. HPP (Hennie) Lötter, *Injustice, Violence, and Peace: The Case of South Africa.* A volume in **Philosophy of Peace.**

57. Lennart Nordenfelt, *Talking About Health: A Philosophical Dialogue.* A volume in **Nordic Value Studies.**

58. Jon Mills and Janusz A. Polanowski, *The Ontology of Prejudice.* A volume in **Philosophy and Psychology.**

59. Leena Vilkka, *The Intrinsic Value of Nature.*

60. Palmer Talbutt, Jr., *Rough Dialectics: Sorokin's Philosophy of Value,* with Contributions by Lawrence T. Nichols and Pitirim A. Sorokin.

61. C. L. Sheng, *A Utilitarian General Theory of Value.*

62. George David Miller, *Negotiating Toward Truth: The Extinction of Teachers and Students.* Epilogue by Mark Roelof Eleveld. A volume in **Philosophy of Education.**

63. William Gerber, *Love, Poetry, and Immortality: Luminous Insights of the World's Great Thinkers.*

64. Dane R. Gordon, Editor, *Philosophy in Post-Communist Europe.* A volume in **Post-Communist European Thought.**

65. Dane R. Gordon and Józef Niznik, Editors, *Criticism and Defense of Rationality in Contemporary Philosophy.* A volume in **Post-Communist European Thought.**

66. John R. Shook, *Pragmatism: An Annotated Bibliography, 1898-1940.* With Contributions by E. Paul Colella, Lesley Friedman, Frank X. Ryan, and Ignas K. Skrupskelis.

67. Lansana Keita, *The Human Project and the Temptations of Science.*

68. Michael M. Kazanjian, *Phenomenology and Education: Cosmology, Co-Being, and Core Curriculum.* A volume in **Philosophy of Education.**

69. James W. Vice, *The Reopening of the American Mind: On Skepticism and Constitutionalism.*

70. Sarah Bishop Merrill, *Defining Personhood: Toward the Ethics of Quality in Clinical Care.*

71. Dane R. Gordon, *Philosophy and Vision.*

72. Alan Milchman and Alan Rosenberg, Editors, *Postmodernism and the Holocaust.* A volume in **Holocaust and Genocide Studies.**

73. Peter A. Redpath, *Masquerade of the Dream Walkers: Prophetic Theology from the Cartesians to Hegel.* A volume in **Studies in the History of Western Philosophy.**

74. Malcolm D. Evans, *Whitehead and Philosophy of Education: The Seamless Coat of Learning.* A volume in **Philosophy of Education.**

75. Warren E. Steinkraus, *Taking Religious Claims Seriously: A Philosophy of Religion,* edited by Michael H. Mitias. A volume in **Universal Justice.**

76. Thomas Magnell, Editor, *Values and Education.*

77. Kenneth A. Bryson, *Persons and Immortality.* A volume in **Natural Law Studies.**

78. Steven V. Hicks, *International Law and the Possibility of a Just World Order: An Essay on Hegel's Universalism.* A volume in **Universal Justice.**

79. E. F. Kaelin, *Texts on Texts and Textuality: A Phenomenology of Literary Art,* edited by Ellen J. Burns.

80. Amihud Gilead, *Saving Possibilities: A Study in Philosophical Psychology.* A volume in **Philosophy and Psychology.**

81. André Mineau, *The Making of the Holocaust: Ideology and Ethics in the Systems Perspective.* A volume in **Holocaust and Genocide Studies.**

82. Howard P. Kainz, *Politically Incorrect Dialogues: Topics Not Discussed in Polite Circles.*

83. Veikko Launis, Juhani Pietarinen, and Juha Räikkä, Editors, *Genes and Morality: New Essays.* A volume in **Nordic Value Studies.**

84. Steven Schroeder, *The Metaphysics of Cooperation: The Case of F. D. Maurice.*

85. Caroline Joan ("Kay") S. Picart, *Thomas Mann and Friedrich Nietzsche: Eroticism, Death, Music, and Laughter.* A volume in **Central-European Value Studies.**

86. G. John M. Abbarno, Editor, *The Ethics of Homelessness: Philosophical Perspectives.*

87. James Giles, Editor, *French Existentialism: Consciousness, Ethics, and Relations with Others.* A volume in **Nordic Value Studies.**

88. Deane Curtin and Robert Litke, Editors, *Institutional Violence.* A volume in **Philosophy of Peace.**

89. Yuval Lurie, *Cultural Beings: Reading the Philosophers of* Genesis.

90. Sandra A. Wawrytko, Editor, *The Problem of Evil: An Intercultural Exploration.* A volume in **Philosophy and Psychology.**

91. Gary J. Acquaviva, *Values, Violence, and Our Future.* A volume in **Hartman Institute Axiology Studies.**

92. Michael R. Rhodes, *Coercion: A Nonevaluative Approach.*

93. Jacques Kriel, *Matter, Mind, and Medicine: Transforming the Clinical Method.*

94. Haim Gordon, *Dwelling Poetically: Educational Challenges in Heidegger's Thinking on Poetry.* A volume in **Philosophy of Education.**

95. Ludwig Grünberg, *The Mystery of Values: Studies in Axiology,* edited by Cornelia Grünberg and Laura Grünberg.

96. Gerhold K. Becker, Editor, *The Moral Status of Persons: Perspectives on Bioethics.* A volume in **Studies in Applied Ethics.**

97. Roxanne Claire Farrar, *Sartrean Dialectics: A Method for Critical Discourse on Aesthetic Experience.*

98. Ugo Spirito, *Memoirs of the Twentieth Century.* Translated from Italian and edited by Anthony G. Costantini. A volume in **Values in Italian Philosophy.**

99. Steven Schroeder, *Between Freedom and Necessity: An Essay on the Place of Value.*

100. Foster N. Walker, *Enjoyment and the Activity of Mind: Dialogues on Whitehead and Education.* A volume in **Philosophy of Education.**

101. Avi Sagi, *Kierkegaard, Religion, and Existence: The Voyage of the Self.* Translated from Hebrew by Batya Stein.

102. Bennie R. Crockett, Jr., Editor, *Addresses of the Mississippi Philosophical Association.* A volume in **Histories and Addresses of Philosophical Societies.**

103. Paul van Dijk, *Anthropology in the Age of Technology: The Philosophical Contribution of Günther Anders.*

104. Giambattista Vico, *Universal Right.* Translated from Latin and edited by Giorgio Pinton and Margaret Diehl. A volume in **Values in Italian Philosophy.**

105. Judith Presler and Sally J. Scholz, Editors, *Peacemaking: Lessons from the Past, Visions for the Future.* A volume in **Philosophy of Peace.**

106. Dennis Bonnette, *Origin of the Human Species.* A volume in **Studies in the History of Western Philosophy.**

107. Phyllis Chiasson, *Peirce's Pragmatism: The Design for Thinking.* A volume in **Studies in Pragmatism and Values.**

108. Dan Stone, Editor, *Theoretical Interpretations of the Holocaust.* A volume in **Holocaust and Genocide Studies.**

109. Raymond Angelo Belliotti, *What Is the Meaning of Human Life?*

110. Lennart Nordenfelt, *Health, Science, and Ordinary Language*, with Contributions by George Khushf and K. W. M. Fulford.

111. Daryl Koehn, *Local Insights, Global Ethics for Business.* A volume in **Studies in Applied Ethics.**

112. Matti Häyry and Tuija Takala, Editors, *The Future of Value Inquiry.* A volume in **Nordic Value Studies.**

113. Conrad P. Pritscher, *Quantum Learning: Beyond Duality.*

114. Thomas M. Dicken and Rem B. Edwards, *Dialogues on Values and Centers of Value: Old Friends, New Thoughts.* A volume in **Hartman Institute Axiology Studies.**

115. Rem B. Edwards, *What Caused the Big Bang?* A volume in **Philosophy and Religion.**

www.ingramcontent.com/pod-product-compliance
Lightning Source LLC
LaVergne TN
LVHW102225300926
849160LV00005B/29